Geronimo! U.S. Airborne Uniforms, Insignia & Equipment in World War II *Bill Rentz.* Explores the uniforms, insignia, and equipment of American Airborne, Glider, Troop Carrier, and Airborne Engineers in World War II. Detailed images of individual items, multi-side views of full combat rig, and over 100 World War II photos, most unpublished, showing the uniforms and equipment as worn by the troops. Also a comparative section with both British and German airborne gear. An important reference work for the airborne collector, reenactor, historian and veteran.
Size: 9"x12" • over 630 color and b/w photos • 192 pp. •
ISBN: 0-7643-0677-4 • hard cover • $59.95

Paramarine! Uniforms and Equipment of Marine Corps Parachute Units in World War II *Chris Mason.* *Paramarine*! is devoted to the Marines and Sailors of the Marine Corps parachute units of World War II. It explores in fascinating detail their unique, specially-designed uniforms, equipment, weapons and insignia. For the first time, over 500 photographs and images from World War II and modern collections have been assembled in one volume and combined with a wealth of facts and information about all aspects of Paramarine gear and training. *Paramarine* is the latest in Schiffer's series of studies on the airborne forces of World War II and is a must for historians and Marine Corps and Airborne collectors around the world.
Size: 9"x12" • over 500 color & b/w photos • 240 pp. •
ISBN: 0-7643-1924-8 • hard cover • $69.95

We are always interested in working with authors or collectors on wonderful book ideas. Contact us at **www.schifferbooks.com**, and page down to "For Authors – Submit a Book Proposal." Or call or write to us about your proposal.

Talk with your local bookseller. Their knowledge is one of the best tools in finding the correct book, and we all need to support them.

* *Peter Schiffer and Team* *

Designed & composed in America.

Printed in China.

Vultures over Israel

Vultures over Israel

The Vautour in Israeli Service
Squadron 110

1957-1971

Shlomo Aloni

Book Design by Stephanie Daugherty.

Library of Congress Control Number: 2011941483

Printed in China.
ISBN: 978-0-7643-3960-8

We are interested in hearing from authors with book ideas on related topics.

Published by Schiffer Publishing Ltd.
4880 Lower Valley Road
Atglen, PA 19310
Phone: (610) 593-1777
FAX: (610) 593-2002
E-mail: Info@schifferbooks.com.
Visit our web site at: www.schifferbooks.com
Please write for a free catalog.
This book may be purchased from the publisher.
Please include $5.00 postage.
Try your bookstore first.

In Europe, Schiffer books are distributed by:
Bushwood Books
6 Marksbury Avenue
Kew Gardens
Surrey TW9 4JF, England
Phone: 44 (0) 20 8392-8585
FAX: 44 (0) 20 8392-9876
E-mail: Info@bushwoodbooks.co.uk.
Visit our website at: www.bushwoodbooks.co.uk
Try your bookstore first.

CONTENTS

ACKNOWLEDGMENTS

THE AUTHOR WISHES TO EXPRESS GRATITUDE TO VAUTOUR AIR AND ground crews, to friends and colleagues who contributed, over a time frame of some 15 years, their personal materials, both memories and photos, for the production of this humble tribute to the IAF Squadron 110 Vautour.

Special thanks to modeling maestro Asher Roth, to Doron Avi-Ad and Shirli Reuveni of the IDFA, to the Israeli Censorship (that reviewed and approved this title for publication), and to Jacob Agassi, Ezra Aharon, Yossi Arazi, Dalya Ash, Eli Bahat, Yigal Bar-Shalom, Yair Ben-Shefer, Herzle Bodinger, Yehuda Borovik (BIAF), David Brisk, Ilan Brunner, Arie Dagan, Yoel Dan, Gideon Dror, Oded Einav, Giora Epstein, Itche Erez, Oded Erez, Achikar Eyal, Jacob Gal, Elisha Galon, Micha Gil, Shabtai Gilboa, Isaac Golan, Ran Goren, Albert Grandolini, Reuben Harel, IDFS, IGPO, David Ivry, Shlomo Kahana, Mati Kaspit, Shabtai Katz, Abraham Kimchi, Samuel Kislev, Judah Koren, Zvi Kreisler, Ruth Livneh, Eldad Maayan, Gideon Magen, Judah Manor, Uri Margalit, Hugo Marom, Ovadia Nachman, Isaac Nir, Reuben Nir, Guri Palter, Amos Peleg, Alain Pelletier, Jean-Jacques Petit, Ilan Ron, Ephraim Ronen, Ran Ronen, Simi Saar, Yossi Sarig, Nathan Segall, Jonathan Shahar, Daniel Shapira, Zeev Sharon, Meir Shefer, Pessach Shraga, Nimrod Tavor, Zeev Tavor, Yoash Tsiddon, Raanan Weiss, Arie Yaakobi, Ran Yahalom, Uri Yarom, Adam Zivoni, Beni Zohar, and Levy Zur.

All photos originated from the collections of the aforementioned; in many cases the same photo originated from the personal collection (PC) of several contributors. To save space, as well as to avoid confusion, the formal military style presentation of [rank] [name] was rejected in favor of the less formal [name] only presentation. IAF code words for radio call signs, equipment, and operations were translated from Hebrew to English in order to ease reading.

FOREWORD

My personal instructor during Flying School advance training was Eitan Karmi, a Vautour pilot. Eitan used to praise this special aircraft and emphasize the professional challenge that the Vautour set for its pilots. Eitan's stories planted in me a desire that someday I would be assigned to the Vautour squadron as a member of a small, selected team of Israeli Air Force Vautour pilots.

To my great rejoice, a short while after I graduated from Flying School, the IAF decided to rejuvenate the ranks of the squadron's Vautour pilots. Four young 1st Lieutenants were assigned to the Vautour squadron, and I was one of them.

The Vautour exceeded my expectations. My first impression was its size; it was significantly larger than any other aircraft that I had experienced until that time. A fighter-bomber with two engines, the Vautour had a wing span and an airframe weight that were almost twice as large and twice as heavy than the dimensions and weights of aircraft that I had previously flown. The Vautour was armed with four 30 mm guns and was designed to lift six 400 kg bombs in its belly bomb-bay, in addition to four stations for external loads under the wings. Thus, the Vautour could be configured in various combinations of bombs and fuel in the fuselage and under the wings that enabled the Vautour to fly further than any other IAF combat aircraft, and turned the Vautour squadron into the long range strategic arm of Israel.

Beyond the awesome power of the Vautour as a combat aircraft, the Vautour also excelled in the introduction of many engineering innovations, including a guns pack that

(PC)

enabled fast turnaround, a bomb-bay elevator that provided the aircraft with the capability to dive bomb, and a bombing computer (intervalometer) that calculated the coverage of the bombs on the ground and optimized attack patterns. It is also important to mention the highly praised French human engineering that resulted in a cockpit with excellent ergonomics; a cockpit that was large and comfortable, with all the controls and instruments readily accessible; and a cockpit that was friendly for the pilot.

Flying the Vautour was a challenge, primarily due to the large size of the aircraft; taking off and landing was especially challenging due to the Vautour's unique four points undercarriage. The Vautour also challenged our dedicated teams of maintainers. The squadron's maintainers worked around the clock, days and nights, to keep the Vautours flying, and to maintain our aircraft in serviceable status. Their work was more difficult because every aircraft was different than the other Vautours in its structure, systems, and its character. The aircraft differed from one another to the point that it was impossible to remove an airframe part from one aircraft and to fit it to another aircraft. Thus, the dictum at Ramat David at the time was that the lights in Squadron 110's hardened aircraft shelters, pens, and hangars were never switched off.

Those special qualities of the Vautour shaped the missions of Squadron 110 and defined its part in the future war for which the IAF was preparing throughout that time. The squadron's reconnaissance aircraft reached every relevant point in the Middle East, no matter how far it was, to collect intelligence information and mapping data that were crucial to the IAF's operational success during the Six Day War. As the long range, heavy bomber arm of the IAF, the squadron was tasked to attack the farthest away airfields where Tupolev Tu-16 bombers were based; those Tu-16 bombers posed the main threat against the population of Israel.

My years of service in the Vautour squadron shaped my qualities as a fighter pilot and influenced my operational perspectives; qualities and perspectives I carried forward with me in all my subsequent assignments during my IAF service. I will mention three personal milestones – during my service as a Vautour pilot – that shaped my qualities as a fighter pilot and influenced my operational perspectives:

- The crash landing that I experienced, my injury, and the return to operational flying matured me to become a better pilot, with better understanding of the human ability to recover from severe blows.

- During the Six Day War I was fortunate to take part in three long range air strikes—against Bani Suef and Luxor in Egypt and against H-3 in Iraq—that neutralized the most dreadful threat against the Israeli population at the time, the threat of the enemy's Tu-16 heavy bombers force; three missions from which I learned the importance of being able to neutralize strategic threats no matter how far away these would be.

- From the three years of the Attrition War I internalized the necessity to be able to maintain routine training, including qualification of new pilots, in parallel with intensive fighting.

Major General Herzle Bodinger, IAF Commander, 1992-1996. (PC)

Finally, I would like to thank Shlomo Aloni for his initiative and research that produced this important book, bringing to light a glorious and important chapter in the history of the Israeli Air Force.

Major General Herzle Bodinger
Squadron 110 pilot, Deputy B and Deputy A; 1963-1971.
Israel Air Force Commander; 1992-1996.

1

ROADMAP TO SQUADRON 110 VAUTOUR

1954-1957

ISRAEL AIR FORCE SQUADRON 110 WAS ACTIVATED IN LATE August 1953 as a Mosquito fighter-bomber unit. Prior to activation of Squadron 110, Air Department Chief Daniel Shimshoni signed on 13 August 1953 the document "IAF Plan for 1953 to 1957". An objective of this plan was to field six squadrons of combat aircraft by 1957, including three squadrons equipped with jet fighters and one squadron flying jet bombers. In line with this plan, and within a year from activation of Squadron 110, the IAF started to evaluate re-equipment options; the objective was to purchase a new jet aircraft that would replace the Mosquito fighter-bombers of Squadron 109 and Squadron 110, the Mosquito reconnaissance aircraft of Flight 115, the Mosquito night-fighters of the planned Squadron 119, and the B-17 bombers of Squadron 69.

The mainstay of the IAF fighter-bomber force from 1953 were the American single-engine P-51 Mustang (left) and the British two-engined Mosquito (right); the former was operated by the Ramat David based Wing 1 "North Air Force" and the latter flying from Hatzor were Wing 4 "South Air Force". The "North Air Force" and "South Air Force" nicknames were evidence of the savage competition between the two IAF fighter-bomber wings. (PC)

Squadron 110 Mosquito pilot Ovadia Nachman inspects the Aircraft Servicing Form 4420 prior to a flight from Hatzor in December 1953; the original badge of Squadron 110 is painted on the door of Mosquito 06. The sign in front of the aircraft reads "DANGER, WEAPONS; AIRCRAFT LOADED". Nachman flew the Vautour from 1958 and was Squadron 110 Commander from 1963 until 1964. (PC)

Acquisition Options

The IAF Air Department evaluated the suitability of IAF fighter-bomber types (Mosquito and Mustang) to expected mission profiles. This early 1954 evaluation was to form a basis for planning of the IAF order of battle. The evaluation concluded that both types of fighter-bombers were essential for the then IAF order of battle. As for the future, the evaluation estimated that re-equipment of the IAF fighter-bomber force should be accomplished in 1957. Air Department Chief Daniel Shimshoni reported the conclusions of this evaluation on 10 February 1954, and presented possible replacements for the Mosquito and the Mustang:

> Other aircraft in the world that can be considered for these roles are: for Mosquito missions – B-26 Invader, Canberra and possibly Vautour.

The British Canberra was a first generation jet aircraft (subsonic design with unswept wings) that entered service from 1951; the prospects of HMG selling the Canberra to Israel were remote. The French Vautour II (from here on simply Vautour) was a second generation jet aircraft (transonic design with swept wings) but was still in development and would not enter service before 1957. Therefore, the Invader was the preferred Mosquito replacement type, as Air Department Chief Daniel Shimshoni explained:

> The piston powered B-26 Invader seems to be more efficient than Mosquito and will last longer.

Regardless of explicit indication that the B-26 Invader was the preferred Mosquito replacement type (and regardless of stating the F-86 Sabre as the preferred Mustang replacement type), the Air Department Chief directed Logistics Department and AIR2 as follows:

> It is essential to immediately explore the purchase prospects of Corsair, Invader, Canberra, Vautour, Corsair [sic] and to ascertain the suitability of these types to IAF service.

On 30 March 1954, AIR2 Chief Adam Shatkai forwarded to the IDF Attachés to the USA, England, and France a request for information concerning Canberra technical data. Nearly half a year later, on 12 September 1954 AIR2 Chief Adam Shatkai delivered to the IDF Attaché to the USA another request for information, this time concerning the Invader, Mitchell, Skyraider, and Corsair.

IAF Plan 1954-1958 – Combat Aircraft

On 30 October 1953, AIR2 Chief Adam Shatkai authorized reserve officer Naftali Kadmon to prepare a "staff study concerning standardization of fighter aircraft in the IAF". The objective of this staff study was defined as follows:

> Consolidation of IAF requirements concerning combat aircraft and recommendation for selection of standard combat aircraft for service in the time frame from 1954 until 1960 (inclusive), from the variety of combat aircraft currently in service but still in an early phase of their program timeline or from aircraft undergoing testing towards future service introduction.

Naftali Kadmon delivered his staff study on 27 September 1954 and AIR2 distributed the study, on 30 October 1954, as a milestone in the process towards preparation of IAF Plan for 1954 to 1958. The main considerations that influenced the study were conditions in theatre of operations, enemy order of battle, military balance in the air, combat aircraft missions and economic limitations. The offensive missions of IAF combat aircraft were defined as:

> A. Strategic attack; bombing of regional targets up to 500 miles away.
>
> B. Air support and operations over the battlefields closer to Israel's borders.

Re-equipment alternatives were single role aircraft, multi role aircraft and an intermediate solution. The single role aircraft alternative recommended re-equipment with four types (strategic attack, air support, interception and patrol), the multi role aircraft alternative promoted purchase of one type for all four missions while the intermediate solution suggested:

> Aircraft designed for more than one mission, in different combinations. None of these aircraft types would be capable of flying all four missions, therefore necessitating another type or types to fly the balance of missions. The advantage of this third method is that the number of aircraft types is higher than one but lower than the number of roles, four in this case.

The B-26 Invader was the IAF's preferred Mosquito replacement type during the early 1954 evaluation of potential successors; this French B-26 was photographed in 1958 during conversion in France of IAF Vautour B crews. The Vautour IIB replaced the B-26 in French service. (PC)

The British Canberra was the yardstick of the light bomber breed during the early 1950s. Since entering RAF service in 1951 as a Mosquito replacement, the Canberra was the best light bomber in the world; so good, that it was also introduced into US service as the B-57, a successor to the B-26. This photo of a British Canberra was taken during the 1956 Suez War. (PC)

The recommended re-equipment path was, obviously, the intermediate solution:

> The third solution for the combat aircraft re-equipment question is a compromise between maximum combat efficiency (single role aircraft solution) and maximum economical efficiency (multi role aircraft solution). Assuming that the average IAF pilot would be superior to the average enemy pilot, it is safe to state that maximum combat efficiency can be compromised in comparison with enemy aircraft. In this case, the third solution will be the preferred alternative since it is relatively cheaper while aircrew superiority will enhance the combat efficiency of aircraft types that are not single role.

The staff study pointed at two classes of combat aircraft as candidates for re-equipment: a lightweight fighter (such as the British Gnat, the French Trident, or the Italian Saggitario) and a heavier fighter-bomber (such as the American F-84 Thunderstreak and F-86 Sabre, the French Mystere, or the Swedish SAAB J29). Acquisition of a third combat aircraft type was suggested with reservation:

> To these combat aircraft types can be added a light bomber such as Canberra, though the difference in bomb load between Canberra and a heavier fighter bomber type is quite small, especially over short ranges, and therefore cannot justify acquisition of a light bomber except, perhaps, in small numbers for long range strategic bombing of targets such as Baghdad (600 miles), Basra (780 miles) and elsewhere.

Light Bomber Availability

Very few types of "a light bomber such as the Canberra" were available worldwide, and only one type would be available to Israel: the French Vautour II. The original SO 4000 Vautour (Vulture) was a first generation jet aircraft that first flew on 15 March 1951, but did not enter production and service. The ultimate SO 4050 Vautour II first flew on 16 October 1952 and would enter production and service in France and in Israel.

The French era of the IAF started in the summer of 1954, when Israel ordered Mystere II fighters and Noratlas airlifters. The proposed mix that the aforementioned staff study promoted was partially realized due to actual circumstances, rather than meticulous planning. Mystere II development ran into troubles, and by the end of 1955, the IAF decided to trade their purchase of Mystere IIs for acquisition of Mystere IVAs, with Ouragans as a stop gap solution; the former as the "heavier fighter-bomber type" and the latter in place of a light weight fighter, as proposed in the staff study. Previously, Israel had started to express an interest in the Vautour for IAF service as a bomber, fighter-bomber, reconnaissance aircraft and night-fighter. It was pure coincidence that the timing of the service introduction of the only type of light jet bomber available for Israel (1957) would overlap, precisely, the planned date for IAF Mosquito retirement and expected introduction of a Mosquito replacement.

The May 1955 statement "We demanded agreement in principle for 15 Vautours" related to the visit to France of IMOD Director Simon Peres. "Our request was presented, for his support to enable us to purchase ... 12 Vautours" was another indication

Ezekiel Somech (center) took over command of Squadron 109 Mosquito from Israel Lahav (right) in the presence of Wing 4 Commander Leslie David Easterman (left) on 15 August 1955. Squadron 110 was amalgamated with Squadron 109 in September 1955 to make room at Hatzor for Squadron 113 Ouragan. (PC)

The three Vautour II prototypes were designated 001 (N for Nuit/Night that first flew on 16 October 1952), 002 (A for Attaque/Attack that first flew on 16 December 1953), and 003 (B for Bombardement/Bomber that first flew on 5 December 1954). The pre-production batch aircraft were designated 04 to 09; this batch included one B (04), two A (05 and 07), and three N (06, 08 and 09) aircraft, with the first flying from 18 March 1955. The first production aircraft - Vautour IIA number 1 - first flew on 30 April 1956. (PC)

of Israeli interest, as reported on 14 May 1955 after a meeting between the IMOD Director and the FAF Commander. Four months later, the Czechoslovak-Egyptian Arms Deal was made public in September 1955 and Egypt introduced a few dozen Il-28 bombers from early 1956; the first Middle East nation to field a light jet bomber force. The Vautour was not merely an appropriate answer to the Il-28, but was actually vastly superior. The snag was that the Vautour was still a long way from service introduction. In his 12 October 1955 reply to a 27 September 1955 request from Acting AIR2 Chief Nathan Eldar to evaluate the feasibility of camera installation in the Vautour "without an adverse impact on the bomb load of the aircraft," the IMOD Ordnance Department Assistant for Air Benjamin Kagan stated:

> In reply to your letter I inform you that the requested inquiry is too early to be discussed. The French did not complete the development of the aircraft and we did not yet enter negotiation concerning acquisition.

Squadron 101 Commander Benjamin Peled (left, with Mae West) with Prime Minister David Ben-Gurion (second from right, gray hair) and IAF Commander Dan Tolkowasky (first from right, back to camera with beret) at Hatzor in April 1956. Peled was the test pilot of the Vautour Evaluation Committee. (PC)

Vautour Evaluation Committee

The Vautour was the only combat aircraft in its class available to Israel. Until the summer of 1956, the Israeli sources of information concerning the Vautour were secondary: published materials, classified documents and background briefs. When general discussions started to give way to detailed negotiations, Israel was allowed access to primary data from evaluation in flight; the IAF Vautour Evaluation Committee was in France from June to July 1956. Chairman of the committee was Air Department Chief Shlomo Lahat, and the committee test pilot was Squadron 101 Commander Benjamin Peled. Initially, the committee summed up all available data concerning the Vautour and its three variants: Vautour IIA attack aircraft, Vautour IIB bomber, and Vautour IIN night-fighter. The following presentation focused on the Vautour IIA, which was the type planned for service with Squadron 110.

Vautour A ordnance options included four 30 mm guns (with 100 rounds per gun), 68 mm rockets (up to 316, including 240 in the belly and 76 in four pods under the wings), T-10 rockets (up to 74, including 50 in the belly and 24 under the wings), bombs (up to ten 400 kg bombs, including six in the belly and four under the wings), and napalm.

The bomb-bay was modular and could be adapted to bombs, rockets or fuel. Adaptation to bombs included installation of a special elevator. Prior to bombing, an electrical motor opened the bomb-bay doors and slid the doors outside. The same motor also lowered the elevator so that the bomb-bay was sealed to airflow and the bombs were lowered below the airframe. Releasing the bombs from this position improved safety, while flying with bombs hanging below the fuselage had no known negative impact concerning speed or Mach limitations; it was possible to release bombs at up to 580 knots or Mach 1.0. While in a successful test, 400 kg bombs were released at sonic speed from an altitude of 40,000 feet.

The bomb-bay elevator could lift two triple bomb racks (for six 750 pound bombs, six 500 pound bombs, or six 400 kg bombs) or three triple bomb racks (for nine bombs up to 260 pounds each, six 120 pound training bombs, or six 54 kg training bombs). To load the bomb-bay, the bombs first had to be attached to the rack on a special trolley, fitted with detonators, and lifted—rack and bombs—into position inside the bomb-bay.

Squadron 109 Mosquito was closed in early June 1956, making room at Hatzor for Squadron 101 Mystere.The IAF planned to reform Squadron 109 at Ramat David as a second Mystere squadron, so Squadron 110 was revived as a reserve Mosquito squadron. Seven of the airmen captured in this photo – taken at Hatzor shortly before 109 closed - would later fly the Vautour: pilots Ezekiel Somech and Simon Ash; and navigators Achikar Eyal, Itche Erez, Oded Erez, Uri Ofer and Ephraim Ronen. (PC)

The Vautour A was to be equipped with a gyro sight "that in the future will enable accurate dive bombing; as for low altitude bombing there is no such mode at this time". An intervalometer controlled release of bombs because it was impossible to drop all the bombs at the same time due to center of gravity considerations. The intervalometer enabled separated release of bombs from belly and wings or combined release of bombs from belly and wings. When combined release was selected, the bombs loaded to the wing stations were always dropped in pairs ahead of the bombs from the bomb-bay.

The guns were installed in a pack that could be replaced in 15 minutes; the manufacturer claimed for "perfect exchangeability" between packs of all types in general and all aircraft in particular. It was possible to separately zero the sight in the aircraft and the pack outside the aircraft. The pilot could select firing the two lower guns, the two upper guns, or all four guns. During firing, all cartridges and links were collected inside the airframe.

Rockets and napalm were supposedly standard weapons in IAF inventory. Despite lack of data, the committee may have had a hunch concerning future difficulties when the compiler of the report, Benjamin Peled, stated:

> I assume that it is possible to drop the napalm bombs in our inventory from both belly and wings – ten 250 kg napalm bombs – at speeds lower than 500 knots.

In due time, the IAF would indeed realize that the Mystere rocket pods, supposedly fit for the Vautour, were not compatible with the Vautour due to the latter's higher speeds of rocketry. The committee did not have data concerning future weapons for the Vautour, such as air-to-air missiles, except for sketchy details about a plan to fit a 1,000 kg air-to-ground missile still in development. Specific ordnance handling equipment developed for the Vautour included:

A. A trolley for transport and load of bombs fitted to racks (full load of six bombs on a trolley).

B. A universal platform for all work under the wings and for the guns pack.

C. Jacks for all ordnance and loading jobs.

D. Small trolley for transport and load of small ordnance, such as training bombs.

The calculated maximum range of the Vautour A—in take-off weight of 20 tons with 10,650 liters of fuel (5,150 internal, 3,000 in the bomb-bay, and 2,500 external) and full ammunition for guns—was 4,200 km in five flight hours. The calculated radius of action of the Vautour A—with full internal fuel load, full ammunition for guns, and 208 rockets—was 550 km in a scenario of high altitude cruise, low level approach to the target, five minutes combat over target and return to base at high altitude.

The Aircraft ... Accelerated at a Phenomenal Rate

The Vautour 09 was a pre-production aircraft in the N configuration; the pre-production aircraft (designated Vautour 04 to 09) were an intermediate development phase between the prototypes (designated Vautour 001 to 003) and the production aircraft (the A were numbered from 1 to 30, the N from 301 to 370, and the B from 601 to 640). The unique feature of Vautour 09 was a pair of Rolls Royce Avon turbojet engines. The Avon produced a thrust of 4,200 kg, compared to only 3,500 kg thrust in the standard SNECMA Atar 101E-3 engine, but was 500 kg heavier than the Atar. Therefore, the aircraft that the Vautour Evaluation Committee evaluated for first hand impression was not quite representative. Nevertheless, the evaluation was conducted and covered three sorties: familiarization flight, evaluation at high altitude, and evaluation at low altitude. Committee chairman Shlomo Lahat accompanied test pilot Benjamin Peled during the first two flights.

The familiarization flight was flown on 21 June 1956. It resulted in major surprise and had to be cut short to only 20 minutes duration. The cause of surprise was the acceleration of the Vautour, as Benjamin Peled reported:

> Acceleration, upon release of brakes, is so fast that I was unable to estimate the distance of the take-off run. The aircraft took-off by itself (the undercarriage sets the airframe at the correct incidence for take-off – six degrees nose up attitude) and accelerated at a phenomenal rate; the thrust to weight ratio is unusual (10 tons thrust to 16 tons weight). I hardly managed to retract the undercarriage below 180 knots. I did not manage to retract flaps in time (below 220 knots). Therefore, the flaps got stuck at 15 degrees position at 350 knots. The circuit breaker of the flaps activation motor failed and I had to land immediately after take-off at a weight of 15.5 tons.

The second flight, also flown on 21 June 1956, covered climb to 40,000 feet, evaluation of high altitude performance, descent, and a taste of aerobatics. Climbing time to 40,000 feet was some six minutes; the pilot commented that "the initial rate of climb is far beyond the presentation scale of the instruments, an estimated 16,000 to 18,000 feet per minute". During high altitude flight, the aircraft had limitations on maximum RPM (due to maximum nozzle temperature) and maximum Mach (because of shuddering). Considering these limitations, Benjamin Peled clocked Mach 0.94 at level flight after shuddering, and rolling down of the right wing started at Mach 0.93. The French reported to the Israeli pilot that in a dive above 30,000 feet, the Vautour clocked Mach 1.19 (faster than the Mystere), but Peled only managed Mach 0.98; shuddering and rolling down of the right wing stopped at Mach 0.97. During descent the evaluation of Peled was that the air brakes were not particularly efficient. Rolling rate in aerobatics—70 to 90 degrees per second—was considered just fine, but Vautour 09 had a private 4.5 G limitation because of the heavier Avon engines and due to the lack of the guns pack. This non-standard configuration also impacted the fuel system; the wing tanks were not filled to preserve the aircraft center of gravity.

Low altitude handling was evaluated in the third flight, flown on 24 June 1956 with Commandant Bigand. A while earlier, on 16 May 1956, Vautour 001 was lost in a spin after a stall during a test flight (the crew ejected safely). Therefore, Benjamin Peled was asked to avoid slowing down below 120 knots. Overall the "aircraft handled well at low altitude". Rate of roll, only "fine" in aerobatics during the previous flight,

Vautour pre-production 09 that Benjamin Peled evaluated in flight. (PC)

was considered adequate for the "aircraft size and weight". Sink during recovery from dives "was not more than could have been expected". As for turns, "the radius of turn is quite significant at 550 knots and this fact should be accounted for in planning of flight and attack patterns". The aircraft behavior during stalls was exemplary; the only surprising feature in this department was that the impact of the elevator "improved as speed slowed down" (the explanation was a shift of the aerodynamic center due to stall of the wingtips). As expected, the powerful Avon and the rather light 16 ton take-off weight contributed to impressive single engine performance: the Vautour climbed 6,000 feet per minute and slowed down to 140 knots. Peled estimated that the performances of the Vautour on a single Atar would not be so good, but "there should not be any problems in single engine flight".

The bottom line of the first, first hand impression of an Israeli pilot of the Vautour was certainly positive. The evaluated aircraft was not fully representative, but French willingness to allow an Israeli pilot to evaluate in flight one of the most advanced French combat aircraft of that era, was a major step forward towards the next phase of the IAF Vautour saga, practical negotiation.

The Contract for 18 Vautours was Signed

At the time of the IAF Vautour Evaluation Committee activity, the French decided to reject the Vautour A fighter-bomber and to delay development of a Vautour reconnaissance aircraft because they were already operating F-84F Thunderstreak fighter-bombers and RF-84F Thunderflash reconnaissance aircraft. The French focused on the service introduction of the Vautour N night-fighter (to replace the Meteor NF.11) and the Vautour B bomber (to replace the B-26) in this order; FAF service introduction was scheduled to start at the end of 1957. The Israeli priorities were exactly the opposite, with reconnaissance first and then fighter-bombers. The bright side for the Israelis was that the French rejection of the Vautour A was decided after manufacture of this model was already underway. IMOD civil servant Benjamin Kagan reported from France on 15 July 1956:

> Today we notified the FAF concerning our request to purchase 18 tactical aircraft [Vautour A] and six bombers [Vautour B] that we will convert into reconnaissance aircraft.

France rejected service introduction of the Vautour IIA attack aircraft due to availability of the F-84F Thunderstreak; the depicted FAF examples – in mixed FAF/IAF markings - were photographed in Israel during the 1956 Suez War, when the FAF Escadre de Chasse (Combat Wing) I operated as Squadron 200 from Lod. Due to this French decision only 30 production standard Vautour IIA attack aircraft were manufactured, and 17 of these were exported to Israel from 1957 to 1958. (PC)

France delayed development of a Vautour reconnaissance variant due to availability of the RF-84F Thunderflash; the depicted FAF example was photographed during the 1956 Suez War. Eventually, France converted 12 Vautour IIB bombers to Vautour IIBR reconnaissance aircraft during the 1960s. (PC)

In light of the French rejection of the Vautour A, it was expected that the first two aircraft would be delivered before the end of 1956, with the balance of 16 aircraft following until the end of 1957. This schedule was based on delivery of "as is" aircraft (aircraft manufactured to FAF specifications), because an IAF requirement for specific modifications was expected to delay deliveries. IMOD Director Simon Peres reported to Prime Minister David Ben-Gurion, who wrote in his diary on 16 July 1956:

> The French promise us two Vautours in November (faster than Ilyushin); can lift 3.5 tons of explosives (and also a B!!). Each Vautour will cost 600,000 dollars (excluding spares and ammunition – these will also cost 600,000 dollars); can fly four hours and can fly all the way from France to Israel nonstop.

Eight days later, Ben-Gurion reported in his diary "Simon [Peres] departs to Paris tomorrow for discussion concerning acquisition of ten Vautours," and indeed, this trip was successful, as Ben-Gurion wrote on 3 August 1956:

> [IDF Chief of Staff] Moshe [Dayan], [IMOD Director] Simon [Peres] ... arrived at 18:00. The contract for 18 Vautours was signed. The [French] Air Force was asked if these can be dispensed and gave a negative answer but the [French] Minister of Defense ordered otherwise; in the knowledge of the [French] Prime Minister. They also agreed to delay payment of 20 million dollars for one year but Simon still states that there will be shortage in cash.

Where Will the Money Come From?

The Vautour was expensive. Purchase of 18 Vautours was an element of the French Israeli HIGH TIDE Arms Deal. Total value of the purchased Vautours was 13,371,300 dollars at a unit cost of 742,850 dollars. For comparison, the HIGH TIDE deal also included 36 Mysteres at a unit cost of 364,000 dollars, or a total value of 13,104,000 dollars. These costs did not cover training, spares, weapons, ammunition, ordnance, facilities, and infrastructure. Therefore, the economical burden to introduce the Vautour into IAF service was expected to exceed 20 million dollars.

Meanwhile difficulties surfaced. The Vautour was still in the development phase and would not enter French service before 1957. The report to Ben-Gurion concerning delivery of two Vautours in November 1956 was unrealistic. So was an Israeli request to add reconnaissance capability to the Vautour, a request that Ben-Gurion mentioned on 22 August 1956:

> They [the French] will not give us Vautours for reconnaissance. There are only eight Vautours in France; these are prototypes. They are good for bombing and reconnaissance but not yet ready for reconnaissance.

Still, the unrealistic report of imminent delivery resurfaced, as Ben-Gurion wrote on 5 October 1956:

> Four Vautours are ready in France but are still undergoing testing; can be delivered to us a month and a half from now, each will cost 742,000 dollars without spares and ammunition.

Vautours were not delivered to Israel before the end of 1956. It was during this time that diplomatic connections between France and Israel developed to such an extent that IMOD Director Simon Peres reported to Prime Minister Ben-Gurion on 11 December 1956 about French willingness to sell to Israel 36 Vautours (on top of the 18 already ordered). Seemingly good news, to which the Prime Minister retorted, "Where will the money come from?"

A week later, on 18 December 1956 the practical Prime Minister again had to "ground" a visitor. This time it was IAF Commander Dan Tolkowsky, who presented to the Prime Minister the IAF purchase plan for 1957:

> At 10:00 a General Staff meeting with the IAF concerning next year's budget. Dan demands to acquire: ... 12 Vautours – 12 million dollars ... The total demand of the IAF is

Squadron 110 was mobilized in October 1956 and operated from Ramat David throughout the Suez War; the two-engine safety of the Mosquito was proven in combat, as all Mosquitoes - even those with severe battle damage – returned to base safely, whereas the IAF Mustang squadrons – flying similar air support missions – lost nine aircraft in combat. (BIAF)

> therefore 168 million pounds. I told the present that this is out of question. I do not believe that there will be war this year (except perhaps for an armed conflict between us and Egypt over Sharem El Sheikh). I do not see a necessity for additional Mysteres; we better buy a new type. There also should be no insistence on 18 Super Mysteres and we can do with less than 12 Vautours. This year there will be immigration from East Europe [that will consume a significant slice of the Israeli budget] and the IAF should re-plan.

Moshe Suggests, Dan Wants

The IAF indeed re-planned in line with the guidelines of the Prime Minister. This resulted in dispute between IDF Chief of Staff Moshe Dayan and IAF Commander Dan Tolkowsky; Ben-Gurion wrote in his diary on 7 January 1957:

> The staff meeting discussed planning. There is a dispute between Moshe and Dan. Moshe suggests investing 15 million dollars in acquisition of 12 Super Mysteres and six Vautours and the same next year [1958]. Dan wants only three Vautours and six Super Mysteres with the balance for spare parts and flight hours.

By then, the Vautour had started to enter squadron service in France, and the IAF requirement for Vautour introduction transformed from vital to urgent. The Vautour was planned to re-equip, gradually, four IAF squadrons: the Vautour A attack aircraft to replace the Mosquito fighter-bombers in Squadron 110 service; the Vautour B bomber to replace the B-17 in Squadron 69 service; the Vautour B reconnaissance aircraft to replace the Mosquito reconnaissance aircraft in Squadron 115 service; and

A Squadron 110 Mosquito – still wearing the 1956 Suez War stripes – inside a hangar at IAI; the caretaker of the reserve Squadron 110 Mosquitoes until 1957. (PC)

the Vautour N all weather fighter to augment the Meteor night-fighters of Squadron 119. The delay in delivery from the end of 1956 resulted in a gap between retirement of old aircraft and the introduction of new airframes. The first to be so impacted was Squadron 69, which was inactivated on 1 March 1957; the AIR2 inactivation order—dated 5 March 1957—ended with the phrase "Squadron 69 will be reactivated only after arrival of its planned complement of Vautour aircraft".

Squadron 110 followed, and was inactivated on 1 April 1957. Unlike Squadron 69—which remained inactive from 1957 until 1969 as a result of the changes in IAF Vautour purchase plans—the continuity of Squadron 110 was already secured at the time of Mosquito fighter-bomber retirement. By then, the named commander of Squadron 110 Vautour—who was also the last commander of Squadron 110 Mosquito—was already in France, undergoing a type conversion course to the Vautour.

Conversion in France

IAF maintainers attended a Vautour conversion course in France from 18 December 1956 to 16 January 1957. Vautour type conversion course flights started in France on 8 February 1957. The students were Daniel Shapira, who was to be Squadron 115's Vautour Commander, and Ezekiel Somech, who was to be Squadron 110's Vautour Commander. The French assigned Vautour A Number 6 to this effort, as well as a NCO and two maintainers who were in charge of the aircraft and supervised the practical training of the Israeli team of maintainers. Vautour A6 was an early production aircraft, and as such was powered by the lower rated Atar 101D-3 with 3,000 kg thrust; the 3,500 kg thrust Atar 101E-3 engines were planned to power the IAF Vautour A.

Somech and Shapira flew 19 sorties (nine and ten, respectively) by 17 February 1957 in a tempo of two to four flights per day. Seven malfunctions were reported during these flights, mostly related to the electrical system. There were six principle systems in the aircraft: hydraulic, fuel, pneumatic, electricity, instruments, and weapons. The hydraulic system was utilized to move the control surfaces in flight and to steer the aircraft on the ground. The fuel system was automatic, except for transfer of fuel from the external tanks. The pneumatic system was utilized to pressurize the fuel tanks, the hydraulic reservoir, the cockpit, and the pilot's G suit, as well as heating the cockpit and the windshield. The electricity system was the "least advanced and most malfunctions were related to electricity". The instruments system included several new features for the IAF, including a radio compass, a radio altimeter (previously only installed in the IAF in the Mystere reconnaissance variant), a gyro for dumping of nose movements, and a new type of attitude indicator with a selector that enabled presentation of additional data so that the pilot would use a single instrument during instruments flying. The weapons system included a gyro sight, a sight camera, the guns pack, the bomb-bay, four hard points under the wings, and special ground equipment.

The Israeli maintenance team performed daily inspection, line maintenance, and repair of malfunctions. Daniel Shapira reported that the Vautour was a maintenance friendly aircraft:

> It is generally felt that the designers invested considerable effort to ease maintenance; as a result of this approach, most accessories and systems are easily accessible for maintenance.

The electrical system was indeed diagnosed as a weakness, as was the coupling between the electrical system and the other systems in the aircraft, especially the coupling between electricity and hydraulics. Engine life was only 200 to 250 hours, but a French objective was to improve this figure to 300. Daniel Shapira also pointed at the fuel tanks as a possible maintenance challenge, because access to repair or to replace the fuselage tanks necessitated dismantling of many systems, and because sealing of wing tanks was with "special sealant (American) and we do not have a clue how this will work". Single point pressure refuel was from the front undercarriage bay. Weapons system tests were discontinued after France rejected service introduction of the Vautour A; the only completed weapon tests covered gunnery, straight and level flight bombing from the belly bomb-bay, and rocketry from the front suite of the bomb-bay. Discontinuation of weapons tests, the relatively heavy load of weapons, and the planned variety of ordnance were all expected to lead to special effort, including testing, acquisition of weapons and equipment, specialization in advance weapon systems, and larger teams of ordnance specialists. Daniel Shapira summed up the issue of weapons:

> We have no additional practical knowledge so I believe that we must experience flights with gunnery, rocketry and bombing. As for the installed sight, my opinion is that such an expensive aircraft – capable of lifting a lot of bombs and rockets – must have a capable sight because in modern warfare it is impossible to attack one target more than once. Therefore, aiming equipment must be advance enough to ensure that rockets or bombs will surely hit and crush the target.

Impression from flying the aircraft was positive. The cockpit was spacious compared to the Mystere. The instruments panel was a lot more complicated than instruments panels of previous IAF jets due to the many systems onboard the aircraft; many of these systems were new to IAF pilots, including the radio compass and intervalometer. An instruments panel innovation was the warning lights panel, which grouped together on a single panel all of the warning lights. Start up procedure was dependent on an external battery that started the right engine; once the pilot noted that hydraulic pressure was in the "green" the left engine was started. Taxiing the aircraft on the ground was with foot activated pedals for steering the nose wheels and with hand operated brakes; Shapira summed up taxiing as "after one gets used to taxiing, there is a sense of safety as though one was driving a brand new taxi". Vautour A6 took-off without any negative aspects, but climbed to 40,000 feet in more than 11 minutes! This was almost double the time it took the Avon powered Vautour 09; it was hoped that the Vautour A with the Atar 101E-3 would be a major improvement over the Vautour A6. Other aspects in flight that Daniel Shapira reported were:

- Prolonged flight at 40,000 feet resulted in stiffer operation of ailerons due to coldness but this does not impact handling while it is recommended to occasionally shake the ailerons.

- The elevator is not extremely efficient at sonic speed ... it is assumed that monobloc efficiency will be bet-

The IAF Vautour conversion team at Mont de Marsan in February 1957, together with their French hosts. Ezekiel Somech – assigned to lead IAF introduction of the Vautour A - is standing fifth from left, and Daniel Shapira – assigned to lead introduction of the Vautour reconnaissance aircraft – is standing seventh from left. Technical Officer Ariel Bar-Lev is standing first from left and Technical Officer Mordechai Alkalai is standing first from right. (PC)

Daniel Shapira was to lead IAF introduction of the Vautour reconnaissance aircraft; initial IAF plans emphasized early introduction of the Vautour reconnaissance, and these aircraft were planned for assignment to Squadron 115. When the IAF plan to activate Squadron 115 Vautour reconnaissance was abandoned, Shapira assumed the role of IAF Chief Test Pilot and was subsequently assigned to the FAF Test Pilots School from December 1958. (PC)

Vautour A Number 6 was delivered to Mont de Marsan during the second half of 1956, equipping a detachment of FAF Night Fighter Wing 30 for training of pilots pending planned introduction of the Vautour N all weather fighter; this aircraft was lost in an accident on 14 February 1963. (PC)

ter but in my opinion the monobloc is unnecessary because it is expensive and, after all, the Vautour is not a fighter.

- Flying on one engine is not a problem.
- Turning rate is a lot slower than in the Mystere but quite good relative to size and weight of the aircraft.
- Immediately after touch down, the braking chute is released with a pleasant braking impact (much like a hydraulic damper) and a squeeze of the brakes handle result in a short stopping distance ... the aircraft taxis to the line with the braking chute that is released at the line. During landing, the pilot should be cautious to avoid accidental activation of the emergency brakes instead of the braking parachute.

The braking parachute was also an innovation for the IAF; actually, the principle lesson from this pioneering conversion was the relatively high number of systems onboard the Vautour, including advance systems, complex systems, and innovative systems that pilots and maintainers were expected to learn and master. Daniel Shapira summed this up with a bit of foresight:

> Training pilots will not pose any special problems but Vautour pilots must be experienced jet pilots and they will have to learn in detail all of the onboard systems, even though all warnings were grouped together, because in some special cases, the pilot is required for some sort of resourcefulness prior to activation of the relevant emergency system. It must be remembered that this aircraft is relatively heavy so regardless of being so fast, the Vautour is not a fighter aircraft but a fighter-bomber and it has to be treated like a fighter-bomber.

Three Vautours will be Ready in a Few Days

Meanwhile, French procurement changes impacted the Israeli Vautour acquisition project. The French rejected re-equipment with the Vautour A and prioritized introduction of the Vautour N over the Vautour B; the former entered French service from the summer of 1957 and the latter followed from July 1958. Some of the already manufactured Vautour A aircraft – only 30 of this variant would ever be built – were issued to Vautour N and Vautour B squadrons, pending introduction of their own definite aircraft, as a stop gap for intermediate training.

Quite amazingly, at such a late point along the program's acquisition and introduction timeline, the IAF lacked a policy concerning the Vautour. AIR2 Chief Nathan Eldar issued the document "IAF Acquisition and Introduction of Vautour Aircraft" on 19 April 1957. This document presented a somewhat confusing state of affairs concerning numbers, types, squadrons, and bases. Contrary to HIGH TIDE numbers, AIR2 acquisition figures were 30 Vautours for purchase during IAF Year 1957 and IAF Year 1958, with "first priority for reconnaissance aircraft," even though France was not developing, at the time, a Vautour variant for reconnaissance. Vautour types, according to AIR2, were 15 Vautour A attack aircraft (including three "fitted for reconnaissance"), eight Vautour B bombers (including two "fitted for reconnaissance"), and seven Vautour N all weather fighters. These were intended for service with Squadron 115 (reconnaissance), Squadron 110 (attack), Squadron 69 (bombing), and

A Vautour pair – each aircraft armed with four napalm bombs under the wings – in flight in 1958; Ezekiel Somech recommended operating the Vautours in 2-ship or 4-ship formations. (PC)

Vautour weapons loads on display at Ramat David in 1958: bombs, napalm or rockets under the wings and bombs in the bomb-bay. (PC)

Squadron 119 (all weather), though this planned assignment was not yet finalized. Likewise, the bases from where the Vautours would operate were not decided yet, so construction of essential infrastructure for the new aircraft did not start.

Squadron 110 – supposedly waiting for its new aircraft since late 1956 and with a designated commander who had already converted to the Vautour – was not officially in existence at the time of issue of the AIR2 document. In an appendix to this document, Ezekiel Somech presented a number of dilemmas concerning IAF Vautour operations:

> Targets: In order to fully exploit the considerable weapons load, the Vautour should attack one large target or several targets. In case of attacking several targets, the aircraft will have to fly for sometime at low altitude in order to attack all targets so its radius of action and vulnerability will be compromised. In case of attacking a single target, it will be advisable to assign secondary targets in case the Vautour will not expend all of its weapons on the primary target.
>
> Attack Patterns: The sight is not of an advance type; it is hardly sufficient for gunnery, flat trajectory rocketry and low altitude bombing. The sight is not fit for curved trajectory rocketry or for high altitude bombing. This should have an impact on attack patterns and selection of weapons.
>
> Formation Size: The aforementioned issues, as well as the fact that the aircraft is fast and depends on its speed to evade enemy anti-aircraft fire while it is heavy and therefore will not be able to engage enemy aircraft at low altitude, only to evade them. Because of this reasons, the aircraft should not operate in large formations but in 2-ship or 4-ship formations. This will not be feasible if it will be decided to operate the aircraft for high altitude level bombing with a Vautour B as leader.

Guests approach the first Vautour to land in Israel; A13 is still in FAF markings immediately after arrival on 1 August 1957 to the Squadron 101 line (Mystere 36 and 80 on the right) at Hatzor. The tall figure below the nose of the Vautour is Air Department Chief Ezer Weizman, and beside him (in beret) are IAF Commander Dan Tolkowsky and (gray hair) Prime Minister David Ben-Gurion. (IDFA)

All Weather Operations: The aircraft has two engines, it is comfortable for flying at night, on instruments and equipped with efficient homing system. These qualities enable the aircraft to operate under all weather conditions at the home base, as well as at night, provided that the home base will have an approach and landing system and a surrounding with no obstacles. Nighttime operations are dependable on illumination of targets and are somewhat compromised due to lack of advance navigation system or a second crew member.

Interception: The aircraft climbs to altitude in a relatively short time and it is fast. Above 30,000 feet the aircraft maneuverability is better than all other aircraft currently in IAF service and the aircraft has long endurance capability. For these reasons, the aircraft can serve, in light of IAF current order of battle, as interceptor against bombers. The drawback of the aircraft as an interceptor is that it has two engines so start up procedure takes more time unless a special device for starting both engines simultaneously will be developed. The aircraft is heavy and may experience difficulties in engagement of enemy escort fighters.

Delivery dates of the first IAF Vautours were meanwhile rescheduled; Ben-Gurion was notified – on 2 May 1957 – that the "first three Vautours will be accepted by the end of this month". Those three Vautours did not arrive before the end of May 1957. A 3 June 1957 report from France indicated that delivery of the first Vautour (A12) was expected between 15 to 30 June 1957, with Vautours A13 and A14 following during July; henceforth, a delivery rate of one Vautour every ten days was expected for the balance of 14 Vautours, all from the production batch A15 to A30. Based on this report, Ben-Gurion was notified on 7 June 1957 that "three Vautours will be ready in a few days". IAF Commander Dan Tolkowsky notified the Minister of Defense on 21 June 1957:

The Vautour is currently being modernized and the next month, probably, we will receive our first three Vautours.

Additionally, the IAF Commander presented to the Minister of Defense the new IAF purchase plan that summed up at 24.65 million dollars, with the lion's share – 15.7 million dollars – allocated to Vautour acquisition: 14.7 million dollars for the purchase of 12 additional Vautours (including spare parts and ground handling equipment) and one million dollars for acquisition of spares for the already ordered Vautours. Two days later, the IAF requirement was accepted as an IDF decision "to purchase nine to 12 Vautours".

Flight V

The first Vautour for Israel departed France on 31 July 1957. The French requested confidentiality, so the ferry flight route was over North Africa. Previous ferry flights from France to Israel of Ouragans and Mysteres were routed via the north sector of the Mediterranean Sea, with stops in Italy and Greece. Moreover, the Vautour was flown under a cover story of a long range training flight from France to North Africa and back to France. Thus, the first leg of the journey east took the Vautour from the Sud Aviation plant at Saint Nazaire to Bizerte, Tunis. Following a night stop at Bizerte, the Vautour departed Tunis on the morning of 1 August 1957, supposedly flying back west to France, but actually heading east to Israel. Prime Minister and Minister of Defense David Ben-Gurion welcomed the first Vautour at Hatzor, as he reported in his diary:

In the Morning I departed to Hatzor where we waited for the arrival of the first two Vautours. When I arrived at Hatzor, and crossed the runway towards the control tower, the lone Vautour to arrive flew right over my head in terrific speed and horrifying sound. The second will follow by the end of next week. A pilot of ours flew the Vautour – Daniel Shapira.

The same day the first Vautour departed France (31 July 1957), AIR1 issued the directive "Squadron 115 Flight V" to put in order the custody of the grounded Vautours:

A Vautour flight will be activated, as an element of Squadron 115, from 1 August 1957. The flight will be known as Flight V. The flight will be based at Hatzor. Wing 4 will support the flight with all necessary services. Squadron 115 Commander will decide the division of squadron's manpower and logistics between the squadron's flights and within the overall standard of the unit. Correspondence to the flight will be addressed to Squadron 115. All correspondence concerning the flight will be classified secret.

2

SQUADRON 110 ACTIVATION

1957-1959

The second Vautour arrived in Israel on 13 August 1957 and was also stored at Hatzor. The secrecy of Israeli Vautour acquisition was well preserved, as a report from the British embassy in Paris dated 13 September 1957 clearly indicated:

> In his letter ... of 11 July the Ambassador referred to a report that nine Israeli pilots had arrived at the FAF base at Dijon. You may like to know that, on 16 August, the nine pilots were still at Dijon. They were all officers, the most senior being a Captain; they were using false names and had been issued with new identity cards by the Israeli Air Attaché in Paris. Although they had been flying Mystere 4As they had not been doing anything worth while with them or learning very much. The source was convinced that they were not at Dijon for this purpose. A possible explanation of the presence of these officers in France has occurred to the Air Attaché. He thinks that they may have come here to ferry back to Israel the ground attack version of the Vautour. About 15 of these have recently been built at St Nazaire, but it seems that the FAF already have no use for them. They might however be extremely useful to the Israelis, for instance in taking action against the forts guarding the Straits of Aqaba. A piece of supporting evidence is that one of the officers (a Lieutenant of German extraction) left for Tours, where the Vautour had started to arrive from the St Nazaire factory, about the middle of August. This is of course only a theory, but it seems sufficiently plausible and interesting to bring to your notice.

The British were correct that IAF pilots were present at Dijon, but they were Squadron 109 Mystere pilots on a six month assignment to the FAF and had nothing to do with the Vautour. There were six Israelis at Tours though, but they were Squadron 119 crews on assignment to the FAF to fly Meteor NF.11 night-fighters. A British report from 19 September 1957 clarified this issue, but was still in the dark concerning Israel's acquisition of the Vautour:

> Eight IAF Lieutenants (four crews) are present at Tours. Two crews are undertaking night-fighter training on the Vautour and the other two crews are doing likewise on Meteor NF.11s. It is rumored on the base that the Israelis will eventually take delivery of some Vautours.

Change of Plans

Originally, the IAF set priority for re-equipment of Squadron 115. For this reason the first two Vautours were delivered to Hatzor. As a matter of fact, the first three Vautour A attackers were allocated for conversion to reconnaissance, and Squadron 115 maintainers started a Vautour Conversion Course at Hatzor on 16 August 1957. Then the IAF revised its plans. The Vautour A reconnaissance conversion scheme was scrapped; instead, it was planned to convert to reconnaissance the Vautour B bombers. Vautour B deliveries were not expected to start before 15 July 1958, so suddenly it was possible to activate a Vautour attack unit sooner than planned.

FAF Tours based All-Weather Fighter Squadron 1/30 LOIRE Meteor NF.11 night-fighters photographed during training of Squadron 119 airmen in France from July 1957. This deployment involved two pilots (Ezra Aharon and Meir Livneh) and four navigators (Shivi Brosh, Benjamin Ashkenazi, Itche Erez and Uri Ofer); all six would become involved in the IAF Vautour program. (PC)

Seven Squadron 109 Mystere pilots were seconded to the squadrons of the Dijon based FAF Fighter Wing 2 (including Squadron I/2 CIGOGNES and Squadron III/2 ALSACE) from May to November 1957. Contrary to British reports they had nothing to do with the Vautour, but two of them – Gideon Magen and Levy Zur – would, eventually, become Squadron 110 Vautour pilots. From left to right: Haim Niv, Gideon Magen, Arie Dagan, Shaike Bareket, an IAF air control officer, Uri Dekel, Assaf Ben-Nun, and Levy Zur. (PC)

Squadron 119 flew Meteor night-fighters from Ramat David since 1956; in August 1957 it was decided to relocate 119 from Ramat David to Tel Nof in order to make room for Squadron 110 Vautour. This photo of two Squadron 119 crews was taken at Ramat David in May 1957. From left to right: David Yariv, Ethan Ayalon, Rahamim Melamed and Oded Erez. (PC)

On 27 August 1957 Air Department Chief Ezer Weizman chaired a discussion concerning revision of Vautour planning. The recommendation was to activate Squadron 110 as a Vautour attack squadron. To be based at Ramat David, Squadron 110 was to take over the facilities of Squadron 119; as a result of this revision of plans, Squadron 119 would move from Ramat David to Tel Nof in February 1958.

AIR1 issued, on 2 September 1957, a schedule for the activation of Squadron 110. It was planned to issue an Activation Order and a temporary standard on 13 September, to ferry the two Vautours from Hatzor to Ramat David on 15 September and to set plans for allocation of equipment and facilities, as well as assignment of pilots and maintainers. This tight timeline was optimistic and unrealistic. Though Vautour deliveries had already started, A-Level and B-Level equipment was not planned for delivery before February and April 1958, respectively. Furthermore, Vautour flights in Israel were still banned.

Hardened Aircraft Shelters

The most complex aspect of infrastructure for the new squadron was HAS. The IAF's ultimate objective was to operate all combat aircraft from HAS; the first such facility was operational at Hatzor since June 1957. Because of initial plans to base the first Vautours at Hatzor, the effort to plan and construct HAS for Squadron 110 at Ramat David started quite late along the activation timeline. The HAS was planned to protect the aircraft against enemy action and to enable continued operations under enemy attack; inside the HAS aircraft were serviced and armed. The advent of the Vautour resulted in a new set of requirements for HAS design, and these were addressed in an AIR2 chaired discussion on 12 September 1957.

The size of the Vautour triggered a new, larger HAS design with a higher ceiling (6.5 meters high to enable the jack up of the aircraft for maintenance) and wider taxiway (to enable maneuvering of the aircraft in and out of the HAS). Each HAS was to have an integral storage facility for rounds and rockets (a stock sufficient for two missions), while bombs were to be securely stored nearby the HAS at a safe distance. Adequate accommodation for maintainers was to be in the HAS so that continuous operations, day and night, would be possible. As in the Mystere HAS design, the Vautour HAS was to be equipped with a fire fighting system and comprehensive communication. A lesson from Mystere HAS design was an AIR3 requirement to equip the Vautour HAS with fuel pipes to enable refueling inside the HAS. A debate developed concerning the size of the HAS and whether it should be specific or universal. The Mystere HAS was designed to accommodate two aircraft, so the debate was whether to construct a specific Vautour HAS for one aircraft or a universal HAS that would be able to accommodate one Vautour or two Mysteres. The latter was to be more expensive and provide lesser protection to a Vautour because of its larger size, but IAF Commander Dan Tolkowsky ruled in favor of the universal HAS on 25 November 1957.

Squadron Standard

AIR1 issued a standard for the new squadron on 21 November 1957. Aircraft allocation was 12 Vautours, and assigned manpower was based on an anticipated activity of 200 flight hours per month. The standard included 14 officers (including 12 pilots) and 102 soldiers. The service persons were assigned to the squadron headquarters and to the Technical Flight, while pilots were divided between Flight A and Flight B.

Three officers and four soldiers were assigned to the squadron headquarters. The three officers were the squadron commander (a pilot), an adjutant (doubling as headquarters commander), and a training officer (a pilot). The four soldiers were a clerk in charge of the squadron office, a typist working in the squadron office, a unit sergeant major, and an intelligence expert.

Squadron 101 maintainers push a Mystere – armed with 250 kg bombs – out of a HAS at Hatzor. The first HAS at Hatzor - in fact, the first IAF HAS - was active from 26 June 1957. The HAS complexes for Squadron 110 were constructed of two cells, and each cell could accommodate one Vautour or two Mysteres. (PC)

The bulk of manpower was assigned to the Technical Flight, which had a standard of one officer (Technical Flight Commander) and 98 enlisted persons. The right hand of the TFC was a NCO in charge of the line and the NCO assistant. The principle bodies of the Technical Flight were maintenance and ordnance. The manpower of these two bodies was built around six teams, a team per two aircraft (actually a team per HAS, since every HAS was to accommodate two Vautours). Every maintenance and ordnance team was composed of seven mechanical maintainers, two electrical maintainers, and six ordnance specialists. Total assignment was therefore six team commanders (mechanical maintainers in trade), 36 mechanical maintainers, and 12 electrical maintainers with a NCO in charge of ordnance, plus 35 ordnance specialists. Four smaller bodies supported the six maintenance and ordnance teams: ground equipment, safety/emergency equipment, electronic maintenance, and logistics. Manpower assignment to these support bodies was quite complicated, with six Squadron 110 service persons and 14 Wing 1 Maintenance Squadron service persons. The former were assigned to ground equipment (four) and logistics (two); the latter were assigned to safety/emergency equipment (two) and electronic maintenance (12).

In those pre-computer days, a numerical difference between lists was common. The difference in the Squadron 110 standard between "manpower detailed list" and "manpower trades list" was 17 soldiers. In the latter list there were 50 mechanical maintainers (in the first list there were 42), 14 electrical maintainers (in the first list there were 12), 37 ordnance specialists (versus 36), and 14 others (two technical clerks, three logistics specialists, five communication technicians, and four ground equipment specialists) versus eight (NCO in charge of aircraft line, his assistant, and the six service persons assigned to ground equipment and logistics).

Ten pilots – all officers – were assigned to the two flying flights, with a flight commander and four pilots in each flight. Every pilot had a specific duty. Flight A Commander (later to be referred to as Deputy A) also functioned as Squadron Deputy Commander, and Flight B Commander (later to be referred to as Deputy B) also functioned as officer in charge of the squadron headquarters flying cell. Ten duties were specified for the eight pilots: readiness officer worked under the squadron commander and the deputy commander; two duties (operations officer and intelligence officer) in the flying cell; and seven duties (air safety officer, gunnery officer, navigation officer, air ordnance officer, air support officer, electronics officer, and physical fitness officer) in the training cell.

Set Up Team

The Vautour A succeeded the Mosquito fighter-bomber in IAF service in general, and in Squadron 110 service in particular. Three of the four senior members of the Squadron 110 Vautour

Eight maintainers service Vautour 12 in the Squadron 110 line at Ramat David during July 1958; the bulk of manpower in the squadron standard was assigned to the Technical Flight. (IDFA)

set up team were former Mosquito pilots: Squadron 110 Commander Ezekiel Somech, Flight B Commander Simon Ash, and Zeev Tavor. Flight A Commander Jacob Agassi was previously a Spitfire, Ouragan, and Mystere fighter pilot. The younger members of the set up team were Gershon Lapid, David Kishon, and Reuben Harel. All members of the set up team were combat veterans, having flown the Mystere (Harel), the Ouragan (Agassi and Kishon), the Meteor (Lapid), or the Mosquito during the 1956 Suez War.

Four Vautours were delivered from France to Ramat David from 26 October to 23 November 1957. The Vautours were well hidden, and the world was unaware that Israel had acquired such advance combat aircraft. The annual report of the British Air Attache, compiled on 1 December and delivered on 9 December 1957, stated:

> There has been no evidence that the IAF has taken delivery of additional aircraft so far this year, but it is believed that they have purchase some 15 Vautour ... aircraft from France. Whether these aircraft will be the bomber/ground attack version or night-fighters is not known definitely, but it is believed they will be bombers and if so they will be a potent weapon in the Middle East. Though these aircraft have not yet been delivered one has been seen in France bearing Israel national markings, so delivery cannot be far off.

There were no more places to hide the Vautours at Ramat David, so from the next ferry operation (SHULA 2 in December 1957) the Vautours were again flown to Hatzor, where adequate concealed storage space was available. Squadron 110 Technical Flight teams periodically ran engines and switched on systems of the brand new Vautours in order to prevent serviceability issues as a result of the imposed grounding. Meanwhile, the combat qualifications of the squadron pilots were preserved through flying in the neighboring Squadron 109 Mystere and Squadron 117 Meteor at Ramat David; Zeev Tavor accumulated 39:35 flight hours in Meteors from November 1957 until March 1958, and Reuben Harel flew 18:25 flight hours in Mysteres from January to March 1958.

Gunnery Test

Ezra Aharon was a Squadron 119 pilot undergoing training in France; a member of the Tours based group that the British Embassy in Paris reported on 19 September 1957. Aharon – who first flew the Vautour at Tours on 4 October 1957 (also in A6) – ferried Vautour A25 from Tours to Cazaux on 25 February 1958. Then Squadron 119 Commander Yoash Tsiddon flew Vautour A25 in a gunnery test from Cazaux on 4 March 1958. The allocation of ammunition for the test was 400 rounds, 100 rounds per gun; the objectives of the test were:

A. Experience ground equipment for handling of guns.

B. Evaluation of methods for zeroing of guns and sight.

C. Assessment of sight features during air-to-air and air-to-ground gunnery.

D. Determine whether abrupt maneuver of aircraft might cause jams due to movement of collected cartridges and links.

E. Assessment of scratches caused to aircraft inner structure from ejection of cartridges inside the airframe.

F. Ascertain optimal patterns for air-to-ground gunnery.

Zeroing and loading were demonstrated to the Israelis by a team from the FAF FTC. Tsiddon then took-off, climbed to 20,000 feet, and fired a short burst from all four guns, while flying straight and level, in order to collect cartridges and links for the first phase of the test. The first phase of the test covered gyro

sight gunnery while maneuvering at 20,000 feet. Then Tsiddon strafed a French FTC target that measured six meters wide and six meters high. Strafing passes were flown at a diving angle of 15 to 25 degrees and a speed of 400 to 520 knots. The structure holding the target was hit during the fifth strafing pass and the target collapsed to the ground.

The aircraft, the guns, and the guns pack were examined after landing. It turned out that the upper two guns only fired 26 rounds each and jammed, while the lower two guns fired all rounds (100 each). The pilot reported severe vibrations of sight during gyro gunnery and the disappearance of the diamonds (dots of light on the sight's projector that created a dynamic circle that represented the wing span of the target, therefore defining range to target) when firing long bursts. Otherwise, the test was extremely successful. Though the collected cartridges and links did not evoke any kind of malfunctions during abrupt maneuvering of the aircraft, the cartridges did scratch the inner structure, but these scratches were "not worrying" and the jamming of the two upper guns was attributed to switchology.

The French guns' zeroing instruments and ordnance handling equipment were considered well suited to task, but it was recommended to modify IAF ordnance handling equipment, to manufacture in Israel special equipment (such as a guns pack electrical test instrument), and to purchase only the French guns zeroing instruments. The Israeli pilot recommended solidifying the base of the sight, a recommendation that "will be explored" in the words of the manufacturer's representative at Cazaux.

Tsiddon recommended to adopt the French zeroing practice and to strafe a target on the ground with the fixed sight at a dive angle of 20 to 25 degrees and a speed of 500 knots. Pilots were advised to open fire from a range of 800 meters to the target, but were warned "one should not get too close to the target since the aircraft has a significant tendency to sink during recovery

Daniel Shapira flew five Vautours from France to Israel from 31 July to 20 December 1957, including three to Hatzor and two to Ramat David. (PC)

Ordnance specialists service the guns pack of Vautour 03 at Ramat David in July 1958. (IDFA)

from dive". As for gunnery results, 85 round hits were counted for a 67% success rate, a definitely impressive achievement that warranted the remark "the hits on the ground target were very good".

Start of Vautour Flights in Israel

There were 13 Vautours secretly stored in Israel on 1 April 1958, including nine at Hatzor and four at Ramat David. The following night, Egyptian Il-28 reconnaissance aircraft penetrated Israel and flash photographed key facilities, including Ramat David. This was a major failure of the IAF command and control system, as well as highlighting the fact that Squadron 119 only fielded three outdated Meteor NF.13 night-fighters. The IAF night-fighter force had to be bolstered, and this was accomplished on 4 April 1958 when the first four Vautour N night-fighters landed at Hatzor. Originally intended for storage in the same manner as the 13 Vautour A fighter-bombers, the Vautour N was now needed immediately to fill a gap in IAF operational qualifications. As a result, the ban on Vautour flights in Israel was lifted.

Vautour flights in Israel started on 7 April 1958, when Squadron 119 pilots Yoash Tsiddon and Ezra Aharon flew the four Vautour N night-fighters from Hatzor to Tel Nof, while Squadron 110 Commander Ezekiel Somech took-off from Ramat David in a Vautour A for a training flight. From 8 to 9 April 1958 the only three IAF Vautour qualified pilots in Israel – Somech, Tsiddon, and Aharon – ferried the nine Vautour A fighter-bombers from Hatzor to Ramat David. Ezra Aharon flew 4314 (Vautour A IAF Number 4314, or simply Vautour 14) with landing gear down and Vautour 17 from Hatzor to Ramat David on 8 April, then Vautour 15 (including a dive to Mach 1.14) on 9 April 1958.

The same day – 9 April 1958 – the members of the Squadron 110 Vautour set up team departed Ramat David on their solo flights: Simon Ash soloed in Vautour 20, Zeev Tavor in Vautour 11, and Reuben Harel in Vautour 08. Jacob Agassi recalled:

> What was our conversion like? This was one of the most beautiful experiences of this era, not good but beautiful was the fact that on many occasions we flew aircraft for the first time alone without previous flights with an instructor. Most astonishing of all was flying the Spitfire without previous dual seat instruction ... [Then] Mustang, Ouragan, Mystere, Vautour, Mirage and Skyhawk. This was one of the greatest experiences of my generation, seven types of aircraft in which our first flight was also our solo flight.

Vautour 11 taxis into the Squadron 110 line on 9 April 1958. (PC)

Five Vautours in the new Squadron 110 line on 9 April 1958. (PC)

Pilots assigned to Squadron 119 were also initially posted to Squadron 110 for qualification as Vautour pilots; the first of them was Ovadia Nachman, who flew his solo in Vautour 15 on 28 April 1958. Last to fly solo in the Vautour during this early time frame was Gideon Magen, who flew his solo in Vautour 11 on 5 May 1958. Magen replaced David Kishon as a member of the Squadron 110 Vautour set up team after the latter decided against three years service commitment as required from the Squadron 110 Vautour pioneers.

Vautour Seminar

In parallel with the start of actual service introduction, the IAF arranged a Vautour study day on 20 April 1958. The objectives of the Vautour seminar were to present the capabilities of the new combat aircraft through a series of lectures that covered various aspects of all three Vautour sub types: Vautour A for attack, Vautour B for bombing and reconnaissance, and Vautour N for all weather interception. The first lecture presented a general description of the Vautour A:

> This aircraft is a single seat, all metal, transonic fighter-bomber. Nearly all systems and fuel are contained inside the fuselage. The wings are swept back and the center section is glued rather than riveted. All controls are servo operated with mechanical back up (practically, only rudder and elevator), the flaps also move backwards – important for low speed – and are electrically operated. The landing gear includes two double main wheels in tandem, plus two balancing wheels under the engines, and it is operated by a hydraulic system and an electrical system.
>
> The hydraulic system composes of two mechanical pumps on the engines and an electrical emergency pump; a common reservoir feeds all pumps.
>
> The internal fuel is contained in flexible rubber tanks inside the top rear section of the fuselage and in two aluminum tanks in the wings, in front of the engines. Total weight of fuel is 4,160 kg; it is fed by compressed air. The bomb bay can contain one or two fuel tanks (1,500 liters each) that are fed by electrical pumps. There is a single reservoir tank with fuel input from all tanks and fuel output to the engines. The external wing tanks are 1,250 liters each.
>
> The electricity system operates all systems in the aircraft. There is a 28 Volt battery and two generators that are independent of the battery and each one can supply all required current.
>
> Two air compressors, one on each engine, are utilized to feed fuel, pressurize the cockpit and for air conditioning.
>
> The ejection seat is semi automatic; the cockpit canopy should be ejected separately. The ejection seat has a double action operation, a pull and then a push. The ejection load is 16 G. Maximum speed for ejection is 350 knots (though it is possible to eject at up to 450 knots). Overall,

Simon Ash flew his Vautour solo in 4320 on 9 April 1958. (PC)

> the seat delivers perfect safety above 3,000 feet; the safety above 800 feet is 70%.
>
> The engines are Atar 101E-3 with 3,600 kg thrust each. There are almost no malfunctions. The throttle is operated via a controller that enables unlimited throttle movements [because the controller mediates – through real time calculation of engine parameters – between the pilot throttle inputs and the actual impact on the engine]. The engines are relatively powerful for the airframe, the aircraft is overpowered.

Compared to the Mystere – the IAF cutting edge fighter at the time – the Vautour was superior, flying faster and turning quicker in altitudes above 30,000 feet. Lower than 30,000 feet the Vautour had a better initial acceleration, thanks to faster responding engines, but ultimately the Mystere was faster. In a very hot day climb contest, the Vautour was vastly superior to the Mystere, climbing 8.5 minutes to 40,000 feet, while it took the Mystere almost twice the time (16 minutes) and double the distance to reach 40,000 feet. The turning performances of the Vautour were also highly praised thanks to its tolerance to relatively high G loads.

The Vautour was very stable in high speed, so it was recommended to open gun fire in the speed range of 450 to 560 knots. Diving was possible at constant speed, without acceleration, so it was possible to fly over a target at 500 knots, pull up, and climb to 10,000 feet within half a minute. Various mission profiles, in different configurations, were presented. These included radius of action from 160 to 690 nautical miles, with configurations ranging from clean aircraft with internal ammunition only flying at low altitude to fully fuel loaded aircraft (two belly tanks and two wing tanks, with the latter jettisoned once empty) flying at high altitude. Thus, the Vautour successfully answered most of the operational requirements envisioned in the "IAF Plan 1954-1958 – Combat Aircraft" staff study. Moreover, the mission profile data presented during the Vautour seminar were based on the aircraft manual, while IAF experts felt that these presented performance on the conservative side of the scale:

> The actual range of the Vautour was the subject of a debate between the FAF, the manufacture and us. Our experience and calculations indicate that full fuel range is up to 1,540 nautical miles and that this can be stretched up to 1,740 nautical miles, if external tanks are dropped after emptied.

As already indicated, the FAF rejected re-equipment with the Vautour A. The few Vautour A aircraft that entered French service were utilized for training of pilots in squadrons destined to operate Vautour N fighters and Vautour B bombers. When the N and B models were introduced, the A aircraft were relegated to other duties. As a result, the IAF had to conduct weapons testing in the Vautour A. These tests were flown in France, and were aimed to qualify the Vautour A for gunnery, rocketry, and bombing. Five issues were evaluated:

> A. Gunnery with all four guns (firepower equivalent to eight Meteors).
>
> B. Vibrations of sight.

Maximum fuel capacity of a Vautour in 1958 was 10,650 liters, hence the IAF requirement for more, larger tankers. (PC)

The Vautour was refueled via a single point in the front undercarriage bay; the ladder belonged to Maintenance Team 4. (PC)

The guns pack weighed 760 kg; the four 30 mm guns could fire some 90 rounds per second. (PC)

The undercarriage configuration - two main tandem units in the fuselage and two balancing wheels retracting into the outer side of the engine bays - is clearly evident in this view, as well as the doors of the fuselage bomb-bay that stretched from the front undercarriage bay to the rear. (PC)

C. Strafing air-to-ground.

D. Rocketry (wings only).

E. High speed bombing from bomb-bay.

Results of air-to-ground strafing were poor at speeds below 450 knots. Results were "a lot better" when strafing at speeds between 500 to 550 knots. The firepower of the guns pack – weighing 760 kg fully loaded – was highly praised, with a rate of fire of 1,400 rounds per minute per gun and at least 2.5 hits per square meter of the target. As for vibrations in the gyro sight, it was reported that "in line with our requirement, the French agreed to produce a new base that will stabilize the sight".

As for rocketry, it turned out that the IAF standard rocket pods were unfit for the Vautour because the front covers were unable to withstand the stresses that were generated in speeds higher than 500 knots.

In a bombing test with six 400 kg bombs in the bomb bay, the Vautour flew with ease even when the bombs were exposed, but the aircraft sunk 80 to 100 feet immediately after release of the bombs, a dangerous phenomenon when bombing from low altitude. A positive property of the Vautour that was assessed to improve bombing results was very good stability while flying through bumpiness, a property attributed to the flexible wings. Maximum speed for closing of bomb-bay doors was 480 knots. Therefore, if bombing was flown at a higher speed – "even up to Mach 1" – the pilot had to slow down to 480 knots and only then could close the bomb-bay doors. In a bombing test with bombs released at a speed of 460 knots from an altitude of 10,000 feet, all bombs hit the ground within a rectangle 130 meters long and 90 meters wide.

From a weaponry point of view, the Vautour was considered equivalent to three Mysteres. The Vautour was armed with four 30 mm cannon with 100 rounds per gun (the Mystere had two 30 mm cannon), and could be loaded with four rocket pods or four 500 kg bombs under its wings (compared to two rocket pods or

Squadron 110 maintainers in front of Vautour 11; Norik Harel was Technical Flight Commander until 1961. (PC)

two bombs in the Mystere), as well as 112 rockets or six 250 kg bombs or nine 120 kg bombs in the bomb-bay (against 55 rockets in a belly magazine in the Mystere).

The size of the Vautour, its capabilities, and systems challenged the IAF maintenance and logistics echelons. Starting the Vautour's engines and taxiing consumed 200 kg of fuel that took 25 nautical miles off the range of a pair leader, and even more from the range of a Vautour leading larger formations. The powerful engines exhausted hot gases that damaged the asphalt coating then in use at IAF air bases. An intermediate solution was the gradual opening of the throttles, but this obviously resulted in a longer take-off run, so the ultimate solution was to re-equip with concrete runways.

Refueling the Vautour required larger tankers than those in service with the IAF. It was recommended to avoid parking the tankers in the sun, because even a small reduction in the density of the fuel resulted in a significant range reduction in the order of 40 nautical miles. While heat impacted fuel density on the ground, exposure to intense cold – down to minus 78 degrees Celsius at altitudes higher than 40,000 feet – could actually freeze fuel; the freezing temperature of the fuel was minus 60 degrees Celsius, but the fuel did not freeze because it was pumped for constant flow.

The impressive weaponry options of the Vautour had an impact on the number of ordnance specialists. Preparing a guns pack for action demanded a team of two ordnance specialists working for over 30 minutes and utilizing special equipment. The Vautour bomb-bay was reckoned more advanced than the Canberra's bomb bay and larger; it measured five meters long, one meter high, and one meter wide. However, unlike IAF standard practice, bombs had to be loaded into the bomb bay after installation of detonators because it was impossible to fit detonators to bombs after loading. Special bomb trailers had to be used to load the bomb-bay. A change of bomb-bay configuration, from bombing to rocketry, took more than 90 minutes, with another 40 minutes plus to load the 112 rockets.

Many of the advance features still suffered from immature technologies, but this fact would only be revealed along the Vautour service timeline. The first hint was a hunch that too much depended on electrical systems:

> The electricians are the key to almost all aspects of Vautour maintenance because nearly all of the aircraft systems are electrical. The mechanics are almost jobless in the maintenance of the engines but they have a lot of work in the fuselage. The French instruments are difficult for maintenance, unlike the American instruments in the Super Mystere, and demand a knowledgeable officer in charge.

Improvements

The Vautour had only recently started flying in Israel, and initial operational capability was yet to be achieved, but discussions concerning improvements had already started. Propositions for improvements were discussed in an AIR2 meeting on 11 May 1958; these focused on fuel, electronics, ordnance, and structure.

Special emphasis was placed on the Vautour's range performance. Therefore, it was recommended to purchase 2,000 liter belly tanks instead of the 1,500 liter belly tanks, for a high altitude range gain of 100 nautical miles or an endurance increase of four minutes in combat at low altitude. Additionally, it was agreed to evaluate adaptation of Super Mystere 1,300 liter external fuel tanks to the Vautour instead of the latter's 1,250 liter tanks; this was calculated to add seven nautical miles to high altitude range or 0.75 minutes to combat endurance at low altitude. It was also recommended to increase the number of tankers to be allocated to the Vautour refueling task.

As for the electronic suit, it was recommended to install indigenous rear hemisphere warning radar in the Vautour A (the French developed an equivalent system, but obviously did not install this in their few, non-operational Vautour A aircraft), to replace the radio compass with an improved unit capable of reception of two stations simultaneously and to evaluate installation of Decca 7 and Decca 10 navigation equipment if the IAF would decide to purchase such systems.

Vautour weaponry was – as already mentioned – a major leap forward compared to all other IAF jet fighters. It turned out that the IAF standard bomb trailers were unfit for the Vautour; utilizing these trailers for loading the Vautours with bombs necessitated the jack-up of the tail section. Therefore, it was recommended to purchase a single purpose built Vautour bomb trailer and to manufacture additional trailers in Israel. To resolve the issue of rocket pods – those in IAF stock could not withstand the stresses in speeds higher than 500 knots – it was decided to manufacture new covers that would be able to withstand such stresses. The impact of gunnery on jet engines was already well known, and the Vautour was no exception. In many cases, hot gases from the guns entered the engine inlet and changed the mixture of air and fuel to the point of the engine cutting out; the most common cure was to automatically cut fuel flow to the engines during gunnery. In the Vautour – while flying at high speed and at high altitude – gunnery created a shock wave that resulted in the engine cutting out. The French had already evaluated – during November 1957 – a modification of a shock wave breaker that was installed over the nozzles of the guns, but this modification was not added to previously manufactured aircraft. Ultimately this was not to be considered a problem for Squadron 110, which was not expected to open fire while flying in high speed at high altitude.

Structural modifications discussed included undercarriage and wing profile. The undercarriage was recommended to be strengthened to support a maximum weight of 21 tons, instead of 18 tons. The explanation to the wing profile change proposal was "currently, the Vautour is unable to efficiently engage the MiG-17 because it is unable to sufficiently slow down". Again, the wing with convex leading edge was a feature that was already tested in France. Tests in France yielded that the convex leading edge modification improved low speed performance – as well as high altitude maneuverability (an improvement of 0.5 G in turns at 40,000 feet) – at the expense of slight degradation in high speed performances (less Mach 0.01 in maximum speed and a minute and a half more in climb to 40,000 feet). Therefore, it was recommended that the Material Department study the issue of the convex leading edge and file a report for forwarding to AIR2.

Vautour 18 and Vautour 03 in the line; some aircraft, like 18, had their number painted higher, on the fuselage above the guns pack, while others, like 03, had their number painted lower, half on the fuselage and half on the guns pack. (PC)

The Vautour was the first IAF aircraft equipped with a braking parachute. (PC)

Vautour B Deployment

Squadron 110 was not planned to operate the Vautour B. The IAF planned that the first three Vautour Bs would be modified locally as reconnaissance aircraft to re-equip Squadron 115 Mosquito. Subsequent deliveries of Vautour B bombers were planned to revive Squadron 69 B-17 Flying Fortress. With deliveries of Vautour B scheduled to start in July 1958, the timing was ideal for Air Department Chief Ezer Weizman to challenge this plan.

The discussion process was launched on 29 May 1958, when Air Department Chief Ezer Weizman issued the document "a platform paper for Vautour B deployment discussion". The document reminded readers that Vautour B introduction was planned in two phases. The first phase was introduction of six Vautour B aircraft – including three for photography – from 1 July 1958 to 1 July 1959. The second phase covered an objective of 12 Vautour B bombers in service, in addition to the Vautour B reconnaissance aircraft. The view of the Air Department Chief was that this procurement policy would justify activation of a Vautour B bomber squadron in due time, when enough aircraft would be available. For the meantime, the Air Department Chief proposed to avoid activation of a separate Vautour B squadron.

Since this was the concept that the Air Department Chief promoted, the discussion paper ignored the question where to base a Vautour B squadron, but presented arguments for and against assignment of the few first phase Vautour B aircraft to either one of the already established Vautour squadrons.

Arguments for assignment of the Vautour B aircraft to Squadron 119 at Tel Nof were:

A. Vautour B aircraft modified for reconnaissance will be permanently based at Tel Nof [this was the IAF dictum until then].

B. Many Vautour B and Vautour N missions are overlapping, especially nighttime interdiction.

C. Better flexibility in control and rotation of aircrews between bombers, reconnaissance aircraft and all weather fighters.

The arguments for assignment of the Vautour B aircraft to Squadron 110 at Ramat David were:

A. Operation of Vautour B aircraft, especially in interdiction missions at night, in cooperation with Vautour N aircraft is not a rational utilization of the two types of aircraft. This is because the main mission of the Vautour N, especially due to the small number available, is interception at night while, for as long as the installed bombing sight is of inadequate quality, the Vautour B will be more efficient in daytime bombing. For this reason there is more mission overlapping between Vautour A and Vautour B with the latter able to improve on the former in navigation and in bombing of runways for neutralization of enemy air bases.

B. Squadron 119 only recently started to introduce the Vautour N so it is not desirable to add to the squadron responsibility for such issues as bombing or nighttime interdiction while the core activity of the unit is still being developed.

C. Assumption that reconnaissance aircraft should be based at Tel Nof, closest to AHQ, is not to be taken for granted. It is reasonable to assume that the efficiency of the Vautour force will be enhanced if all Vautour aircraft will be based in a single air base, namely Ramat David. This will also benefit the Ramat David photo lab and it should be remembered that even today there are reconnaissance aircraft based at Ramat David: one of the recce Mysteres and all recce Meteors.

The shear weight and volume of the three arguments for basing the Vautour B aircraft at Ramat David, against the telegraphic nature of the three arguments for basing the Vautour B aircraft at Tel Nof, obviously derived an expected bottom line:

Vautour 17 initially had its number painted in a smaller, different style. (IDFA)

Vautour 17 after application of the Ramat David standard font. (IDFA)

> Following consideration of the two options, it is recommended to adopt the option of basing the Vautour B aircraft at Ramat David as an element of Squadron 110. This deployment should also include the aircraft planned for assignment to Squadron 115.

Vautour B Conversion

The Vautour B was developed as a bomber "capable of delivering a nuclear bomb", as explained in the aforementioned Vautour seminar (and as hinted in the aforementioned Prime Minister Ben-Gurion diary entry on 16 July 1956). The principle difference between the Vautour B and the Vautour A was the configuration of the nose section, which contained the guns pack in the Vautour A and a navigator/bombardier in the Vautour B. Additionally, the Vautour B was equipped with a new design of an all flying, monobloc elevator. Other differences between the Vautour A and Vautour B were the inability of the Vautour B to operate under wing rocket pods and the lack of a gyro sight in the Vautour B, which only had a fixed sight for the pilot. Electronic systems also differed; the Vautour B was equipped with a bombing sight, a radar altimeter, and a Decca navigation system. The then current Vautour B bombing sight, the Norden 14, was an intermediate solution. The Norden 14 limited the Vautour B to bombing from altitudes ranging from 10,000 to 25,000 feet and in a certain speed range. Obviously, the French were developing a new bombing sight for the Vautour B, but at the time it was not known when this new bombing sight would enter service.

Three experienced navigators were assigned to the Vautour B project as early as 5 February 1958: Achikar Eyal, Samuel Kislev, and Uri Talmor. All three of them served as Mosquito navigators with relevant background in photography; Eyal even commanded Squadron 115 Mosquito reconnaissance prior to his assignment to the Vautour B project.

The first phase in training navigators for the Vautour B was a bombardier course in the FAF Navigation School at Training Base 702, Avord. The navigators flew in the Dassault 311 – a navigator/bombardier training aircraft – and flying started on 10 April 1958. Achikar Eyal accumulated 56:40 flight hours until 4 June 1958.

The second phase in Vautour B training was assignment to Bombing Training Center 328 at Cognac. Bombing Training Center 328 was officially inaugurated on 26 September 1957 – as IAF Attaché in France Josef Keidar reported on 21 October 1957 – to initiate FAF Vautour B service introduction. The training center included two squadrons and operated B-26 Invader, Vautour A, Vautour B, and Fouga Magister aircraft for training of Vautour pilots and navigators selected for assignment to French frontline Vautour B units, the first of which was activated on 1 March 1958.

Pilot Zeev Tavor joined the three navigators during their transition from Avord to Cognac. Zeev Tavor was also a veteran Mosquito aircrewman with considerable experience as a reconnaissance pilot. Tavor departed Israel for France in June 1958, and first flew at Cognac on 30 June 1958 (navigation exercise number 1, 2:15 flight hours in a B-26 Invader). First flight for Tavor at Cognac in a Vautour A followed on 2 July 1958, and his first flight in a Vautour B was accomplished on 18 July 1958. Obviously Achikar Eyal did not fly the Vautour A, so his first Vautour flight at Cognac was in a Vautour B on 6 August 1958. Zeev Tavor recalled:

> I was deployed to France for a bombing course with navigators; our objective was to master the Vautour B. I departed Israel with very little knowledge in French and I improved my French during my stay in France until I spoke fluent French.
>
> The navigators did not travel with me, they were already in France. I travelled alone to Cognac and was told that the navigators were on vacation; they finished their course and split up, every one of them travelled with his family for a vacation in a different region. I waited for them at Cognac and I was worried, how will they find such a remote place? They arrived right on time.
>
> I was already a qualified Vautour pilot but I joined the course with new Vautour pilots so the first phase was ground school and type conversion. During this time the navigators flew with other pilots. I also attended an instruments flying course in the Fouga because they regularly flew in clouds and I had no instruments flying rating. A little bit of exercises in the B-26, a few flights in the Vautour A and only then I started flying the Vautour B. The Vautour A and the Vautour B were

The line of training B-26 bombers at Cognac in the summer of 1958. (PC)

Achikar Eyal flew his first flight in a Vautour B on 6 August 1958 in 609 with French pilot Lt Raimbault; Zeev Tavor first flew Vautour B609 on 29 August 1958 with Israeli navigator Samuel Kislev. (PC)

Navigator Samuel Kislev (left) and pilot Zeev Tavor (right) flew three flights together in Vautour B607 from 25 August to 23 September 1958. The badge of Bombing Training Center 328 was a bomber vulture, a bird of prey loaded with two bombs. (PC)

Achikar Eyal flew eight sorties in Vautour B607 from 14 August to 11 September 1958; overall Achikar Eyal accumulated 35:20 Vautour flight hours, 9:05 Fouga flight hours, and 24:30 B-26 flight hours at Cognac. (PC)

The four IAF Vautour B conversion students on the day of graduation, from left to right: Achikar Eyal, Samuel Kislev, Zeev Tavor and Uri Talmor. (PC)

> almost exactly the same aircraft except for the B having a navigator in the nose; otherwise, there was almost no difference. We exercised bombing but the aircraft was outdated and the bombing sight was antiquated. Small smoke bombs were loaded onto two wing stations; we flew straight and level at high altitude and dropped these bombs.

The Israeli pilot only started flying with the Israeli navigators during the advance stages of the course: with Samuel Kislev from 29 July, with Uri Talmor from 19 August, and with Achikar Eyal from 8 September 1958. Most of Tavor's sorties during the advanced stages of the course were flown with Kislev: 26 out of 33 Vautour B sorties that Tavor flew from 29 July to 14 October 1958. Overall, Zeev Tavor accumulated 53:45 flight hours in a Vautour, 7:10 flight hours in a Fouga, and 11:00 flight hours in a B-26 during the conversion to the Vautour B at Cognac.

REBECCA 3 and ARIELA 1

Pending achievement of Squadron 110 Initial Operational Capability, the Vautour pilots continued to preserve their operational qualifications through flying Mysteres: Squadron 109 allocated – on 23 March 1958 – 70 flight hours in April, 150 flight hours in May, 150 flight hours in June, and 40 flight hours in July 1958 to the Squadron 110 pilots.

The qualification of the Squadron 110 set up team members as Vautour pilots enabled their participation in the ferrying of new aircraft from France to Israel. Such participation was viewed as excellent training in long range operations. First participation of Squadron 110 set up team pilots in the ferry of Vautours from France to Israel was during Operation REBECCA 3. The objective of REBECCA 3 was to deliver to Israel two Vautour N night-fighters and two Vautour A fighter-bombers. Assigned pilots were Squadron 119 Commander Yoash Tsiddon and Squadron 110 pilots Ezekiel Somech, Simon Ash, and Reuben Harel. The four flew from Tours to Tebessa, Algeria, on 20 June 1958. Tsiddon and Somech flew the N fighters (with Squadron 119 navigators Shivi Brosh and Itche Erez, respectively), while Ash and Harel flew the A attackers. Reuben Harel recalled:

> Tebessa is at an altitude of 2,500 feet so we had to take-off very early in the morning when it was supposed to be relatively cool. But it was very hot so we were not sure if the [heavier] Vautour N could make it [take-off safely]. There were no manuals so Yoash Tsiddon decided that he would stand near the end of the runway and watch the take-off of the two Vautour A aircraft. From our lift off point he would evaluate if the Vautour N could follow. Today, after 36 years of flying as an El Al airline pilot, this approach seems to be so amateur but at the time we simply did not have adequate data.
>
> Unfortunately there was a sirocco at Tebessa. Sirocco is a hot air blowing north from the Sahara and there we were at three o'clock in the morning facing a temperature of 35 degrees Celsius. Yoash stood by the end of the runway, we took-off and he decided that they would not make it. We pressed

Yoash Tsiddon (right) and Shivi Brosh (center) prior to departing from France in Vautour B616, IAF Number 4231. (PC)

Vautour B616 immediately after the accident at Tel Nof, still with FAF markings. (PC)

Vautour B616 some time after the accident at Tel Nof, with FAF markings already removed. (PC)

Not a pleasant position to be in during a crash landing! The navigator's compartment in the nose section of Vautour B616 after the accident at Tel Nof on 26 July 1959. (PC)

> ahead while they had their aircraft fuel tanks not fully fueled. Then they flew to Bizerte on the coast at Tunis, a lower elevated airport a bit closer to Israel and not impacted by the sirocco. They refueled to the maximum and followed us to Israel.

The next ferry operation (ARIELA 1) also included two Vautour A aircraft for Squadron 110, as well as the seventh and final Vautour N for Squadron 119 and the first Vautour B, with its assignment still undecided. The four Vautours flew from Tours to Bizerte on 25 July 1958 and continued to Israel the following day. Daniel Shapira and Reuben Harel landed the Vautour A at Ramat David, while the Vautour B and the Vautour N headed to Tel Nof. The Vautour N – flown by pilot Jacob Agassi and navigator Itche Erez – landed safely, while the Vautour B – flown by Yoash Tsiddon and Shivi Brosh – was damaged in a landing accident due to a technical malfunction that was related to the monobloc elevator. This unfortunate accident actually sealed the prospects of the Vautour B entering service with Squadron 115, as originally planned.

Lifting the Curtain of Secrecy

Initial Operational Capability was accomplished towards the end of the first training term. At the time, the IAF Year started in April and was divided into three training terms: Term 1 from April to July, Term 2 from August to November, and Term 3 from December to March. All training terms ended in a combat day exercise; Squadron 110 Vautour's first combat day exercise was on 9 July 1958. The combat day exercise started as a squadron level simulation of full scale operations, evolving over the years into a wing or base level exercise and eventually into a full force IAF level maneuver. Combat day exercise placed special emphasis on very fast turnaround. Turnaround is the time frame from return to base at the end of a mission to take-off on the next mission; during this time frame the aircraft is serviced, fueled, and armed. The fast turnaround was the IAF's answer to the enemy's numerical superiority. As a force multiplier, the fast turnaround enabled a small force to generate more missions per day than a much larger force. During their first combat day exercise, the Squadron 110 Vautour pilots flew four missions each; the average duration of a mission was 0:55 flight hours.

The curtain of secrecy that shrouded the IAF's introduction of the Vautour was lifted in the wake of Operation ARIELA 1, during which deliveries to Israel of Vautour A and Vautour N aircraft were completed. The day before the arrival of the ARIELA 1 Vautours, the Vautours took part in the IAF command handover parade fly-past on 25 July 1958. A day after the arrival of the ARIELA 1 Vautours, the new IAF combat aircraft was exposed to military attachés during their visit to Hatzor on 27 July 1958. The official public debut of the Israeli Vautours was during the IAF Day fly-past over Beer Sheba on 12 August 1958. Six Vautours (including Reuben Harel in 4303 and Gideon Magen in 4312) flew in formation, while Daniel Shapira demonstrated aerobatics in a lone Vautour.

Fuel Leaks

Aerobatics demonstration in the Vautour was an indication of how the IAF perceived its new attack aircraft. The Vautour was the most advanced combat aircraft in IAF service from arrival in Israel during August 1957 until the start of Super Mystere deliveries in December 1958. Still, IAF fighter pilots viewed an assignment to the Vautour as beneath one's dignity when compared with assignment to the Mystere or the Super Mystere. Obviously, this attitude only spurred Vautour pilots to squeeze

The 17th and last Vautour A for Squadron 110 was A26 – IAF Number 4326 – which arrived at Ramat David on 26 July 1958; the aircraft was photographed shortly afterwards with its French style "26" number painted over and prior to painting of the Ramat David style "26". (IGPO)

The Squadron 110 line in July 1958; present Vautours are 12, 09, 07, 03 and 18. Reuben Harel flew 03 and Gideon Magen flew 12 during the IAF Day fly-past on 12 August 1958. (IDFA)

the maximum, and more than the maximum, out of their aircraft during training air combats versus Mystere and Super Mystere pilots. Jacob Agassi recalled:

> The Vautour was a big aircraft with two powerful engines and we loved it a lot; in a way we loved it like one loves a retarded child. We, the veteran fighter pilots assigned to the Vautour from fighter squadrons, initially sensed inferiority in comparison with fighters. Some of us were former Mosquito pilots, but we, fighter pilots assigned from the Mystere, strove to transform the Vautour into a fighter. As a result, we maneuvered in an extremely aggressive manner. We exceeded, not once and not twice, the maximum permissible G load and fuel leaks started almost immediately.
>
> The sealing of the fuel tanks was not perfect. The wings contained integral fuel tanks and, since sealing was not good, fuel leaks started to the point of fires and a few very serious safety incidents. We had to close fuel tank number 7 that were in the wings above the engines so we had aircraft with tanks 7 and without tanks 7. When we practiced air combat, a Vau-

Vautour A18 – IAF Number 4308 – in the Squadron 110 line during August 1958; fuel tanks 7 (right and left) were integral tanks in the wings between the engines and the fuselage. (PC)

Squadron 110 Technical Flight Commander Norik Harel (in cockpit) and crew chief Gdalya Admon (on the wing) service a Vautour on the line at Ramat David. (PC)

> tour without tanks 7 had an edge over a Vautour with tanks 7. To play fair fight, the Vautour with tanks 7 took-off first to burn fuel from tanks 7.

Wing 1 Maintenance Squadron Commander Johanan Bar-Sela reported the fuel leak issue as early as 19 August 1958. The integral tanks 7 were in between the wing roots and the engines, and suffering from excessive bending and torsion loads. The inner surfaces of these tanks were covered with sealant, but the combination of excessive loads and immature technologies of integral fuel tanks and sealing techniques resulted in a major problem of chronic fuel leaks.

Corrosion

Engineering Research Unit 230 employee Immanuel Bracha visited Ramat David on 16 April 1958 to inspect corrosion on the outer skins of the Vautours. The aircraft were almost new, yet they suffered from quite substantial corrosion. Fitting corrosion was noted in several aircraft, but the inspector was unable to determine if the skin was fully penetrated to the extent of actual holes. The proposed treatment – in light of previous experience with the Mystere – was cleaning and repainting the aircraft. The recommendation was accepted, as MAT2 Chief Peretz Posner reported on 12 June 1958, but repainting was expected to be a prolonged process that did not start immediately:

> For the meantime, and because these aircraft are especially vulnerable to corrosion, we emphasized many times, during our visits to the units, the necessity of often cleaning of the aircraft skin. The objective of cleaning is to minimize skin corrosion due to dust, dirt and the impact of dew.

For the time being, pending repainting, various anti-corrosion treatments were evaluated. MAT2 Materials Officer Adam Siton reported, on 15 October 1958, an experiment that started at Ramat David on 13 October 1958, when Vautour 14 was treated with a coating "that provides good anti corrosion protection (galvanized corrosion included) and is an excellent primer for painting. The protective coating has a golden hue and its thickness is 0.00001". Its weight is negligible and it has no impact on the aerodynamic smoothness of the aircraft surfaces".

The treatment process included covering of all openings, covering of surfaces that may be adversely impacted by the coating (such as wings' gluing materials), thorough cleaning of outer surfaces, washing of cleaning materials, treatment with corrosion removal materials, and application of the protective coating. MAT2 evaluation was that a team of four could treat an aircraft in two days.

Following this experiment, it was decided to similarly treat all Vautour A aircraft. Wing 1 Maintenance Squadron Commander Avner Halivni reported to Wing 1 Administration Squadron Commander on 23 October 1958:

> The Vautour A aircraft are suffering from severe corrosion. In the wake of several experiments, MAT2 ordered to remove corrosion with chemical fluids. This work requires two actions each consuming three working days per aircraft (overall, six working days). Since we do not have sufficient manpower for this work that is outside the scope of normal B-Level maintenance, I request an assignment of four workers (they can be non professional workers) for two months.

Vautour Conversion Course 2

The intake of Vautour Conversion Course 2 was six Mystere pilots: three from Squadron 101 at Hatzor (Jonathan Shahar, Maoz Poraz, and Oded Marom) and three from Squadron 109 at Ramat David (Eli Bahat, Oded Einav, and Simi Saar). The objective of this intake was to double the number of Squadron 110 regular pilots and to achieve the squadron standard of 12 regular pilots. Additionally, two Squadron 119 pilots (Adam Zivoni and Meir Livneh) also attended the IAF Year 1958 Term 2 Vautour Conversion Course. Jonathan Shahar recalled:

> I was in mourning after being informed of my assignment to Squadron 110. At the time, I was a Squadron 101 pilot and Squadron 101 was THE interception squadron of the IAF. During this time, the Vautours used to practice air base attack missions against Hatzor and we used to watch these simulated attacks. Among the Squadron 101 pilots watching the Vautours was Benjamin Peled, at the time an engineering student in a university and a Squadron 101 emergency posting pilot, and he told us that he was the first IAF pilot to fly the Vautour and what an awful aircraft it was. Indeed, the Vautours pulled up [from low altitude approach] to 8,000 feet or so – the Mystere was unable to pull up so high and usually pulled up to 3,000 feet – and started to pull out of the [subsequent simulated bombing dive] much earlier than the Mystere because, Benjamin Peled told us, of its handling difficulties.
>
> I was so happy in Squadron 101 and I did not understand this assignment, why me? But an order is an order so I reported to Ramat David. I was shocked from our reception at Squadron 110; it was a totally different attitude than my previous experience at Hatzor; an absolutely different air force. Squadron 110 Commander [Ezekiel] Somech greeted us 'welcome to 110' and then he told us what we were going to do.
>
> The first surprise was two nighttime flights every week; at Hatzor we used to fly two nighttime flights in a term! The first of these two flights per term was refreshment and the second was high altitude navigation. The objective that Somech set for Squadron 110 was to fly at night exactly like flying during daytime. At the time, this was a revolution in the IAF.
>
> Then Somech detailed the missions of the squadron. Back in Squadron 101 we practiced air combats, navigation and air base attack. Here, in Squadron 110, we were to practice air base attacks, low altitude navigation, long range flights, air support over land, maritime strike at sea, nighttime attacks and low altitude navigation at night. This was all new to me.
>
> Then we converted to the Vautour. Deputy B Simon Ash was in charge of training and every morning briefing started with an examination; this was also new in the IAF. We learnt all of the emergency procedures so every morning he entered the briefing room, wrote in a chalk on the board

Meir Livneh flew his Vautour solo in 23 on 8 August 1958. (PC)

Jonathan Shahar in the cockpit of a Vautour; Shahar flew his Vautour solo in 03 on 8 August 1958. (PC)

Perfect aiming: the aiming dot (pipper) in the center is right on target and the diameter of the "diamonds" circle – representing range to the target – exactly matches the wingspan. (PC)

the title of one emergency and we had to write down the full procedure, this was the morning test. Thus we started the working day that ended exactly at 17:00. This was also an innovation that Somech set. The Squadron 110 working day started at 07:30 when we were picked up from the mess to the squadron and ended at 17:00. We did not work at night except for nighttime flights. In other squadrons work continued into the night but Somech set limitations that normalized our routine. For me, this was a totally different approach and I really liked it.

The aircraft itself was huge but it was fun to fly. All you had to do to take-off was to open the throttles and to maintain heading on the runway; the Vautour took-off on its own because it was already set on its undercarriage in the correct incidence for take-off, the pilot did not have to pull up the nose. Handling was very pleasant and to our surprise Somech, Agassi and the rest of the pilots treated the Vautour in the same manner as a Mystere, this bomber was flying air combats and was fighting very well! Obviously, the Vautour was unable to fight a Mystere on equal terms because the

A Vautour in air combat training as photographed via a sight camera; the Vautour's sight camera was used for assessment of air combat training, but not for BDA, because of severe vibrations when the guns fired. (PC)

Mystere's radius of turn was smaller. Thus we learnt to fight air combats in pairs and it was Squadron 110 that developed the IAF concept of fighting air combats in pairs; coordination of pairs compensated for our disadvantage in one versus one air combat against a Mystere.

As a guns platform, the Vautour was the best aircraft I had flown; it was like a sniper rifle. The aiming point of the guns was along the roll axis of the aircraft so the Vautour made it easy to hit the target if you flew it correctly and Squadron 110 turned into a unit of snipers. Likewise concerning bombing, the Vautour was an extraordinary attack platform. Within a short while we all became addicted to the Vautour.

One of the subjects that we learnt during conversion was tracking and ranging. One aircraft was turning at a bank of 45 degrees and the practicing pilot was supposed fly passes, starting to photograph from 800 meters range and closing down to 200 meters range. The pilot had to utilize the stick in order to maintain the pipper, the aiming dot, right on target while at the same time to constantly change the range selector so that the circle of diamonds would match the wing span of the target. This was a basic maneuver but in the Vautour we practiced tracking and ranging at an altitude of 40,000 feet. Nobody else practiced tracking and ranging at an altitude of 40,000 feet, the Mystere could hardly climb to 40,000 feet, but for us this was evidence to the potential of the Vautour that could practice air combats at 40,000 feet!

A unique conversion course flight was maximum range with internal fuel; the flight plan was from Ramat David to a point nearby Rhodes and back. The technique was constant RPM so the aircraft climbed and slowed down as fuel was consumed and the aircraft became lighter and lighter. We started to cruise at 45,000 feet and I reached a maximum altitude of 58,000 feet! We were unaware that a pressure suit was a must for safe flying above 50,000 feet but the experience was overwhelming, it was a different world up there, the sky were black and we could see Turkey on one side and the Nile Delta on the other side. As lack of pressure suit was not enough, I used to fly without gloves, it was more comfortable. In this flight my right hand held the stick but my left hand was idle until it was time to change RPM for the descend to landing. I then placed my left hand on the throttle and my hand was glued to the throttle, it was so cold. I suffered severe frostbite and ever since then I always flew with gloves.

Riveting Tool

Vautour Conversion Course 2 flights started with two flights in Squadron 117 Meteor trainers to practice general handling of a multi-engine jet aircraft in the first sortie, and to master single engine handling characteristics of such an aircraft during the second Meteor sortie. Exercise 1 was the first flight in the Vautour, and Exercise 2 to 14 followed; those 13 flights covered practice of general handling, ground control approach to landing, single engine handling, navigation, and formation. Tactical training then started with a series of tracking and ranging exercises. It was during this phase that Vautour Conversion Course

Vautour 05 that Eli Bahat landed with a jammed elevator on 1 September 1958. (IDFA)

A Vautour – 4309 – under tow along the line, past Vautour A22. (IDFA)

2 student Eli Bahat was forced to declare an emergency while flying Vautour 05 on 1 September 1958. Eli Bahat recalled:

> I flew my Vautour solo flight on 8 August 1958, exactly three years after my day of enlistment, and accumulated some 20 flight hours in the Vautour by 1 September 1958 when I took-off to practice air combat versus my friend Jonathan Shahar. It was in the middle of this practice that my elevator jammed. Luckily, I was flying at medium altitude and I managed, very gently, to level the aircraft with the ailerons and the elevator trim tab. I declared an emergency and started to receive a lot of advises over the radio. The Vautour's stall characteristics were problematic [and for this reason Squadron 110 pilots were banned from flying slower than 160 knots] and everybody were worried if the elevator trim tab would be effective down to landing speed, someone even suggested that I should eject.
>
> I pressed ahead with my approach to Ramat David, descending and slowing in a very gentle manner. I planned and flew a long and flat final, cautiously controlling speed with the elevator trim tab while extending flaps and undercarriage. I flew over the edge of the runway a bit fast but my landing was perfectly normal. Emergency vehicles waited for me but were not needed; I deployed the braking chute and taxied to the line where everybody waited. Then Technical Flight Commander Norik [Harel] arrived with a wooden hammer and started to pound, gently, over the skin of the empennage until, suddenly, we heard the sound of a blow and the elevator was freed! The aircraft was towed to the [Wing 1 Maintenance Squadron] hanger where someone, somehow managed to pull out of the rear fuselage a manufacturer riveting tool [that was forgotten in the fuselage during production].

At first maintainers baptized pilots after a solo with a bucket of water; then a bath was added and pure water evolved into a sticky and stinking mixture of water and oils. The depicted fortunate maintainers are baptizing IAF Commander Ezer Weizman after his Vautour solo on 11 December 1958. (PC)

There is No Room for a Seperate Reconnaissance Squadron <u>At All</u>

The first Vautour B for Israel was flown from France to Tel Nof because at the time, the IAF planned to assign the first three Vautour B aircraft to the Tel Nof based Squadron 115. The fact that this Vautour B was damaged during landing at Tel Nof actually torpedoed its assignment to Squadron 115. This unfortunate incident played into the hands of the Air Department Chief, whose proposition to assign the Vautour B aircraft to either Squadron 110 at Ramat David or Squadron 119 at Tel Nof was not discussed yet. Additionally unfortunate for the plan to re-equip Squadron 115 with the Vautour B was the fact that the very day that 4231 was damaged was also the first full day in office of the new IAF Commander, Ezer Weizman, who took over from Dan Tolkowsky on 25 July 1958.

IAF Commander Ezer Weizman promoted the proposition of Air Department Chief Ezer Weizman concerning the assignment of the Vautour B aircraft. Since 4231 was damaged upon arrival, it did not enter Squadron 115 service as planned. The new deadline for a decision concerning assignment of the Vautour B aircraft was therefore the return from France of the pilot and three navigators undergoing Vautour B conversion. Naturally, IAF Commander Ezer Weizman accepted the recommendation of Air Department Chief Ezer Weizman. The decision to assign the Vautour B aircraft to

IAF Commander Ezer Weizman right after his Vautour solo flight on 11 December 1958. Weizman initiated the process to assign the Vautour B to Squadron 110. (PC)

The Vautours operated from a line pending completion of construction of HAS complexes; the "wartime" drill – prior to introduction of HAS – was to disperse the aircraft around the base, and this was practiced during combat day exercises. (PC)

Regrettably, at the time, tight Israeli security included removal of aircraft numbers from censored photos; this Vautour is possibly 05. (PC)

Squadron 110 at Ramat David – as well as construction of two HAS at Ramat David instead of at Tel Nof as previously planned – was made on 25 September 1958.

AIR3 Chief Moshe Peled saw fit to challenge this decision on 9 October 1958. Peled presented three heavyweight arguments to support his preference of the original IAF plan to re-equip the Tel Nof based Squadron 115 with the Vautour B aircraft:

> There is no debate that those Vautour B aircraft that will be modified for reconnaissance will be all assigned to a single unit. To achieve maximum efficiency at the shortest time frame, it is a must to activate a special unit with a commander fully committed to the promotion of this one subject. We do not know yet enough about how to operate the Vautour B but I assume that operating the Vautour B will be different than operating the Vautour A. If we will add Vautour B aircraft to a Vautour A unit then it is only natural that the attention of the commander will not be the same as in a dedicated Vautour B unit.
>
> Our planned emergency deployment (excluding Vautour B) is 62 jet aircraft at Ramat David (Squadron 109, Squadron 110 and Squadron 117) and 57 aircraft at Tel Nof (Squadron 116, Squadron 119 and Squadron 140) but only 15 of the latter are jet aircraft. Therefore, while Ramat David will be quite crowded during emergency deployment, Tel Nof will be relatively roomy. From this perspective it will be preferable to base the Vautour B aircraft at Tel Nof.
>
> The current number of HAS at Ramat David is five with three more under construction. There are four HAS under construction at Tel Nof. This situation – significantly more HAS at Ramat David than at Tel Nof – impacts adversely our flexibility in deployment and the rational of dispersal between our three air bases that – for promotion of our requirement for a fourth air base – we present as already suffering from significant crowding.

The arguments of the AIR3 Chief were indeed significant and concerned the heart of the matter: promotion, density, and dispersal. However, the analysis of AIR3 Chief was based on a past plan that enough Vautour B aircraft would be purchased to re-equip Squadron 115 (reconnaissance) and Squadron 69 (bombers). This plan was no longer valid once the new IAF Commander entered office; Ezer Weizman was a staunch supporter of multi-role combat aircraft, which he viewed as superior to single role aircraft.

In his answer to AIR3 Chief – dated 11 October 1958 – the IAF Commander stated "I do not believe that there will be budget available for acquisition of additional Vautour B aircraft". This statement scrapped the prospects of separate Vautour B squadrons – one for bombing and one for reconnaissance – or even a single separate Vautour B squadron for bombing and reconnaissance. Moreover, the IAF Commander made it clear that his view of an independent reconnaissance squadron was:

> There is no room in the IAF for a separate reconnaissance squadron AT ALL.

In line with this directive of the IAF Commander, pilot Zeev Tavor and navigators Achikar Eyal and Uri Talmor were assigned to Squadron 110 upon their return from France in November 1958, while navigator Samuel Kislev was assigned to AHQ with emergency posting to Squadron 110. Then the next delivery of Vautour B aircraft was also impacted from the change, with the second and third Vautour B aircraft planned for landing at Ramat David rather than at Tel Nof, as did the first Vautour B for Israel.

AIR3 issued an order for Operation ARIELA 2 on 27 November 1958. The order assigned to the ferry operation two pilots (Ezekiel Somech and Zeev Tavor) and two navigators (Samuel Kislev and Uri Talmor). As usual, preparations for the ferry included a long endurance flight to test the functionality of the fuel system; Zeev Tavor and Achikar Eyal flew 3:35 flight hours in Vautour B626 on 20 December 1958. Three days later – on 23 December 1958 – the two aircraft (Ezekiel Somech and Samuel Kislev in Vautour B624 and Zeev Tavor and Achikar Eyal in Vautour B626) were flown from Cognac to Istres (0:45 flight hours) prior to a direct ferry flight from France to Israel; 3:20 flight hours from Istres to Ramat David. Zeev Tavor recalled:

> We flew along a straight line and over Greece we experienced an interception attempt. I looked around while we were cruising side by side; suddenly I saw two aircraft down below on our left side. These were, I think, F-86 Sabres that climbed towards us. However, this aircraft flew really high and we were cruising at high altitude. I reported to Somech, we opened the throttles, started to climb even higher and they were unable to reach us.

Operational Capability

The first operational border crossing mission of Squadron 110 Vautour was flown on 29 October 1958. It was a visual reconnaissance over Lebanon and Syria, and Reuben Harel flew Vautour 20 for 1:30 flight hours in line with Mission Order M74 (the 74th Mission Order that AIR3 issued during month M, October 1958). At the time of this flight, the Vautour was the most advanced combat aircraft in the Middle East; in certain flight regimes the Vautour was actually superior to the IAF Mystere or the Arab MiG-17, the two types of fighters then forming the cutting edge of the Egyptian, Israeli, and Syrian air forces.

The squadron's level of activity was also accelerated during Exercise FOCUS, a wing level combat day exercise on 12 November 1958. The objectives of FOCUS were to exercise:

> Transition to Readiness State C and to emergency deployment.
>
> Mobilization of reserves.
>
> Maximum tempo of operations with armed aircraft.
>
> Commanding the wing from the wing command post.
>
> Delivery of fuel, ordnance and services to dispersal locations around base.
>
> Transmissions of intelligence reports from the air.

Three ranges were assigned to the exercise and ammunition allocation for Squadron 110, which covered 2,600 rounds, 52 250 kg explosive bombs, and 52 250 kg sand bombs. A total of 210 reserve days were to cover mobilization of 70 Squadron 110 reserves (mostly maintainers) for three days, and the flying day was in the middle of this time frame. Flying was to start in a preplanned wave after the activation signal of FOCUS; subsequent missions were to be flown in line with mission orders. All sorties were to be flown with ordnance, and in addition to live ordnance attacks over the ranges, the pilots were authorized to fly dummy attacks – on their way back from the ranges – against Megiddo field, St Jean field, and Plugot field. Hatzor based Squadron 101 Mysteres were to be on readiness for scramble to intercept the Ramat David aircraft as an exercise element.

The core of the combat day exercise was still fast turnarounds, and the Squadron 110 pilots flew up to five missions during FOCUS. This was a timely preparation, as four days later fire erupted along the Israeli-Syrian border. The exchange of fire peaked on 3 December 1958 and the IAF prepared for war. AIR3 issued Operation Order RUTH, aimed at achievement of air superiority through air base attack missions over Egypt and over Syria; both were elements of the United Arab Republic. Although the full complement of 17 Vautour A aircraft was delivered to Squadron 110, there were only 12 Vautour combat qualified pilots in the squadron. Therefore, AIR3 included only eight Vautours in the Operation RUTH task assignment program.

Tension along the Israeli-Syrian border did not deteriorate to war, but emphasized the importance of the IAF as a readily available regular force with a very short time frame for transition from peacetime to wartime, as well as constant readiness for execution of an air superiority campaign. The Vautour was the most qualitative attack aircraft of the IAF, and still the most modern combat aircraft in the region. This enabled the IAF to offer pilots a unique opportunity for first hand acquaintance with enemy air bases that they were expected to attack in wartime. Based on the performance edge of the Vautour, the IDF authorized to the IAF a series of full moon night flights over enemy air bases in Egypt and Syria. Gideon Magen flew such a mission on the evening of 31 December 1958, flying along the Suez Canal in Vautour 20 for accumulation of intimate knowledge of potential targets prior to return to Ramat David after 1:40 flight hours.

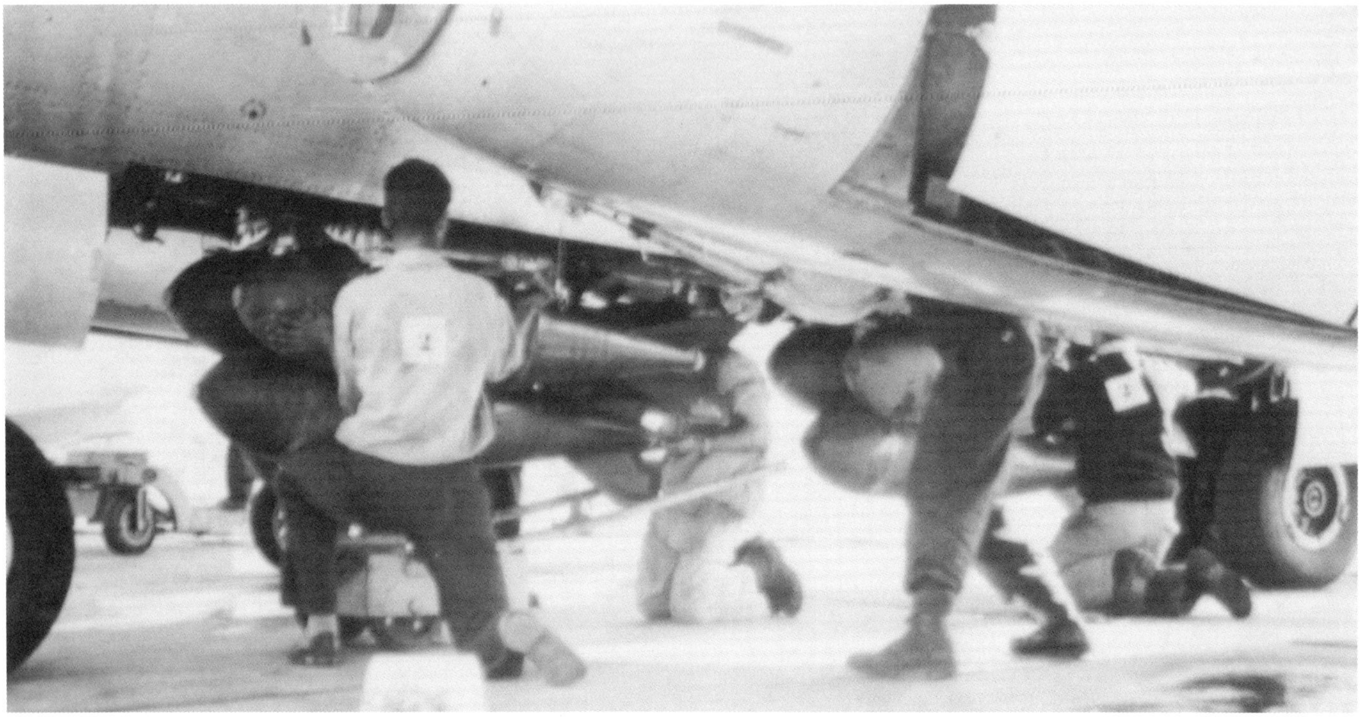

Squadron 110 maintainers load the bomb-bay of a Vautour during fast turnaround, a cornerstone of the combat day exercise. (PC)

Ordnance specialists prepare a guns pack to action; watching the drill, left to right, are Yossi Sarig and Zeev Tavor. (PC)

New Badge

The motif of the original Squadron 110 Mosquito badge was a bat superimposed on a triangular cloud hovering over a group of small stars. The bat represented fighting at night with radar equipped aircraft, for at the time of activation, the planned wartime mission of Squadron 110 was as a night-fighter unit. The cloud stood for flying in all-weather conditions, and the stars – representing navigation aids – were a tribute to the navigators.

The badge of Squadron 110 Mosquito, which Squadron 119 Meteor adopted so it was no longer available to Squadron 110 Vautour. (PC)

Oded Einav in his workshop, welding a metal structure for a sculpture inspired by the badge that he designed for Squadron 110 Vautour back in 1958. (PC)

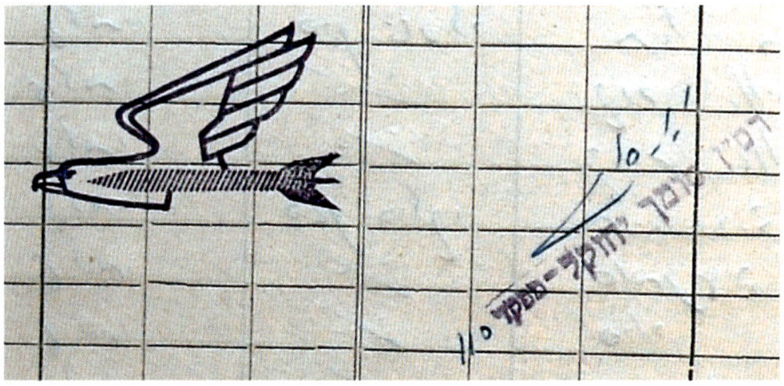

The original version of the Squadron 110 Vautour as stamped in a 1958 logbook beside the signature of Squadron 110 Commander Ezekiel Somech, who added "110" in handwriting; a "Squadron 110" stamp was later added. (PC)

Oded Einav finishes the concrete sculpture that he prepared to represent the badge of Squadron 110 Vautour. (PC)

The finished sculpture - that Oded Einav prepared, to represent the badge of Squadron 110 Vautour - at Ramat David, opposite Vautour 26. (PC)

The regular Squadron 110 Mosquito was inactivated in September 1955, and the reserve Squadron 110 Mosquito was closed in April 1957. During this time frame, Squadron 119 was activated as the IAF dedicated night-fighter unit. In an unprecedented act – that has not been repeated ever since – an active (regular) IAF squadron (Squadron 119) took over the badge of a semi-active (reserve) IAF squadron (Squadron 110).

When Squadron 110 Vautour was activated it no longer had a badge, since Squadron 119 was using the badge of Squadron 110 Mosquito. Oded Finav recalled:

> Squadron 110 Commander Ezekiel Somech asked the artists among us to propose a badge. My proposal was selected. It depicted a vulture carrying a bomb.

A new badge was produced; the motif of this new badge was a white bird of prey "loaded" with a red bomb over a blue circle as a background. The new badge may have been inspired by the badge of the FAF Bombing Training Center 328, which was a bird of prey "armed" with two bombs. The IAF approved the new Squadron 110 badge on 17 November 1958.

Reconnaissance Gap

The Vautour B aircraft arrived from France without reconnaissance gear, except for a K-24 BDA camera with 12" lens. Photography Technical Services Unit 121 expert Arie Yaakobi inspected Vautour 31 after arrival at Tel Nof and concluded that the K-24 BDA camera had no practical value as an imagery intelligence collection tool. Yaakobi estimated that from an altitude of 30,000 feet, the camera would produce photos to cover an area 3.8 km wide and 150 km long, but at a scale of 1:30,000, which was considered inadequate for recognition of aircraft types. From 40,000 feet the expected coverage was 5.0 km wide and 200 km long, but at an even lower scale of 1:40,000.

The damage that Vautour 31 sustained during landing at Tel Nof resulted in a major IAF reconnaissance gap. The Squadron 115 Mosquito reconnaissance aircraft were retired in August 1958, and the IAF reconnaissance squadron was deactivated in line with the aforementioned view of the new IAF Commander. To bridge this operational gap, IAF Commander Ezer Weizman issued a requirement on 1 October 1958 to fit Vautour 1,300 liter external fuel tanks to a reconnaissance Mystere in order to enable the short legged Mystere to venture as far away as Luxor, Egypt. This proposed stop gap solution was problematic. Technically, the ground clearance between the surface and the lowest point of the Vautour 1,300 liter tanks, when fitted to the Mystere, was only 18 cm. From an operational perspective, the reconnaissance gear installed in the reconnaissance Mystere was inferior to the larger and heavier cameras that were installed in the reconnaissance Mosquito and planned for installation in the reconnaissance Vautour. Finally, the overall performance of the reconnaissance Vautour were superior to that of the reconnaissance Mystere; especially in long range mission profiles.

Originally, the IAF planned to acquire enough Vautour B aircraft to support the existence of Squadron 115 for reconnaissance and Squadron 69 for bombing. As already mentioned, the new IAF Commander challenged this planning, probably in line with his ambition to equip the IAF with 100 Mirage interceptors as multi-role combat aircraft.

Initially, on 11 October 1958 the IAF Commander ruled "there is no room in the IAF for a separate reconnaissance squadron AT ALL," while also stating "I do not believe that there will be budget available for acquisition of additional Vautour B aircraft". Four months later, on 3 February 1959, the IAF Commander finished off the brilliant future foreseen for the Vautour B in IAF service in a letter to the IDF Chief of Staff:

> The aircrews returned from conversion to the Vautour B in France. They report that the efficiency of the Vautour B, with the currently installed bombing sight, is in doubt. Bombing accuracy from 20,000 feet is 200-500 meters. Such ac-

Yossi Sarig flew his Vautour solo in 23 on 24 November 1958; this Vautour Conversion Course was the first for Flying School instructors assigned to Squadron 110 in emergency posting, and that conversion course also included Yoel Dan and Zeev Sharon. (PC)

A child looking at the high Vautour 05 in the squadron line. The "05" has been repainted higher, on the fuselage only, whereas previously the "05" was painted lower on both fuselage and guns pack, as seen in the July 1958 photo mentioned above. Gideon Magen flew Vautour 05 during the squadron's 8-ship fly-past on 26 January 1959. (PC)

curacy is useless for our operational plans on which we based our decision to purchase this aircraft.

The French continue in the development and testing of a proper bombing sight but we do not know when this process will be completed. There can be little doubt that this sight will produce good results.

Therefore, I suggest that we will not go ahead with acquisition of the balance of two Vautour B aircraft. We will focus in the first four Vautour B aircraft as reconnaissance platforms and will trade the two Vautour B aircraft with one Vautour A and one Vautour N. In case such a trade will result in a budget gap, it is possible to add to the transaction our French reconnaissance Mystere and to return it to France as it is non standard equipment for the IAF.

If, sometime in the future, a radar bombing sight will fulfill our requirements, then it will be possible to reconsider planning of the Vautour B element in the IAF.

The IDF Chief of Staff endorsed the recommendation of the IAF Commander and authorized conversion of all four Vautour B aircraft as reconnaissance platforms against previous plans to modify the first three Vautour B aircraft for reconnaissance and operate the follow up aircraft as bombers.

Convex Leading Edge Wing

Super Mystere deliveries to Israel started in December 1958, and the Vautour lost its status as the most modern combat aircraft in IAF service. The Super Mystere, a third generation (supersonic) jet fighter, entered service with Squadron 105 at Hatzor as the new IAF interceptor. Squadron 105 initiated a process towards achievement of Initial Operational Capability as the IAF top interception unit, while Squadron 110 was already approaching Full Operational Capability as the IAF's leading attack element. Regardless of this seemingly absolute difference, the inner contest between the two most capable combat aircraft squadrons was indeed intense. Jacob Agassi recalled:

The Super Mystere entered IAF service in December 1958 with actual flying from February 1959; this photo grouped together the IAF frontline at the time, from left to right: Meteor (operated by Squadron 117 based at Ramat David), Ouragan (Squadron 113 at Hatzor), Mystere (Squadron 101 at Hatzor and Squadron 109 at Ramat David), Super Mystere (Squadron 105 at Hatzor), and Vautour. (IDFA)

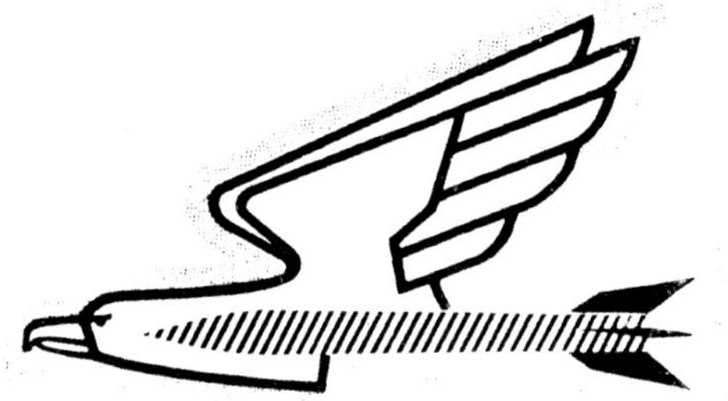

The invitation to the first anniversary evening of Squadron 110 Vautour on 26 January 1959; the program included "speeches; artists; surprises; and the IAF Orchestra".

In our struggle to be fighter pilots, we used to our advantage the Vautour's Fowler flaps; flaps that extended aft of the wing's trailing edge as well as dropping down. In air combat training, when the aircraft slowed down and started to shudder, we used to deploy the flaps. The aircraft stopped shuddering; we gained extra lift and we had a huge advantage in climbing at slow speeds. On the other hand, the French never concluded the Vautour's spin tests so we banned our pilots from slowing down to below 160 knots; a limitation that had an adverse impact on the ability of the aircraft to fully exploit its maneuverability potential. Undisciplined pilots who used to slow down below 160 knots immediately had an edge over disciplined pilot who did not slow down to below 160 knots. When training air combats were against other types of aircraft, such as the Super Mystere, we used to ignore the 160 knots limitation and we slowed down to very dangerous speeds.

In an attempt to improve maneuverability and performance, France employed Vautour A4 to test a couple of innovations: the all flying monobloc elevator and a wing with convex leading edge.

The monobloc elevator was aimed to rectify lack of handling sensitivity in high Mach numbers and shuddering in transonic flight. The phenomenon resulted, respectively, in abrupt nose pitch up during recovery from dives and in high G loads that adversely impacted the structure of the aircraft. Therefore, the Vautour – with the standard non monobloc elevator – was limited to Mach 1 at 30,000 feet and to Mach 0.95 at altitudes below 30,000 feet. The monobloc elevator resolved these issues and was installed during manufacture in all Vautour B aircraft from Vautour B611 onwards and in all Vautour N aircraft from Vautour N346 onwards.

The convex leading edge was aimed to improve the stability of the Vautour in slow speed flight. As indicated, the convex leading edge was proposed for evaluation to enhance the maneuverability of the Vautour – in order to match the slow speed maneuverability of the MiG-17 – during the Vautour seminar on 20 April 1958. Prior to this, on 13 March 1958 Squadron 119 Commander Yoash Tsiddon reported from France that "Mr Fischel [French test pilot] stated that Vautour A4 equipped with convex leading edge and monobloc elevator has no limitations at all in maneuverability".

Almost exactly a year later, the French authorized IAF test pilot Daniel Shapira to fly a single test flight in Vautour A4, the monobloc elevator and convex leading edge wing test aircraft.

Repainted Vautour 25 at IAI, the Vautour D-Level contractor throughout IAF service. Zeev Tavor flew 25 from Ramat David to Lod on 25 February 1959 and the aircraft was back at Ramat David by 12 March 1959. (PC)

Repainted Vautour 20 at IAI; Zeev Tavor returned 20 from Lod to Ramat David on 5 May 1959. (PC)

Daniel Shapira reported on 8 April 1959 eight subjects that he evaluated during his flight in Vautour A4:

A. Climb to 40,000 feet in nine minutes.

B. Testing of maximum G and gunnery limitation (the speed in which the aircraft starts to be unstable for accurate gunnery). This test included four turning dives in varying speeds from Mach 0.93 to 0.76. Shapira concluded that an aircraft with convex leading edge has a slight advantage in maneuverability and a slight improvement in stability during roll.

C. The aircraft accelerated from Mach 0.75 to 0.91 – at an altitude of 40,180 feet – in two minutes and 16 seconds. Shapira figured that this acceleration test yielded a loss of Mach 0.15 to 0.2 compared to Vautour with regular wing.

D. Engine performance was tested, at certain altitude and given weight, in a set of speeds and G loads, without a change of altitude. Shapira felt this test yielded no difference between a Vautour with regular wing and a Vautour with convex leading edge.

E. Ailerons efficiency was tested in a dive from 40,000 to 31,000 feet, with acceleration. The right wing started to drop at Mach 0.93 and at Mach 0.95 air brakes were deployed to end the test. Shapira re-

Vautour 15 was photographed at Hatzor during a Wing 4 combat day exercise on 2 March 1959, hence why the Squadron 101 Mystere 92 in the background is wrapped in a camouflage net. (PC)

Superb view of Vautour 20 at IAI on 1 May 1959 with an amazing selection of aircraft in the background: Beech 18, Noratlas, Mosquito, Harvard, Spitfire, C-47, Consul, B-17 and Instructor with, just behind the Vautour, a Noratlas airlifter unloading an Alouette helicopter. (PC)

ported that the wing drop was less significant in an aircraft with a regular wing.

F. Stability in slow flight was tested with landing gear up and down. With undercarriage in the up position, shuddering started at 168 knots but control remained efficient; severe shuddering started at 140 knots as well as instability in roll; vertical nose oscillations started at 139 knots at which point the French asked Shapira to stop, in order not to risk the aircraft, though the Israeli pilot was told that French pilots managed to slow down to 125 knots in such a test. With undercarriage in the down position, slight shuddering started at 128 knots, severe shuddering started at 120 knots and instability in pitch and roll, coupled with shuddering, started at 115 knots at which point the French asked Shapira to stop, again informing the Israeli pilot that French pilots managed to clock 107 knots during such a test. Shapira summarized that 'this facts indicate a considerable improvement in the handling qualities of the aircraft'.

G. Attack was tested in three passes at 540 knots. Shapira had the impression that the monobloc improved handling but stability was not improved in comparison to a Vautour with regular wing. Later on, Shapira tested acceleration in dive from 10,000 to 1,000 feet up to 570 knots; a speed that was ten to 15 knots slower than achieved in a Vautour with regular wing.

H. Landing was accomplished at 140 knots over the runway edge with comfortable handling and stable aircraft. Shapira evaluated that it was possible to land at 135 knots over the runway edge in order to minimize the landing roll. However, Shapira was unable to detect differences in landing behavior between a Vautour with convex leading edge and a Vautour with a regular wing that landed at 145 to 150 knots over the runway edge.

After this flight, Daniel Shapira was told that the French planned to install the convex leading edge in all of their Vautour aircraft, including aircraft that were already manufactured. As for the monobloc elevator, the French plan was to equip all aircraft still to be produced, but not to modify already manufactured airframes. Additionally, the French presented to Shapira their own flight test data concerning the monobloc elevator and convex leading edge. The bottom line, from both French flight test data and the impression of the Israeli test pilot, was generally similar: improved stability at slow speeds, enhancement in high altitude (above 45,000 feet) performances and stability, and a minor maximum G load improvement. Shapira summarized:

> The major improvements are: better maneuverability at sonic speed and at slow speeds with forward center of gravity, correct longitudinal stability throughout the speed range, good control over elevator movements that can – in certain cases at high speed – prevent oscillations, slight gain in maximum G load. These benefits degrade when the aircraft is loaded with external stores due to shift backwards of the center of gravity.

First Air Combat

The IAF discussed air combat tactics versus the MiG-17 on 12 February 1959. This discussion was in the wake of several Mystere versus MiG-17 engagements and the ongoing introduction of the Super Mystere. The MiG-17 was treated as a respectable foe, superior to the Mystere and equipped with a number of advanced systems, such as an afterburner and a rear hemisphere radar warning receiver. Compared to the Vautour, the MiG-17 had a lower wing load and a higher thrust to weight ratio (though only with afterburner on and against a Vautour weighing more than 12 tons); as already mentioned, the Vautour climbed to 40,000 feet in six minutes or so, while it took the MiG-17 only four minutes to reach a similar altitude. The weaknesses of the MiG-17 were instability at high speeds and a fuel limitation of two to four minutes engagement with afterburner on in local conditions. Therefore, the IAF recommendation was to avoid low and slow engagements, to avoid competition in the vertical dimension, and to avoid combat at high altitude. The only remaining envelope region was high speed, low altitude engagement, but this was not an IAF official recommendation because of the following reservation:

> The alleged difficulties of operating the MiG-17 at high speeds are not proven and should not be trusted. Therefore, it is recommended to engage the MiG-17 under such conditions in which flying qualities will have the most decisive impact.

The IAF was confident that the quality of the Israeli fighter pilot would prevail, even against a superior adversary, as the MiG-17 was. Yet another factor was the balance of forces in the air during an engagement. The Arabs flew in 4-ship formations, including scrambles for interception, while the IAF standard formation was a pair. Most previous engagements involved an IAF pair versus an Arab 4-ship, but the Arab standard procedure was to split the 4-ship formation into two pairs at the start of the engagement; this tactic actually enabled the IAF to achieve a local numerical balance of a pair versus a pair. The IAF's dilemma for future engagements was whether to switch to a 4-ship formation or depend on the Arabs to continue this split tactic. In case of the latter, the next dilemma was to split the IAF pair for two parallel single versus pair engagements, or to preserve the IAF pair for a single pair versus pair engagement. The IAF preference was to allow leaders to decide in light of "the number of enemy aircraft, the tactical situation, the enemy's behavior, the weather and the proficiency of the Israeli wingman". Nevertheless, the IAF recommendation was to avoid a split of the pair:

> In other circumstances the pair should not split and engage in close formation or in open formation. The consideration in the selection of the formation type will remain in the hands of the leader according to the development of the combat but mostly in light of the proficiency of his wingman. Anyway, the basic formation will remain close formation until the leader will decide otherwise.

Reuben Harel flew Vautour 03 during the 7 June 1959 engagement versus Egyptian MiG-17s, when he fired 60 rounds; note the Super Mystere dummy in the background. (PC)

This discussion was not academic. Air space violations were commonplace in those days. Ground control systems lacked in coverage, and most violations – be it IAF aircraft venturing into Arab territories or Arab aircraft incursions into Israeli air space – did not result in any kind of engagement. Time and space issues mostly denied IAF interceptors even the slightest chance of an interception. Arab incursions were almost invariably so brief that flying time of scrambled interceptors from Hatzor or Ramat David to the sector of violation was longer than the actual endurance of enemy fighters over Israeli territory.

Such a scenario surfaced on 7 June 1959. Ground control tracked an incursion of Egyptian aircraft, but the scrambled Super Mysteres were unable to intercept the intruders. At the same time, a Squadron 110 pair – Reuben Harel in Vautour 03 and Moshe Ofer in Vautour 25 – were flying an interdiction training sortie. Reuben Harel recalled:

> We were in the sector of Halutza Gvulut during a routine training flight but even during such training flights we had our guns loaded and our pack of four guns had an awesome firepower. Suddenly, ground control notified us about a possible target, enemy aircraft that intruded Israel from the Gaza Strip. We were vectored north-west, low altitude, 1,000 or 2,000 feet.
>
> And so it was. From quite a distance we saw four aircraft, possibly MiG-15, and we engaged them; trying to open fire and shoot them down. First we tried to get into position right behind their tails but this was not a piece of cake; the Vautour maneuvered very well – the official maximum G was 5.5 – but probably not enough for the MiG, they managed to accelerate to the west and increased the distance from us. The combat was quite short, we chased them away but we did not harm them.
>
> We actually opened fire but this was from long range, they accelerated faster than us, escaping back to the Gaza Strip in a shallow dive at low altitude. Then we disengaged and we returned to Ramat David; we still had sufficient fuel left.

The MiGs also opened fire. At the start of the engagement, the MiGs activated their afterburners and split into two pairs. One MiG-17 disengaged and retreated across the border, one MiG-17 climbed to 4,000 feet, and two MiG-17s remained in a pair at low altitude. Reuben Harel climbed to engage the lone higher MiG-17, but it also retreated across the border. Moshe Ofer engaged the MiG-17 pair at low altitude and exchanged fire until he ran out of ammunition, at which point Harel, who had returned from his pursuit after the lone higher MiG-17, took his

Air combat acts 1 to 3 as related by the Egyptian MiG-17 pilot. (IDFA)

place. The MiG-17 pair then seized the opportunity to disengage and accelerated towards the border. Harel could only attack their rear with a short burst – 60 rounds, 30 each from the two lower guns – from a range of 1,500 meters.

Only the lower right gun of Vautour 25 fired all of its 100 rounds; the other three guns had all jammed. The lower left gun fired 68 rounds and jammed due to link separation. The upper right gun fired only five rounds and jammed due to improper cocking. The upper left gun fired 42 rounds and jammed due to link separation as a result of high G load; both Vautours exceeded 8 G during the combat. Finally, Vautour 25 returned to Ramat David with two minor hits in its nose section; the Israelis attributed these hits to fragments from either a damaged enemy aircraft or self detonation of rounds.

It is possible, though, that damage to Vautour 25 was the result of Egyptian gunfire. The MiG-17 formation leader recounted the engagement for the benefit of contemporary Egyptian press:

> It was circa 10:00 when we flew over the Gaza Strip on a routine training mission to familiarize new pilots with this sector. We were four and I did not anticipate combat but I did expect the birth of our first child as my wife entered hospital that same morning. While we were flying over Khan Yunes, I suddenly spotted two Vautours, the most advance Israeli combat aircraft. The two Vautours were right behind me and I had to react quickly as I was flying at sonic speed and

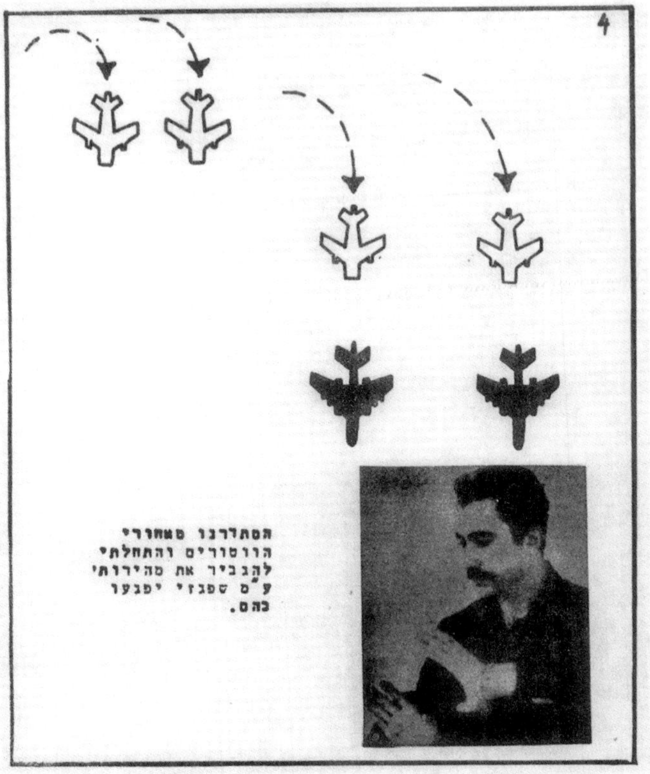

Air combat acts 4 to 6 from the perspective of the MiG-17 leader. (IDFA)

so were they; every second counted. We flew at 800 meters and I ordered two aircraft to climb to 1,000 meters in order to watch over the combat and to safeguard against possible Israeli reinforcements.

So we entered a two versus two air combat, two MiGs versus two Vautours. Then, I took a very daring decision; a tight turn and penetration in between the two Vautours. At this moment, they both opened fire at me but they both missed. I finished this maneuver, flying in between the two Vautours, they were flying at a sonic speed and so did I and no one opened fire. This was pure maneuver aimed at achievement of a position right behind the Vautours; it is a well known fact that an aircraft has to trail the enemy in order to launch an effective strike. My wingman followed me and the other two aircraft reached their altitude. When I finished this maneuver, we were getting further away from each other, flying in opposite directions, tail to tail. I, and my wingman, turned tightly until we were in position right behind the Vautours and I started to approach in order to secure an efficient strike. Suddenly, the two Vautours broke away and then I knew that we had the upper hand. I followed them in their turn and when they were within gunfire range, I opened fire, aiming at the left Vautour. I saw fire erupting form its wing and it turned towards Israel. The other Vautour remained right in front but also turned towards Israel. I could pursue this Vautour but my orders were to avoid flying over Israel; the whole

> engagement took place over the Gaza Strip. Overall, I fired 28 rounds and combat lasted for ten minutes. I was elated, having shot down a Vautour for the first time. Back at base, I was greeted with great rejoicing; my friends suggested that I will name my just born son Vautour.

IAF Intelligence dismissed the recollection of the Egyptian MiG-17 formation leader, even though his story did generally overlap the details given in the IAF debrief:

> It is only right that the story of the formation leader will be read in good humor and in understanding that it was prepared to be agreeable to the palate of the average Egyptian. It is doubtful whether this narrative actually expresses the mentality of Egyptian pilots and in any case should not result in disregard.

Largest Exercise Since the Sinai Campaign

The activation process of Squadron 110 was practically completed. The squadron was issued its full complement of aircraft: 17 Vautour A and four Vautour B (including 4231, which Air Maintenance Unit 22 repaired and handed over to Squadron 110 during June 1959). Operational capability was achieved, missions across the borders were flown, and Vautour B reconnaissance tests were initiated. Vautour conversion courses produced pilots to fill the ranks of Flying Flight that started to have the characteristics typical of an established squadron. At the time, the air crew population of IAF combat aircraft squadrons included executives (including the squadron commander and his two deputies), regulars (including conversion course students), emergency postings (AHQ staff officers and Flying School instructors), and reserves. Inevitable personnel changes impacted the management of the squadron: Deputy B Simon Ash was promoted to Squadron 117 Commander from 1 April 1959; Deputy A Jacob Agassi was assigned to the IDF Staff School; Zeev Tavor entered office as Deputy A; Reuben Harel entered office as Deputy B; and future Deputy A and Deputy B, respectively, Reuben Harel and Yossi Sarig, would be assigned to the IAF Senior Leaders Course that Squadron 101 opened on 16 August 1959.

An Achilles' heel was aircraft serviceability. In addition to corrosion and leaks, the Vautours suffered from a severe shortage of spare parts. Only seven Squadron 110 Vautours were serviceable on 8 July 1959: five Vautour A (4305, 4307, 4309, 4315 and 4320) and two Vautour B (4233 and 4237); 11 aircraft were grounded while awaiting spare parts, and the balance of three aircraft was 4231 (undergoing engines repair), 4235 (being modified for reconnaissance), and 4314 (undergoing repainting in a camouflage scheme).

Regardless of spares shortage, the squadron was in excellent shape at the start of Exercise LANCE, a full scale IAF maneuver that Chief of the General Staff Branch Isaac Rabin described to newspaper editors in a background brief on 7 August 1959:

> Training; there were several major exercises lately. The IAF conducted a large scale exercise; as a matter of fact this was the largest scale exercise since the Sinai Campaign. The objective was that, unlike in the Sinai Campaign, the IAF would operate in full force without mobilization of reserves. The scenario was a surprise, the State of Israel was surprised and the IAF was ordered to immediately operate in full force while, at the same time, mobilizing its reserves to enable continuous operations over a longer time frame.
>
> This exercise proved that our air space is very small, even for the number of aircraft in IAF service. The duration of the exercise was similar to the Sinai Campaign but the number of sorties was higher. There were positive aspects as well as mishaps; we learn from every exercise in order to improve our standard in the next exercise.

Vautour 14 was grounded for repainting on 8 July 1959; this photo was obviously taken after repainting. (PC)

Gershon Lapid was photographed, standing on the ladder beside the cockpit of Vautour 15, in March 1959; Lapid was killed in Vautour 22 on 21 July 1959. (PC)

Gideon Magen on the ladder beside a Vautour; Magen flew 11 sorties during Exercise LANCE, and these included an ejection from Vautour 25. (PC)

Chief of the General Staff Branch's reference to "mishaps" probably pointed at Squadron 110, which started participation in Exercise LANCE on the night of 20 to 21 July 1959 when the Vautour pilots flew at least two missions: Mission Order L111 (including Reuben Harel in Vautour 14) and Mission Order L113 (including Gideon Magen in Vautour 03). At the time, the exercise Mission Order's number prefix letter was also the first letter (or letters) of the exercise name, in this case L for LANCE.

Emergency posting pilots joined the regular pilots, and the squadron started full pace flying from the morning of 21 July 1959. At least four Mission Order missions were flown prior to GUITAR take-off at 09:25 on Mission Order L19, an air base attack mission against a dummy air base at the Matred Heights range with Time on Target at 10:00. GUITAR leader was Gershon Lapid, and his wingman, GUITAR 2, was emergency pilot Jacob Agassi, who – as an IDF Staff School student – functioned during the exercise as an exercise umpire. Gideon Magen was sub leader as GUITAR 3, and Eli Mor, who had just finished his Vautour conversion course, flew as GUITAR 4.

The 4-ship formation entered bombing along the runway axis, from north to south, in two pairs flying in line astern, the second pair flying behind the first pair at proper safety distance. The drill for strafing was to split the 4-ship formation into pairs in order to pose a more difficult challenge to the enemy's anti-aircraft fire. In line with this practice, the first pair turned 360 degrees and entered a first strafing pass also from north to south. The second pair turned 270 degrees and entered a first strafing pass from east to west. The pairs coordinated their strafing attacks, with the leaders transmitting "in" and the wingmen radioing "out".

While the second pair was flying its first strafing pass, the first pair turned right 270 degrees to enter a second strafing pass from west to east. GUITAR 4 reported "out" and GUITAR 1 transmitted "in", but apparently saw GUITAR 4 right in front of him, so he radioed "pass" to indicate an aborted attack and started to climb. Immediately afterwards GUITAR 1 and GUITAR 3 collided completely by surprise; GUITAR 1 was climbing out of the aborted west to east pass and GUITAR 3 was climbing after the east to west strafing pass. Collision was head on at an altitude of 300 to 500 feet above ground level, and with the Vautours travelling at high speeds (GUITAR 3 clocked 400 knots). Vautour 22 exploded and its pilot, Gershon Lapid, was killed. The right wing of Vautour 25 was torn off and the burning machine rolled to the right.

The impact of the collision was so violent that Vautour 25 rolled several times at such a rate that pilot Gideon Magen could not read the instruments and could hardly distinguish between ground and sky. Magen jettisoned the canopy of his cockpit and heard Jacob Agassi exclaiming "Gideon, JUMP!" This shout spurred Magen to speed up his actions; unable to grab both ejection handles on both sides of the seat – as in a sterile ejection drill – Magen ejected himself by pulling only the right handle. Ejection was perfect, despite the low altitude and the rolling of the aircraft. The seat was released, the parachute was opened automatically, and Magen landed on the ground, safe and sound, after only six to eight seconds in the air. Gideon Magen recalled:

The final photo of Vautour 25 was taken on Israel independence Day, 13 May 1959, with Jonathan Shahar on the wing. At the time a Flying School instructor and a Squadron 110 emergency posting pilot, Shahar flew nine sorties during Exercise LANCE. (PC)

The crash site of Vautour 25 at Matred Heights. (PC)

> Right at the moment of this huge impact, I realized that I was involved in a collision and that I have to eject; this instant understanding, of what happened and how I have to react, saved my life. Agassi's shout 'Gideon JUMP, Gideon JUMP' still rings in my ears and undoubtedly stimulated me to react even faster. During the short time frame from collision to ejection I had a very clear vision of my wife Shulamit holding in her arms our two nine months old baby daughter twins Hila and Anat; a vision that still lives in my mind to this day.
>
> The ejection actually took place at an estimated altitude of 400 feet [above ground level at Matred Heights] in a roll; the manufacturer minimum altitude for safe ejection was 800 feet [above ground level] with the aircraft's wings leveled and the nose pointing upwards. I was lucky to eject over Stream Lavan so when the canopy of my parachute opened I was probably at the level of Matred Heights but still a few dozens of meters high above the center of the stream.

The range control team, with pilot Reuben Eyal in command, immediately linked up with Magen. Emanuel Shiber flew Magen from Matred Heights to Beer Sheba in Piper 54, and Davidesko flew him from Beer Sheba to Ramat David in Piper 51. Unlike today, when the IAF is grounded after an accident, Exercise LANCE continued regardless of the accident, and the Vautours kept on taking off from Ramat David in the usual high tempo typical of exercises. Gideon Magen himself returned to fly LANCE missions in the Vautour that same day, and overall he flew 11 Vautour sorties by the time Exercise LANCE ended on 23 July 1959. Magen flew an average of more than three missions per day, a high tempo of flying but not exceptional. Reuben Harel flew ten LANCE missions, Yossi Sarig flew nine, and emergency pilot Jonathan Shahar also flew nine LANCE missions.

Chief of the General Staff Branch Isaac Rabin elaborated on 26 July 1959 about the impact of the collision at Matred Heights in a letter to the Israeli Military Attaché in France, Uzi Narkis:

> The recent major exercise of the IAF ended with several mishaps including one severe [probably referring to the Vautours collision] ... This malfunction resulted in deterioration of relationships between a section of the General Staff members and the IAF. Though I am not part of these sections, or the section that was dragged into this tension, there are surely serious lessons concerning operational training standards in the IAF ... [IAF Commander] Ezer Weizman was in a state of tear out for a while after the malfunction in the IAF exercise but with the help of his friends he already recovered.

150 Attack Aircraft

The IAF's original plan was to purchase some 50 Vautours for four squadrons. This acquisition plan was to be accomplished over a time frame of several fiscal years, and started with the HIGH TIDE contract in the summer of 1956. The naming of Ezer Weizman as IAF Commander changed plans. From the days of Ramat David versus Hatzor, single engine fighters versus two engine fighter bombers, Spitfire and Mustang versus Mosquito, fighter pilot Ezer Weizman was the most prominent representative of the single engine fighter force. Immediately after entering office as IAF Commander on 25 July 1958, Weizman cancelled the plan to re-equip Squadron 115 with Vautour B reconnaissance aircraft, reassigned the Vautour B reconnaissance aircraft to Squadron 110, and postponed purchase of additional Vautour B bombers to equip Squadron 69.

A year after entering office, Weizman finished off acquisition of additional Vautours. Previously, the IDF Chief of Staff had approved purchase of 12 Vautour B bombers and two Vautour B reconnaissance aircraft as an element of the three years plan, April 1959 to March 1962. This time frame was expected to be peaceful, and for this reason the IDF was willing to accept the risk of a temporary reduction in the IAF order of battle until planned introduction of the Mirage from January 1962. The

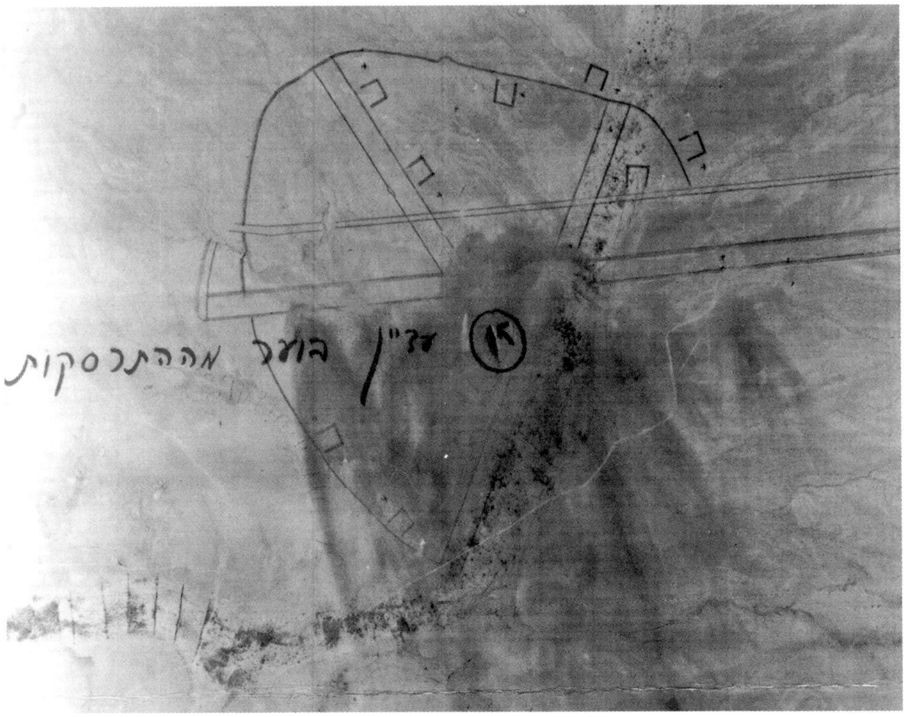

Matred Heights as photographed from 30,000 feet at 11:15 on 21 July 1959; the Vautour wreckage (arrow in circle) is still burning. (IDFA)

The Squadron 110 line in December 1959, a month after the IDF Chief of Staff approved the IAF Commander's recommendation to purchase, as a stopgap solution pending introduction of the Mirage, Super Mysteres instead of Vautours. (PC)

question was what would be the low point of the IAF order of battle during this time frame.

In a 30 July 1959 document addressed to the IDF Chief of Staff, the IAF Commander stated that "150 attack aircraft is the minimum number for fighting the air forces of Arab nations". At the time the IAF fielded 150 attack aircraft, but this number was planned to be cut down – due to attrition and retirement versus small scale acquisition – until the introduction of the Mirage. To maximize the impact of the forwarded forecast, the IAF Commander presented an extreme scenario, claiming that "due to attrition and retirement, the attack aircraft order of battle will drop by [1 April] 1962 (without additional acquisition) to just 81 aircraft". This presentation was an extreme case because acquisition of 14 attack aircraft was already approved, and the first Mirages to enter service from January 1962 were excluded from the presentation; the planned replacement of the Mustang and Meteor (the Fouga Magister armed trainer) was not included in the forecast, and the forecasted attrition was higher than the eventual attrition (the IAF Commander's forecast was that attrition would reduce the number of Mysteres to 40 by 1 April 1962, while the actual number was 43, to 15 Ouragans versus 18 in reality, etc).

In light of this gloomy presentation, the IAF Commander foresaw "a need to acquire 24 modern attack aircraft in the current fiscal year. The recommended type, from performance and availability perspectives, is the Super Mystere". The IAF Commander suggested that this proposed purchase would not be on top of the already approved acquisition of 14 attack aircraft (12 Vautour B and two Ouragans), but in place of this planned purchase.

The IDF Chief of Staff approved the IAF Commander's recommendation on 4 November 1959, thus summing up a meeting to discuss the IAF Year 1960 (April 1960 to March 1961). Actually, the IAF Commander's suggestion was not fully approved. The purchase of Super Mysteres instead of Vautours was accepted, but the approval only covered 12 Super Mysteres instead of 16 aircraft (12 Vautour B bombers, two Vautour B reconnaissance aircraft, and two Ouragans). Thus, the original IAF Vautour purchase plan was terminated, and the IAF would actually field 122 attack aircraft by 1 April 1962 (excluding 38 Fouga Magisters as Mustang replacements and including 12 additional Super Mysteres), or 50% more than the gloomy forecast that the IAF Commander presented in July 1959!

Sitting on the wing of Vautour 09, left to right: Oded Marom, Eli Bahat and Yossi Sarig. (PC)

Frigates to Ceylon

The Israeli Navy Flotilla 5 inducted two destroyers from July 1956 to initiate replacement of three frigates. Shortly after the end of the Sinai Campaign, two of these frigates scheduled for retirement (Israeli Navy Ship F-28 *Mivtach* and INS F-32 *Miznak*) sailed from the Mediterranean Sea, around Africa, and to the Red Sea. Then Israel withdrew from Sinai and the two redundant frigates were offered for sale.

Ceylon purchased the two frigates in May 1959 after a prolonged and exhaustive negotiation. The contract covered training of Ceylon naval crews at Eilat, and the plan was that these crews would deliver the frigates from Eilat to Ceylon; first *Mivtach* and then *Miznak*, as the latter was in need of repairs. Meanwhile, the deal went public and Egypt threatened to prevent the frigates – "seized" at Eilat was the Egyptian narrative – from passing through the Straits of Tiran.

Ceylon preferred to avoid a diplomatic mess and asked Israel to deliver the frigates to Ceylon. At first Israel refused because of the contract terms, the question of passage via the Straits of Tiran, and the fear that any change would be attributed to an

A Vautour in flight over the sea; the Vautour was the IAF's preferred maritime strike aircraft due to its long range, heavy offensive load, and two engine safety. (PC)

Reuben Harel flew Vautour 21 during Operation GALIT; Vautour 21 was photographed departing Ramat David with two 1,250 liter drop tanks. (PC)

Egyptian diplomatic victory. This jeopardized the whole contract, the cancellation of which would have been a true Egyptian diplomatic victory, so Israel agreed to deliver the frigates to Djibouti, where the ships would be handed over to Ceylon.

INS *Mivtach* departed Eilat on the evening of 30 August 1959 and sailed through the Straits of Tiran on the morning of 31 August 1959 under the protection of a Squadron 110 aerial umbrella. The frigate arrived at Djibouti on 7 September 1959 and was handed over to the Ceylonese. Sailing onwards to Ceylon and arriving there on 21 September 1959, the former Israeli frigate was made the flagship of the Ceylonese navy.

Squadron 110 also protected the passage via the Straits of Tiran of the INS *Miznak* on 8 April 1960 during Operation GALIT; Reuben Harel in Vautour 21 and Yossi Sarig in Vautour 17 flew the 1:15 hour mission. Reuben Harel recalled:

> We escorted ships, frigates that sailed from Eilat, all along Sinai until the limit of our range and to the point where the ships were at open sea because there were fears that the Egyptians or the Saudis might try to stop the ships. The reason why the Vautour was selected for this mission was that it was possible to present the Vautour as unarmed. Six bombs were loaded into the bomb bay and with external fuel tanks we still had sufficient fuel to fly to Sharem El Sheikh and back.

Balancing Wheels

The next Squadron 110 end of term exercise (Exercise SHOULDER from 25 to 26 November 1959) was completed without any unusual events. Fast turnaround was emphasized as usual. Reuben Harel flew seven sorties, including one at night; Gideon Magen flew six sorties, including one at night; Yossi Sarig flew five sorties; navigator Achikar Eyal flew five sorties, including one at night; and emergency posting pilot Jonathan Shahar flew four sorties.

Four days later, on 30 November 1959, Ezekiel Somech handed over command of the squadron to Jacob Agassi. IAF Comptroller Israel Lahav issued the IAF accidents report for 1959 a short while after Agassi entered office as Squadron 110 Commander. Squadron 110 suffered 24 accidents of all sorts, or one accident per 98.9 flight hours. This was quite a high rate of accidents, but not unusual in those days, as the vast majority of accidents were Category 1 and resulted in very light damage. Jacob Agassi recalled:

> There was a number of balancing wheels related incidents. The Vautour had a bicycle undercarriage – there were two wheels in each of the two main gears – and the purpose of the outer balancing wheels was to safeguard from a wing hitting the ground. Sometimes, one of the balancing wheels did not detract so we had a few scary landings; in one of these I was personally involved. It was a night flight and one of the balancing wheels did not detract; because it was at night there were not enough service persons to call – in daytime we used to call service persons that held up the wing after the aircraft slowed down so it would not hit the ground – but I performed a perfect landing and the aircraft was not damaged at all... Overall, the Vautour was a big aircraft with two powerful engines and we loved it very much.

Squadron 110 Commander Jacob Agassi handling the hoist that lifts the Vautour 03 guns pack into place. (PC)

Vautour 26 after veering off the runway on 31 May 1959. (PC)

3

SQUADRON 110 VAUTOUR

1959 - 1963

COMMAND HAND-OVER FROM FOUNDING FATHER Ezekiel Somech to his successor Jacob Agassi signaled the end of the infancy phase in the history of Squadron 110 Vautour and the start of maturation. Attack qualifications had achieved Full Operational Capability, and the next challenge for Squadron 110 was to close down the aforementioned reconnaissance gap.

First Vautour B Reconnaissance Conversion

Slightly more than a year had passed since Vautour 33 arrived at Ramat David on 23 December 1958 until MAT4 recommended, on 5 January 1960, to qualify the first reconnaissance Vautour as operational. While Vautour 31 was being repaired, Vautour 33 was selected to serve as the first IAF reconnaissance Vautour B. Zeev Tavor ferried Vautour 33 from Ramat David to IAI on 28 January 1959. IAI was contracted with all IAF Vautour D-Level work, including the conversion of the Vautour B into a reconnaissance aircraft.

Vautour 33 was modified with three 36" lens K-38 cameras, each capable of photographing an area 4.5 km wide and 200 km long from 20,000 feet with sufficient quality for recognition of tank types. The three big cameras were installed in the bomb-bay, one vertically and two angled 30 degrees to either side. Design and modification of the Vautour B bomber as a reconnaissance platform was a challenging project. A special truss was designed, manufactured, and installed in the bomb-bay to support the three cameras. A heating system was fitted to warm up the bomb-bay. A lower partition – with special windows for photography – was installed to seal the bomb-bay. The bomb-bay doors had to be modified to suit the photography system. All of these major issues, as well as many other minor modifications, had to be designed, manufactured, installed, and tested.

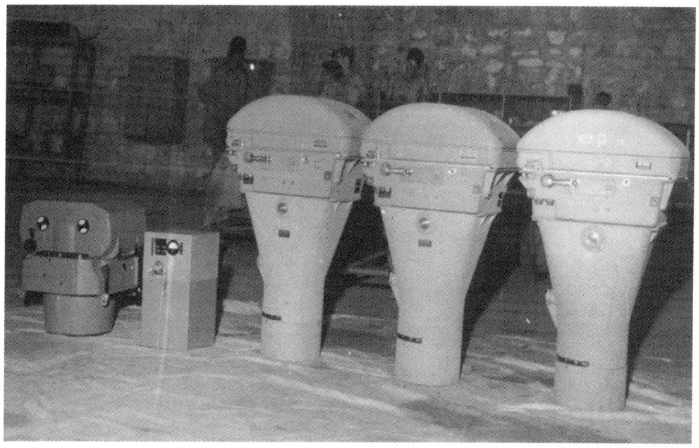

The reconnaissance gear in the Squadron 110 Reconnaissance Pen. (BIAF)

Vautour 33 was back at Ramat David by May 1959. Test flights revealed a number of deficiencies, with consequent modifications and additional testing. The concluding round of three test flights started on 15 December 1959. Maoz Poraz and Achikar Eyal flew the first test flight, which evaluated photography from 35,000 feet, from 48,000 feet, and again from 35,000 feet during a minimum endurance mission profile that lasted 0:55

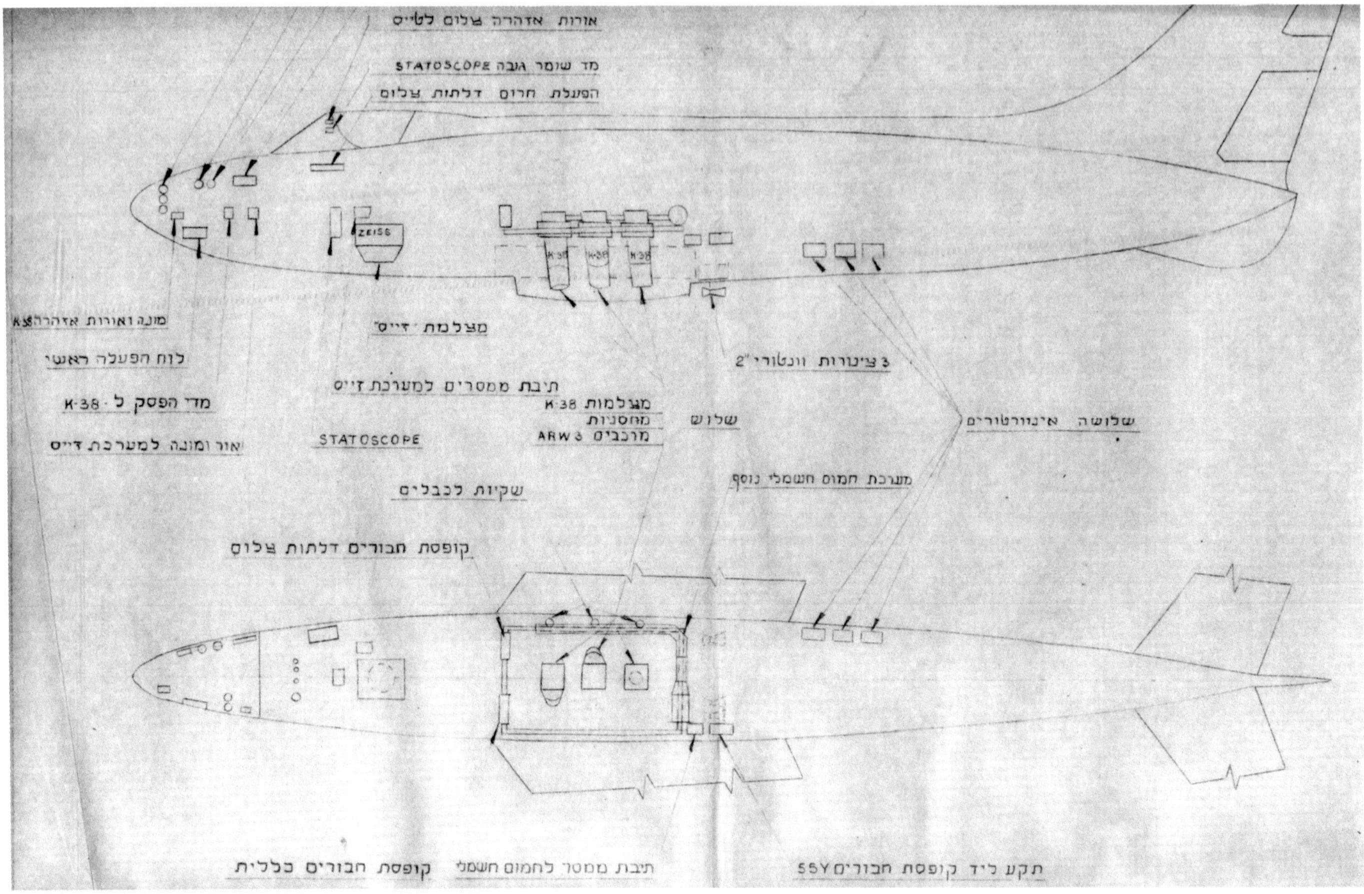

A drawing of the Vautour 33 reconnaissance modification with a Zeiss camera in the front camera bay and three K-38 cameras in the front suite of the bomb-bay. (IDFA)

The rearmost K-38 camera attached to its support frame prior to installation in its vertical position. (BIAF)

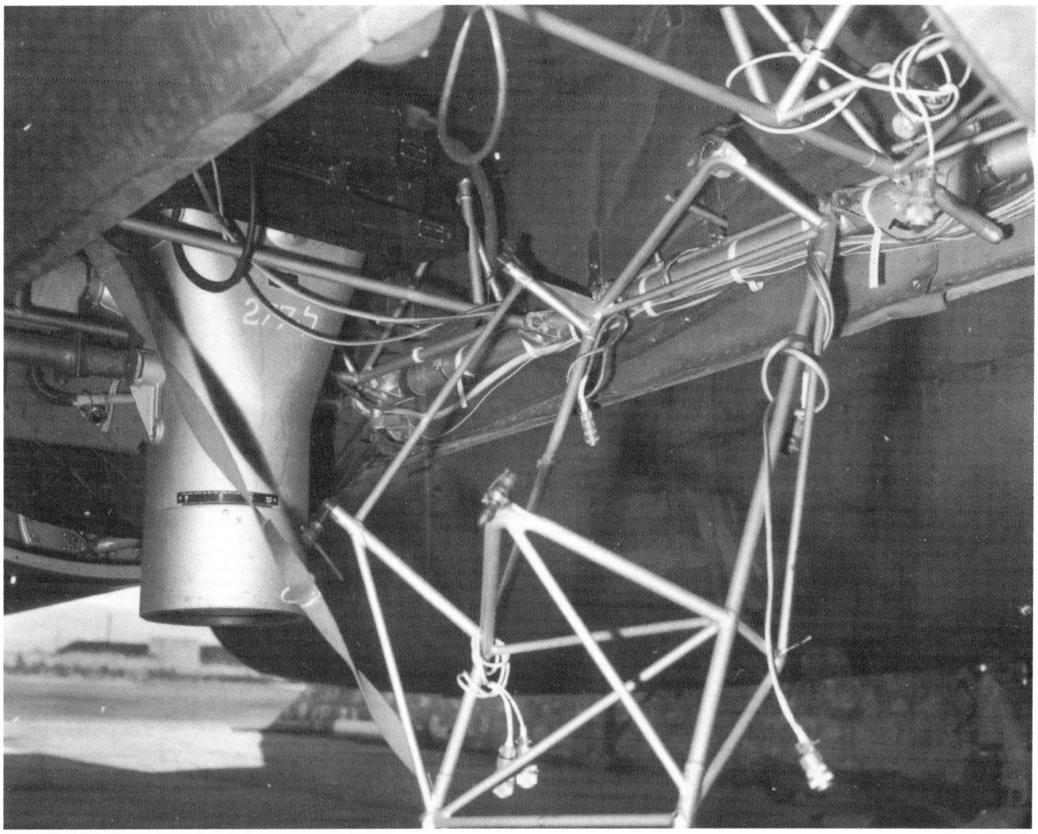

The modified bomb-bay, looking aft towards the vertical camera. (BIAF)

flight hours. The system functioned faultlessly during this first flight. The second test was flown on 16 December 1959, when Vautour 33 climbed to 48,000 feet, flew 63 minutes, descended to 35,000 feet and took photos before climbing back to 48,000 feet and taking more photos. The second test was also successful, even though the weather was hazy. Problems appeared on 17 December 1959, when two flights were attempted but both had to be aborted due to malfunctions in the aircrafts' electrical systems. The first third test was thus flown on 18 December 1959, with Reuben Harel and Yossi Sarig as camera operator. The profile of the 1:50 flight hours sortie was similar to the 16 December test flight, but weather conditions prevented accomplishment of satisfactory photos. A second third test flight was flown on 24 December with Zeev Tavor in the cockpit and Sarig as camera operator. The results of the 24 December 1959 test flight were satisfactory and adequate for operational use.

At the conclusion of those test flights, MAT4 recommended to declare Vautour 33 operational for mission profiles as flown during the tests. However, AIR3 rejected the MAT4 recommendation and set a requirement for five consecutive trouble-free reconnaissance test flights as a condition for declaration of Vautour 33 as an operational reconnaissance aircraft. In line with this requirement, two more reconnaissance test flights were flown on 15 and 16 February 1960. The mission profiles of these two test flights were similar: photography from 30,000 feet, photography from 40,000 feet, photography from 48,000 feet, and prolonged mapping photography from 40,000 feet over the Negev Desert. Yossi Sarig and Uri Talmor flew the first 0:45 flight hours test and both flights were considered successful. As a result, MAT4 again recommended, on 26 February 1960, to declare Vautour 33 as operational.

Start of Shafrir Tests

RAFAEL, the IMOD Weapon Systems Development Authority, started to develop a simple short range, heat seeking air-to-air missile during 1958. The IDF authorized, on 28 February 1959, IAF involvement in the development phase, with future acquisition only if the development would produce a successful air-to-air missile. An IDF Development Order was issued on 30 December 1959 along with basic requirements, including interception of aircraft flying at up to Mach 2.2, launch from aircraft flying at Mach 0.5 to 1.7, operating altitudes from 1,000 to 60,000 feet, a range of at least 1.5 kilometers at all altitudes, launch envelope from a 30 degree cone behind the target, and the ability to pursue a maneuvering target at up to 2 G. The following month the first missile was ready for the start of flight tests, and it was then that Minister of Defense David Ben-Gurion selected the name Shafrir (Royal Canopy from Jeremiah 43:10).

The IAF considered the most suitable aircraft to test the new air-to-air missile would be a Vautour B or N. Eventually the Vautour B was selected, and the first flight with Shafrir was accomplished on 6 January 1960. Five days later the Vautour B launched a Shafrir for the first time; Ezekiel Somech was the pilot and MAT4 Project Officer Josef Shenar (a Mosquito navigator from 1953) flew as observer.

Second Vautour B Reconnaissance Conversion

Upon conclusion of Vautour 33 test flights, MAT4 also suggested on 5 January 1960 that a test flight program for Vautour 35 should be prepared. Vautour 35 was to be modified with an improved reconnaissance system that included the three K-38 cameras like in Vautour 33, but also a 6" lens K-17 camera that replaced the original Zeiss BDA camera in

A Vautour 33 K-38 photo of Hatzor taken from 40,000 feet on 24 December 1959. (IDFA)

its bay, in front of the front wheel, and an improved electrical heating system.

Vautour 35 test flights started on 7 February 1960, when Yossi Sarig and Achikar Eyal flew a 1:40 flight hours sortie. The objective of the first three test flights was to collect data, especially concerning temperatures in the cameras' bay, at several altitude levels from 20,000 to 48,000 feet. The fourth test flight, on 11 February 1960, was the first with all four cameras installed. The cameras were operated separately and together. Mission profile was photography from 30,000 feet, a climb to 48,000 feet, 60 minutes flight at 48,000 feet, photography from 48,000 feet, a descent to 30,000 feet, and photography from 30,000 feet. The system functioned properly, so five more tests were flown from 12 to 19 February 1960, all to the same mission profile as in the 11 February 1960 flight, in order to achieve the requirement for five consecutive faultless reconnaissance test flights. This objective was indeed achieved, and MAT4 recommended, on 26 February 1960, to declare Vautour 35 operational.

RETAMA

Armed clashes along the Israeli-Syrian border resulted in an Egyptian evaluation that Israel might attack Syria, then the north district of the United Arab Republic. Egypt, at the time the south district of the United Arab Republic, started to roll forces forward into Sinai from 17 February 1960. The Egyptian objective was to deter Israel from attacking Syria, but Israel was unaware of the Egyptian deployment. Israel became aware of this development only on 22 February, when the USA alerted Israel concerning this Egyptian action. Israel figured that if this USA report was correct, then the IDF was at a major disadvantage, with up to 500 Egyptian tanks in Sinai facing only some 30 IDF tanks in the Negev Desert. The Israeli Government realized that if this was the case then the IDF was surprised, and Egypt had an opportunity to attack Israel at favorably relative strength until the IDF reinforced Command South. A reconnaissance mission was authorized for the following day in order to verify this information. Although Vautour 33 was not yet declared operational, it was surely the most advanced reconnaissance aircraft in IAF inventory, and as such was selected to fly this crucial mission.

The objective of the 23 February 1960 mission was to photograph Egyptian army camps along the Suez Canal in order to ascertain if the camps of the Egyptian armored units were occupied or empty. If the camps were empty then this was confirmation of the report from the USA. Additionally, EAF air bases along the Suez Canal were to be photographed. The mission profile included an indirect approach. Vautour 33 was to fly south along the Gulf of Aqaba to the Red Sea and then north over the Gulf of Suez towards the Suez Canal. Time on target for the first photography pass was 12:30. The IAF evaluated

A February 1960 24" K-38 camera view – from 30,000' - of the northeast half of Ramat David air base, with runway 15-33 at the bottom of the photo. The three visible lines belonged to, left to right: Squadron 117 Meteor, Maintenance Squadron beside the two hangars (near the edge of runway 27), and Squadron 109 Mystere (going down to the start of runway 33). The Squadron 110 line was northwest of the junction of runways 09-27 and 15-33. (IDFA)

that EAF interceptors might be able to intercept the reconnaissance Vautour over the north sector of the Suez Canal or over the Mediterranean Sea, so a Squadron 105 4-ship formation was tasked to fly a combat air patrol north of Sinai. The Super Mysteres were to be vectored to engage the EAF interceptors only if they endangered the Vautours.

Yossi Sarig and Achikar Eyal flew Vautour 33 with Jacob Agassi escorting them in an armed Vautour A. The Vautour crew was briefed to photograph from 45,000 feet, but contrails developed at this altitude, so they descended to 42,000 feet. Prior to photographing Kabrit, the Vautour crews were notified that Egyptian MiG-17s had been scrambled from Abu Sueir. Over Kabrit the Vautour crew noted, and photographed, the scramble of additional MiG-17s. The Vautours pressed ahead with their mission, but also climbed to make things more difficult for the MiG-17 pilots. The Vautour B climbed faster than the Vautour A, and within a short while Agassi was trailing some ten miles behind Vautour 33. By then Vautour 33 had completed its mission and was over Port Said. Due to this climb the contrails re-appeared, and this enabled the MiG-17 pilots to home in on Vautour 33. The lagging Vautour A did not produce contrails and was thus invisible to the MiG-17 drivers. This situation offered Agassi an opportunity to engage the MiG-17s; Agassi actually asked permission to open fire but was denied. Meanwhile, the Vautours continued to climb and their superior high altitude performances began to show, as the MiG-17s were unable to close the distance.

The photographic evidence that Vautour 33 brought from the Suez Canal was indeed conclusive; the camps of the Egyptian armored units were empty. The IDF issued Operation RETAMA order for immediate reinforcement of Command South and for preparations to war against Egypt. Two more reconnaissance missions were authorized to be flown on 24 February 1960. The objective of these missions was to verify that the Egyptian armored units, no longer in their Suez Canal camps, were indeed in Sinai. A Squadron 101 Mystere was tasked to photograph the Gaza to El Arish sector, and Vautour 33 was tasked to photograph the Abu Ageila to El Arish sector. Reuben Harel and Achikar Eyal flew Vautour 33, and the photographs verified that the Egyptian armed forces from the Suez Canal were indeed deployed to Sinai.

Until the IDF would complete reinforcement of Command South, the IAF was set on Readiness State C (one level below war) from 16:30 on 24 February 1960. As an element of the IAF readiness scheme, Squadron 110 was to be prepared to launch up to 12 Vautour A aircraft to attack enemy targets and up to two Vautour B aircraft to photograph enemy targets. In order to avoid escalation, no more reconnaissance missions over Egypt were approved until 3 March, when the situation was starting to calm down. Zeev Tavor and Achikar Eyal flew the 3 March 1960 mission in Vautour 33 to confirm that the Egyptians were starting to return their forces from Sinai to the Suez Canal Zone. The RETAMA emergency was all over by the end of March, though its lessons left an everlasting impression in IDF heritage. Achikar Eyal recalled:

> Following the [RETAMA reconnaissance] flights I sent a letter to Prime Minister Ben-Gurion. I indicated that although our Intelligence promised a pre-warning of a whole week prior to an Egyptian attack, we were caught by surprise "with our pants down". Therefore, I suggested that the

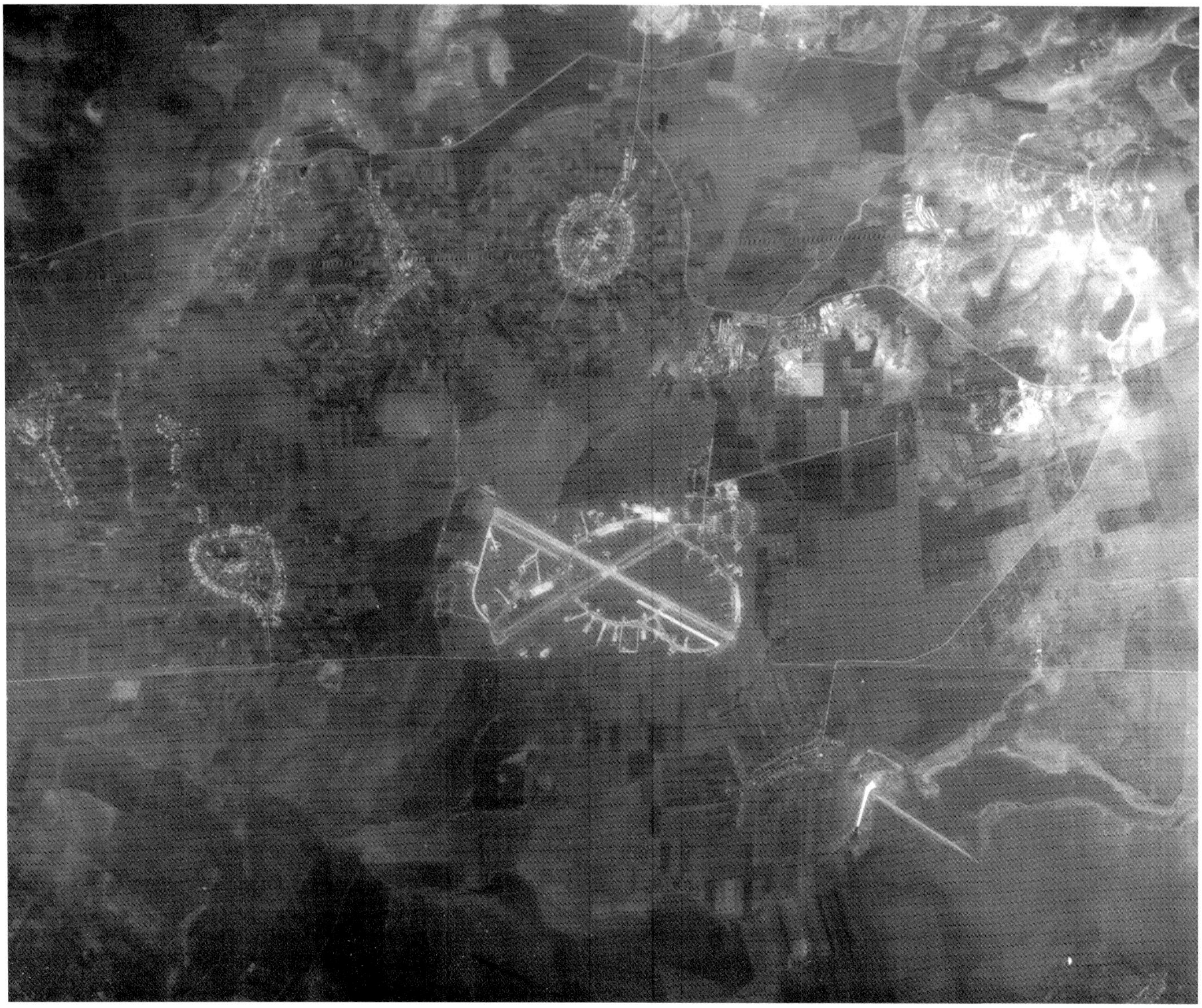

A 6" K-17 mapping photo taken from Vautour 35, while flying at 30,000 feet, on 15 February 1960. Ramat David is right in the center with the four lines clearly visible, left to right: Squadron 110 Vautour, Squadron 117 Meteor, Maintenance Squadron, and Squadron 109 Vautour. The circular village is Nahalal, while the village left of Ramat David is Kfar Yehoshoa. (IDFA)

> IDF would be restructured for readiness against enemy attack within a time frame of less than 24 hours. Ben-Gurion replied to my letter, stating that although the Captain [me] was way out of line, he was right.

Reconnaissance Repercussions

While the Vautour B reconnaissance conversion project was approaching operational status, two related issues had to be addressed. The first issue was assignment of reconnaissance qualified navigators to Squadron 110; subsequent practice would be one or two regular navigators, plus a group of emergency posting navigators. The few emergency posting navigators assigned to Squadron 110 in the wake of the operational qualification of the Vautour as a reconnaissance platform included veteran Mosquito navigators Arie Raviv and Ephraim Ronen. At the time, the Flying School NBR Squadron Commander, Ronen, first flew in the Vautour on 29 February 1960 and experienced one of his nine incidents/accidents during his fourth flight in the Vautour, a training sortie on 4 March to practice photography from 48,000 feet in Vautour 35 with pilot Maoz Poraz. Ephraim Ronen recalled:

> In the Vautour B the pilot and navigator were separated so we had a procedure that ensured constant communication [over the intercom] between the pilot and the navigator. Routine reports from the navigator to the pilot – such as 'opening doors', 'photographing', 'closing doors' – were an element of this communication procedure that ensured audio contact because there was no visual contact. Then, during this flight, I suddenly felt as though I was awakening, the horizon was rolling all around me and I heard Poraz shouting 'Ronen', 'Ronen', 'Ronen', again and again. I woke up and apparently, as we initiated the photography run, I reported 'opening doors', 'opening doors', 'opening doors' and I did not stop reporting 'opening doors' until my voice faded. Poraz realized that something was wrong, he called me but I did not answer so he dived and I regained my

Squadron 110 Reconnaissance Officer Yossi Sarig was the pilot of Vautour 33 during the 23 February 1960 mission over the Suez Canal. (PC)

Reconnaissance Vautour 33, which flew the 23 February 1960 mission, inside the Squadron 110 Reconnaissance Pen at Ramat David. (PC)

consciousness at an altitude of circa 20,000 feet. The post flight inspection yielded that my oxygen regulator malfunctioned. Obviously, an oxygen system malfunction at high altitude is critical. Poraz swift action probably saved my life.

The second issue was whether to repaint the Vautour B reconnaissance aircraft in the same camouflage scheme as applied to the Vautour A attack aircraft. A test was devised, and AIR2 reported on 26 January 1960:

> Since doubts were expressed concerning painting of reconnaissance aircraft (Vautour B) it was agreed with Major [Ezekiel] Somech that a test will be conducted with painted and unpainted aircraft flying at high altitude. The test revealed that in most cases it was easier to spot an unpainted aircraft than a painted aircraft. Therefore, the Vautour B aircraft should be painted in the same scheme as Vautour A aircraft.

AIR2 authorized the repainting of Vautour B (and N) aircraft by IAI on 29 February 1960, but on 16 March, AIR3 indicated that repainting of Vautour B aircraft would only start after the conversion to reconnaissance configuration of the third Vautour B was completed. Repainting would then have to be planned in a staggered schedule so that two Vautour B reconnaissance aircraft would be operational while the third was being painted.

A Vautour A inside a British heritage pen, which were used for dispersal of aircraft from lines during war games and emergency situations; the Ramat David pens were covered with roofs after the rainy winter of 1959/1960. (PC)

A 4-ship formation that took part in the 2 May 1960 Independence Day fly-past: Jacob Agassi with Achikar Eyal flew lead in 37, Yossi Sarig flew 05 (the right wingman in this photo), Jonathan Shahar flew 18, Zeev Tavor flew 20, and Reuben Harel flew 26. (PC)

The guys in the background are anonymous Squadron 110 maintainers on a camouflaged Vautour A. Discussions concerning the camouflage of Vautour B and N aircraft started in 1960. (PC)

A Vautour 31 Zeiss camera photo taken from 30,000 to 40,000 feet on 22 June 1960; the depicted air base is Hatzor, with the Mediterranean Sea coastline on the left, while the white adjacent to the sea are sand dunes. (IDFA)

Third Vautour B Reconnaissance Conversion

IAI converted the third Vautour B reconnaissance aircraft (4231) with three K-38 cameras in the front suite of the bomb-bay and one Zeiss RMK-15 camera in the forward compartment. A fifth camera (type N-9) was installed in the cockpit for the duration of testing, along with a special test panel that included a clock, an air suction meter, and eight temperature meters (seven temperature gauges were fitted in the bomb-bay and one temperature gauge was installed in the forward camera compartment). The N-9 camera took photos of the test panel, although the navigator was also to write down times, altitudes, and temperature readings. Thus, the first phase in the test program was to test the test equipment, with the N-9 camera photographing the test panel at preset intervals during climb to 48,000 feet, 60 minutes cruise at 48,000 feet, and descent back to landing. The K-38 and RMK-15 cameras, plus an altimeter, were to be installed for the second phase of the test program, which included photography missions in simi-

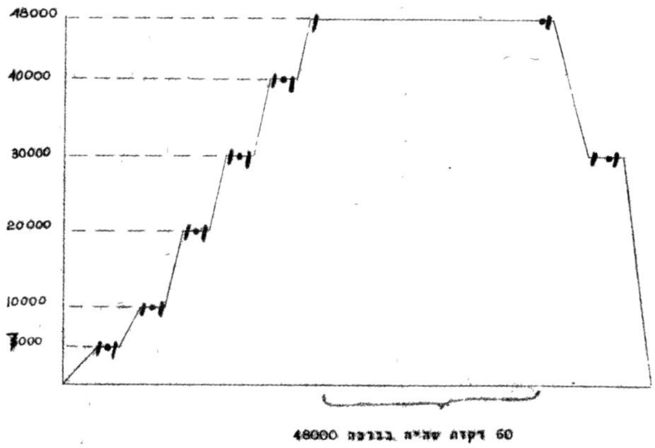

Graphic description of a Vautour 31 photography test flight; the lines denoted the opening and closing of camera doors, while the dot denoted photography with the K-38 and Zeiss cameras. (IDFA)

A Vautour 31 Zeiss camera photo – taken from 30,000 to 40,000 feet on 22 June 1960 – with Tel Nof air base right in the center. (IDFA)

lar flight profile as in the first phase. The third phase covered photography from 30,000 to 48,000 feet in flight profile that simulated an operational mission, with the Vautour flying at 500 knots (Mach 0.85 at 30,000 feet and Mach 0.87 at 48,000 feet).

MAT4 suggested on 26 May 1960 that the test program should start on 30 May and would be completed by 10 June. However, AIR3 only issued the relevant Mission Order on 2 June, and the series of ten test flights was accomplished from 7 to 24 June; navigator Achikar Eyal flew in eight of the test flights, pilot Reuben Harel flew five sorties, pilot Yossi Sarig flew two sorties, and pilot Gideon Magen flew one sortie, with pilots Zeev Tavor and Maoz Poraz and navigator Arie Raviv also taking part in the test.

The heating system included electrical heating and hot air from the engines; the output of this system was doubled in 4231 as a lesson from 4233 and 4235. Still, the temperature of the "hot" air from the engines – after 60 minutes flight at 48,000 feet – was minus 12 degrees Celsius, and it was actually cooling the cameras in the bomb-bay! Thus, the heating with bleed air from the engines had to be disconnected at high altitude. The output of the electrical heating component was insufficient to maintain proper temperature in the camera bay. The K-38 cameras still managed to function, so this issue was left for future improvement. As for the front camera compartment, the heating system managed to produce adequate temperatures (always higher than 15 degrees Celsius). The RMK-15 camera functioned extremely well during the second phase of the test program, producing sharper images than the K-17. However, it turned out during the third phase of the test program (simulation of a mission profile) that the RMK-15 camera was unable to tolerate aerodynamic loads higher than 3 G.

The MAT4 report of the flight test program was issued on 12 July 1960, and called for further exploration and development of RMK-15 tolerance to G, heating system improvement, and research concerning factors that influenced the quality of photography from high altitude. For the meantime, MAT4 evaluated the photographic qualifications of 4231 as adequate and recommended to qualify for operations the third Vautour B converted to a reconnaissance aircraft.

The original IAF objective of three Vautour reconnaissance aircraft was at last achieved, but the quality of the photographs was considered mediocre. AIR2 chaired a discussion on 4 August 1960 that addressed this issue. Apparently, Air Maintenance Unit 22 zeroed the cameras to operate from an altitude of 30,000 feet and a temperature of 20 degrees Celsius, but the temperature in the aircraft was only 15 degrees Celsius at 30,000 feet, while the IAF aspired to photograph from 45,000 feet. The cameras were zeroed based on data from 1945, because no newer data was available and, unlike in the Squadron 115 days, the cameras were not returned to Air Maintenance Unit 22 for servicing after unsatisfactory photographic missions. This was attributed to the distance from Air Maintenance Unit 22 at Tel Nof – where Squadron 115 was based in the past – to Squadron 110 at Ramat David. The product of this discussion was agreement that the quality of the cameras should be addressed prior to any additional improvements in the installation of the cameras inside the aircraft.

Yossi Sarig flew Vautour 14 during the test of a nighttime attack mission profile on the evening of 22 July 1960. (PC)

Reconnaissance Bombers

Versatility was an IAF key buzzword at the time. Operating a single role type of aircraft was out of fashion, and in April 1959 the Air Department Chief ordered utilization of the Vautour B aircraft for additional missions, such as bombing and illumination. The Vautour B aircraft that were still unmodified for reconnaissance were utilized to illuminate targets for Vautour A attackers from Exercise SHOULDER, a November 1959 combat day.

The conversion of the Vautour B aircraft to reconnaissance challenged the versatility dictum. The conversion practically changed the combat capabilities of the Vautour B. The bomb-bay elevator was removed, electrical systems essential for activation of ordnance from bomb-bay and from wing stations were removed, and the critical intervalometer was also removed. The issue of Vautour B versatility had to be re-evaluated in light of these changes.

Since the April 1959 Air Department Chief order was still standing, AIR2 recommended on 22 June 1960 to allow the Vautour B to be loaded with ordnance, bombs, or flares in the bomb-bay or under the wings, but not in the bomb-bay and under the wings. To enable bombing from the wing stations, it was suggested to install a device – in place of the removed intervalometer – that would release the bombs at proper intervals. As for bombing from the belly, AIR2 estimated that it would take 48 hours to remove the reconnaissance system and up to 12 hours to re-install the bomb-bay gear. In light of these findings, AIR2 suggested to Air Department Chief to re-consider the versatility requirement of the Vautour B.

For the meanwhile, Squadron 110 started to practice daylight bomb aiming with the Norden sight; Achikar Eyal flew five such sorties from 1 to 7 July 1960 in the reconnaissance modified Vautour 33 and the yet to be converted Vautour 37. Yossi Sarig and Eyal then flew three sorties in Vautour 37 – from 10 to 12 July – for actual bombing of targets in a range. Then, on the evening of 22 July, Maoz Poraz and Eyal in Vautour 37 illuminated to Reuben Harel (in Vautour 18), to Sarig (in Vautour 14), and to Jonathan Shahar (in Vautour 15) during a test of a nighttime attack mission profile.

Shafrir Tests from B to A

The next round of Shafrir tests started on 8 July 1960. The Shafrir was fitted with a heat seeker and was to be launched against a hot target on the ground. By then the Vautour B was operational as a reconnaissance aircraft. The task of testing the Shafrir was shifted to a Vautour A, and the assignment of project test pilot also changed from Ezekiel Somech to Jacob Agassi. The objective of the 8 July test was to select a suitable hot target, from several types of targets placed in a range, on which the Shafrir would easily home. The selected target was a metal grid that was heated by gas burners. This target was placed in range for the actual launch test on 15 July. The Shafrir acquired the target from a range of three kilometers during the first two "dry" runs, so Agassi was authorized to launch. Agassi launched the Shafrir from a range of 1,200 meters. The seeker acquired the hot target, but the control system did not function properly and the missile missed the target.

Reuben Harel flew Vautour 18 during the test of a nighttime attack mission profile on the evening of 22 July 1960. (PC)

Convex Wing for Vautour B

The aforementioned evaluation of the convex leading edge wing resulted in an IAF decision – formally taken on 8 December 1959 – to equip only the four Vautour B aircraft with the convex leading edge wing. In line with this decision, Israel purchased four convex leading edge wing kits, but none was fitted by July 1960, when Squadron 110 Commander Jacob Agassi challenged this decision and suggested to apply the convex leading edge wing modification to the Vautour A as well.

Jacob Agassi emphasized improved handling at slow speeds. Fitted with the convex leading edge wing, the Vautour was expected to be able to better engage enemy aircraft and to be safer due to improved ability to recover from a spin. Agassi estimated that the total cost to modify the squadron's Vautour A aircraft with the convex leading edge wing would not exceed half the cost of one aircraft. If the loss of one Vautour due to a spin would be avoided then the modification would be more than worthwhile.

Apart from safety and low speed handling, Squadron 110 also concluded that a Vautour with the convex leading edge wing was generally superior to a Vautour with a regular wing. This evaluation was based on comparison of the French manual for a Vautour B with the convex leading edge wing and the IAF manual for a Vautour A with a regular wing. The major advantages of a Vautour with the convex leading edge wing as concluded from this detailed comparison were: smaller fuel consumption (and longer range) at all altitudes, faster optimum speed for maximum range at low altitude, lower minimum controllable speed, smaller radius of turn at altitudes higher than 20,000 feet, shorter take-off distances, and higher service ceilings.

Take-off roll was cut from 1,660 to 1,450 meters at a weight of 21 tons. Climb to 40,000 feet in a clean 19 ton aircraft took 12 minutes for a Vautour A with a regular wing and nine-and-a-half minutes for a Vautour B with the convex leading edge wing; the former covering 90 miles and consuming 1,040 liters of fuel, while the latter covered only 70 miles and consumed only 870 liters of fuel. Climb to 48,000 feet in a clean 19 ton aircraft took up to 43 minutes for a Vautour A with a regular wing and up to 45 minutes for a Vautour B with a convex leading edge wing; the

The original non convex leading edge wing of the Vautour is clearly visible in this view of a Vautour taxiing at Ramat David; in a convex leading edge wing, the leading edge was dropped out of the wing fence so the wing's leading edge was not continuous from root to tip. (PC)

A Vautour landing at Ramat David; the continuous leading edge of the non-convex leading edge wing is clearly illustrated. (PC)

former covering 118 miles and consuming 1,175 liters of fuel, while the latter covered 108 miles and consumed only 1,040 liters of fuel. Range at 50,000 feet (with full internal fuel and no external or internal loads) for a Vautour A with regular wing and Vautour B with convex leading edge wing was 900 miles (endurance of 1:55 flight hours) and 1,000 miles (endurance of 2:10 flight hours), respectively. Range at 48,000 feet with maximum fuel (internal, fuel tanks in bomb-bay, and two external tanks) was 1,680 miles at 3:40 flight hours and 1,925 miles at 4:10 flight hours, respectively.

All of this aforementioned data comparison actually confirmed the logic of the IAF's decision to modify the Vautour B with the convex leading edge wing. The problematic data for the convex leading edge wing in the Vautour A case was in the radius of turn below 20,000 feet. In this regime, the Vautour B with convex leading edge wing was at a distinct disadvantage compared to the Vautour A with a regular wing. At 12 tons weight, the turn radius of a Vautour A with a regular wing was 1,000 meters (with external fuel tanks) and 1,100 meters (without external fuel tanks) against 1,700 and 1,550 meters for the Vautour B with the convex leading edge wing, despite respective minimum speeds of 124 and 112 knots. Since the Vautour A was expected to operate at low to medium altitudes, while the Vautour B was planned for high altitude operations, it was logical to adhere to the original IAF decision, regardless of the Squadron 110 appeal.

Vautour A Limitations

Aggressive flying was encouraged among the ranks of IAF combat aircraft squadrons. Squadron 110 pilots surely flew their Vautours in an aggressive manner. In addition to the aforementioned fuel leaks (more of this subject hereafter), the Vautours suffered from structural damage; mostly cracks, loose riveting and broken supporters. A fine example was the discovery of broken flaps supports in 4311 as reported on 22 June 1960. Subsequent examination revealed that 13 more Vautour A aircraft (out of a total fleet of 15 aircraft) suffered from the same problem, which was related to excessive loads that the rear sections of the engine housings were subjected to and/or to incorrect adjustment of the flaps.

Wing 1 Commander Menahem Bar responded on 4 August 1960 to an IAF Commander enquiry concerning possible high G abuse of the Vautour airframes; an issue that endangered safety and was expected to cut short the service career of the IAF heavy attack aircraft:

> There is a view that these malfunctions result from our manner of operating those aircraft at higher than maximum G loads and this view is probably partially founded.

Ramat David Commander Menahem Bar (left) responded on 4 August 1960 to an inquiry by IAF Commander Ezer Weizman (right) concerning high G in the Vautour (background). (IDFA)

Squadron 110 flew three reconnaissance missions over Jordan from 18 to 24 October 1960; each mission included a reconnaissance Vautour B and an escort Vautour A. Reuben Harel flew Vautour 14 to escort Yossi Sarig with Achikar Eyal, in Vautour 31, on the first of these missions. (PC)

> As for G loads, the current squadron limitation of 5 G is less than the limitation of the manufacturer but even this limitation is sometimes breached, regardless of discipline and punishment, because it is so easy to exceed the G limit during recovery from strafing and bombing [dives] and also during high speed air combat [training].
>
> Considering the aforementioned, it is unlikely to be possible to operate the Vautour A as a fighter-bomber with G limit lower than the current limitation but it is very important to be strict concerning the current G limit.
>
> Concerning the aforementioned, it is likely that the service life of the Vautour A will be shorter than the service life of other combat aircraft in IAF service.
>
> It is possible to extend the service life through extra maintenance and servicing: repairing cracks on time, strict inspections of riveting, periodic inspections of structural segments where Araldite adhesive was used instead of riveting. Obviously these works would require assignment of additional maintainers but will extend the service life of those aircraft and will ensure that no aircraft will unexpectedly disintegrate in flight.

The IAF accepted the view that further limitations on the G limit of the Vautours was impractical. Instead of the internal squadron G limit, Material Department Chief Judah Rabin issued an official IAF set of G limitations on 1 September 1960, a limit per every approved configuration. The official G limit ranged from a maximum of 5 G (for an aircraft with up to 2,000 kg of fuel and any external load) to only 3 G (for an aircraft with maximum internal fuel and load but without any external load).

Nighttime Reconnaissance

Photography from high altitude in a subsonic aircraft was quite safe at the time of the service introduction of the Vautour B. However, deeper incursions into enemy territory were less immune, as was evident during the 23 February 1960 mission over the Suez Canal. A short while later, a Soviet SA-2 SAM shot down the USA U-2 spy plane on 1 May 1960, and this incident signaled the future for subsonic high altitude photography. It was only natural for Israel to expect that Arab enemy defenses would improve. AIR4 reported that the first MiG-19 fighters were delivered to Egypt in September 1960, and initial operational capability of this supersonic interceptor was expected during 1961. The Mach 2 MiG-21 and the SA-2 SAM were expected to follow.

The ultimate IAF solution was a plan to purchase a Mach 2 Mirage reconnaissance aircraft, a variant of the Mirage III fighter, or a version of the Mirage IV bomber. Meanwhile, it was time to explore a different approach in order to revitalize the Vautour B for deep penetration missions. Nighttime reconnaissance from low altitude seemed to be the answer, and by September 1960 it was planned to return 4231 to IAI for modifications. To enable nighttime reconnaissance from low altitude, the Vautour B had to be installed with a flash magazine and with a different type of camera.

Fuel Tanks Work

Wing 1 Maintenance Squadron was unable to cope with the challenge of Vautour fuel leaks. Material Department specialists visited Ramat David on 15 March 1960 to discuss the issue of fuel leaks and to devise methodologies for proper repair work. It was impossible to remove the old sealant with chemicals because available chemicals damaged the Araldite adhesive that was used in the wing structure instead of rivets. Manual removal

Isaac Golan flew the depicted Vautour 08 to escort Reuben Harel with Achikar Eyal in Vautour 31 during a reconnaissance mission over Jordan on 24 October 1960; three days later, Reuben Harel flew Vautour 08 during the Suez War Day fly-past. (PC)

Vautour 12 at IAI; MAT2 reported on 18 January 1961 that "Vautour 12 has been at IAI for months". (PC)

and cleaning was a tedious job that involved a lot of scratching and took quite some time to accomplish.

By year end the issue of leaks from the Vautour's integral wing fuel tanks was not solved. On the contrary, the leaks got worse, with 300 liters of fuel leaking from a Vautour during one night and several incidents of local fires. IAI Chief Executive Officer Al Schwimmer met Material Department officers on 13 December 1960 and undertook to deliver three repaired Vautours per month from February 1961, provided that the IAF would undertake that six Vautours would be at IAI for integral wing fuel tanks sealant work at any given time during the duration of the project; the IAF was also to undertake to deliver sealant materials to IAI.

IAF Commander Ezer Weizman stepped into the arena in order to speed up resolution and ordered, on 14 December, to transfer the fuel leaks repair work to IAI. Work at IAI, in three shifts, was to start immediately, and was to be completed by July 1961. In order to accomplish repair work within this time frame five to six aircraft were to be grounded simultaneously, and the IAF Commander ordered to deliver five Vautours (three A, one B, and one N) to IAI so that work would be initiated. AIR3 was instructed to update the Squadron 110 readiness objectives for the planned duration of this repair work.

Troubles surfaced almost immediately. The work involved dismantling, removal of old sealant, cleaning, application of new sealant, and reassembly. Cleaning consumed 2,000 working hours over a time frame of three weeks in three shifts. Total working hours per aircraft was 4,000. MAT3 reported on 18 December 1960 that, due to simultaneous work on several aircraft, the IAF had run out of sealant stock. A special IAF transport aircraft was to fly to France to collect sealant. MAT3 also suggested to inquire if Vautours in FAF service had also suffered from such severe fuel leaks, and to explore the possibility of replacing the integral wing fuel tanks with conventional old fashion tanks even at the expense of smaller capacity.

Sharfrir Test Refinements

The scenario of the 15 July 1960 Shafrir test was repeated on 16 December. This time the missile acquired the hot target on the ground from a range of 3.5 km, but again the control system was far from perfect and the missile missed. Regardless of this setback, it was decided to progress to tests against an aerial target. The selected aerial target was a T-10 rocket that was modified with extra heat generators. Homing tests started on 23 January 1961, when Reuben Harel flew Vautour 26 as a test target and test rocket launch platform for the Shafrir Vautour piloted by Jacob Agassi. In subsequent flights – this phase lasted until 12 March 1961 – it was proved that the Shafrir was able to acquire the test target Vautour, as well as the test rocket, from ranges of up to four to five kilometers at altitudes of 10,000 to 30,000 feet. The Shafrir was thus deemed ready to progress to the next phase of its test program: launch against a simple aerial target.

OMERA 31 Camera Test

The K-38 cameras produced satisfactory, but not superb, images. A possible alternative was explored when an OMERA 31 camera, the only one of its kind at the time, was loaned to the IAF in January 1961. MAT4 issued a request on 16 January for flight test of the camera and indicated that this request was urgent, because the OMERA 31 was loaned to the IAF for only 30 days.

Equipped with a 600 mm lens, the OMERA 31 was a modern, compact camera with a mirror that minimized its external dimensions. Therefore, it was possible to fit this camera in the forward camera bay of a Vautour B. MAT4 suggested that the camera would be fitted in Vautour 31, replacing the Zeiss camera, and that a single (vertical) K-38 camera would be installed in the bomb-bay for comparison of photographs.

AIR3 issued the relevant Mission Order on 17 January 1961; the objective was to compare the 600 mm OMERA 31 with the 36" (914 mm) K-38, and four flights were authorized for the test. Navigator Achikar Eyal flew two of these test flights in 4231, the first (on 11 February) with pilot Abraham Kimchi and the second with pilot Isaac Golan.

Fuel Tanks Work Problems

IAI undertook to deliver three Vautours per month, fully cured from fuel leaks, from February 1961, but this objective proved difficult to accomplish. Both IAI and IAF failed to achieve objectives set in December 1960. An IAI obligation was to assign to this project 54 workers, working around the clock in three shifts. Recruiting skilled workers for this kind of job proved troublesome, and by 5 February 1961 IAI was able to maintain only 18 hours of full work, rather than the promised 24 hours. The IAF failed to deliver to IAI the sealant material it had undertook to obtain; the first shipment reached IAI on 5 February and

Reuben Harel flew Vautour 26 as infrared target and as a T-10 rocket launcher during the Shafrir test in January 1961. (PC)

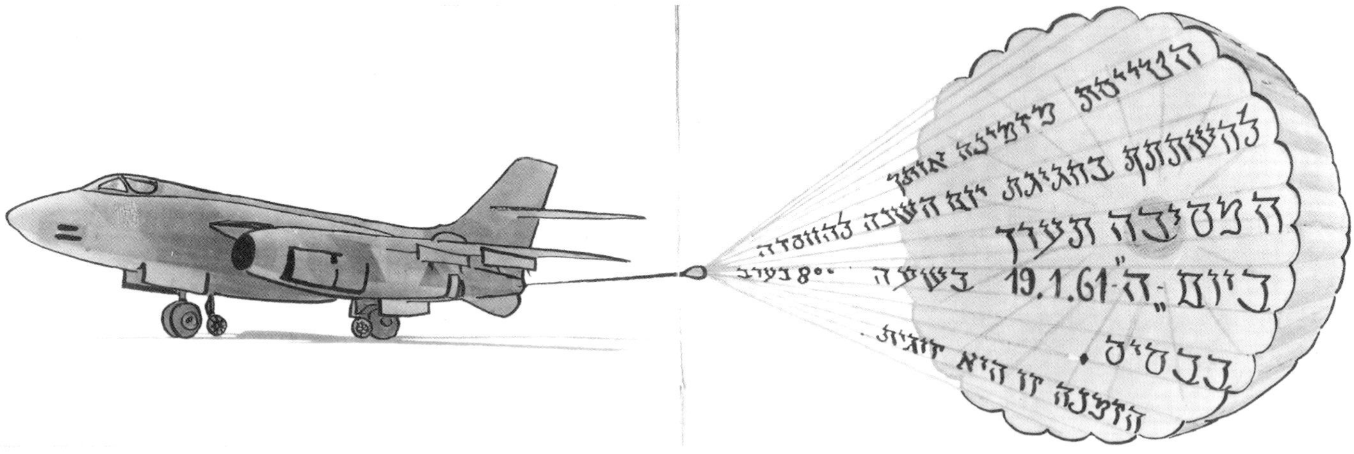

Invitation to the Squadron 110 anniversary evening on 19 January 1961. (PC)

Both Vautour 12 and Vautour 17 were among the first group to attend the IAI fuel tanks repair work. (PC)

was sufficient for only three aircraft. IAI planning of this work was based on the assumption that six aircraft would be at its facility at any given time during the duration of this project, but the sixth Vautour (4318) was only expected to arrive at IAI on 12 February. Of the five Vautours present at IAI on 5 February, only one (Vautour 12) was in the final phase of the work. Another Vautour was undergoing sealant application, while Vautour 17 was completely cleaned of old sealant. Vautour 07 was undergoing cleaning and Vautour 11 arrived from Ramat David in a poor state of previous sealant work, to the point that IAI estimated that an extra month of work would be needed.

Reuben Harel flew Vautour 31 to IAI on 21 February 1961 and returned to Ramat David in Vautour 12; the first to emerge from the IAI fuel tank 7 work. The following day – 22 February – MAT3 telegraphed Wing 1:

> The Squadron 110 aircraft are leaking. The sealing of fuel feed to tank 7 is authorized.

Exercise SCYTHE

With some six Vautours at IAI for fuel tanks work, on top of the normal periodic maintenance schedule, the squadron suffered from availability of a smaller than usual number of aircraft. Exercise SCYTHE, from 27 to 28 March 1961, was a Ramat David wing level combat day that ended Year 1960 Term 3. Emergency posting pilots and navigators arrived at Ramat David on the morning of 27 March, and as usual, fast turnaround was a cornerstone of the combat day exercise. Air crews flew up to three missions per day, so the flying tempo during SCYTHE was less intense than in previous combat days; perhaps this was the impact of the lower availability of aircraft.

It was during the first day of the exercise that Vautour 37 – the only Vautour B that had not been converted to a reconnaissance aircraft – was planned to fly a SCYTHE mission in the lead of a 4-ship formation. Since combat day was a simulation of wartime, the aircraft were loaded in realistic ordnance configurations. The pilot who flew Vautour 37 in a previous SCYTHE mission barely managed to control the aircraft during take-off because of distinct pitch up, but the crew did not report this mishap and Vautour 37 was quickly prepared for the next SCYTHE mission. Ephraim Ronen recalled:

> Every Friday morning a Dakota flew emergency posting air crews from the Flying School [at Tel Nof] to Ramat David where we flew our training flights. We enjoyed these Fridays

Jonathan Shahar – photographed in the cockpit of a Vautour back in March 1959 – was a Flying School instructor and a Squadron 110 emergency posting pilot until November 1961, when he returned to the regular cadre of 110. (PC)

> because the mess at Ramat David was well known in the IAF [for its quality]. Once every Term of four months there was a combat day and we deployed to Ramat David for the duration of the exercise.
>
> It was a combat day and we were in the tent outside the squadron building; tents were erected during combat days to accommodate all of the emergency posting and reserve air crews. I knew that I was assigned to fly with [Moshe] Papo, take-off at 10:10, and for the meanwhile we played backgammon. I looked at my watch and it was 09:00 but the next time I had a look the time was still 09:00. Then I realized that something was wrong with my watch. I asked [Abraham] Kimchi 'what is the time?' and he retorted '10:00' so what shall I do? I have to fly with Papo. Kimchi had a motorcycle and he drove me to the HAS but it was too late. It was a usual flight for a combat day that could have been flown without a navigator so by the time we arrived at the HAS we saw the aircraft taxiing out. We remained there, nearby the HAS, and watched the take-off. Papo was the leader so he took-off first and maybe at 300 or 400 feet the aircraft pitched up and he crashed in an awesome fireball. We rushed with the motorcycle to the crash site but there was nothing we could do.

Squadron 110 pilots felt that if a navigator was in the nose of Vautour 37 then the aircraft may have succeeded in taking off, as was the case in its previous SCYTHE mission. The IAF concluded that Vautour 37 was out of center of gravity limits in any case. Jonathan Shahar recalled:

> I was standing on the lawn outside the squadron building – some 200 meters north-west of the junction of runway 09/27 and runway 15/33 – when Ephraim Ronen approached me and asked 'what is the time?' I answered 14:00 and he retorted 'oh my, I have to take-off right now' so we remained there to watch the take-off. We saw Papo's Vautour roll along runway 15. Right at the junction, where vehicles used to cross the runway, the Vautour lifted its nose, pitched up and crashed.
>
> The [integral wing] fuel tank [7], ahead of the center of gravity, was not filled with fuel ... The aircraft was loaded with two 500 kg bombs on the outer wing stations, aft of the center of gravity, but lacked fuel in the [integral wing] fuel tanks between the engines and the fuselage. These fuel tanks were sealed because of fuel leaks but the impact of this action on the center of gravity was not explored. The Vautour B was more affected because it lacked the [heavy] guns pack in the nose section so even in normal situation its center of gravity was aft in comparison to a Vautour A.

Nighttime K-37 Photography Test

The expected improvement of enemy air defense systems motivated exploration of long term and short term solutions. An optional long term solution surfaced on 6 June 1960, when Israeli Prime Minister David Ben Gurion notified French President Charles de Gaulle, in a summit meeting, that Israel was interested in acquisition of three or four Mirage IV Mach 2 bombers modified for reconnaissance. The Mirage IV would only enter French service in 1964 and would never enter Israeli service, so interim solutions were sought in order to preserve the credibility of the IAF reconnaissance force.

One such short term solution was a switch from high altitude daytime photography to low-medium altitude nighttime photography. Vautour 35 was fitted with a K-37 camera in the front compartment and a flash ejector system in the front suite of the bomb-bay. The K-37 was a camera that was optimized for nighttime photography from medium to high altitude, and even equipped the famous U-2 spy plane.

The first series of test flights covered five sorties from 31 May to 5 June 1961. The objectives of these flights were to test the installation in flight, to test the functionality of the flash system, to test the operation of the camera, and to evaluate the quality of the photos as taken in various combinations of altitude and speed. The installation of the system was tested at up to 520 knots and at up to 5 G and was sound. The flash system functioned properly, did not impact handling, and did not damage the airframe. Photography was tested from 8,000 feet (at 360 and 420 knots) and from 5,000 feet (at 360 and 435 knots); the quality of the photographs was better when taken at higher speed and lower altitude.

Minor modifications were applied, and another test was flown on the evening of 11 June 1961, when Vautour 35 took-off with 25 flashes and flew seven photographic runs over Megiddo airfield, taking two to five photos at every run for a total of 25 photos at a time interval of three seconds between photos. The test was not successful. The aircraft took the photos from 5,800 and 10,000 feet, while the test order specified 5,000, 8,000 and 8,500 feet (at 430, 470, and 500 knots); the camera did not function properly (including improper feed of film, so that only 22 photos were taken); the aircraft was not stable enough, the camera vibrated, and the photos were not sharp.

The process of modifications and tests therefore continued. On 18 June 1961 AIR3 issued a Mission Order covering two test flights of a K-37 camera as installed in Vautour 35. Mission Order brief was for the first flight to take-off from Tel Nof at 01:00 on 21 June, with the second to follow from 02:00. Specified crew was Jacob Agassi and Giora Bar-Nir, but the actual crew was Yossi Sarig and Bar-Nir. Interestingly, after flying those two nighttime test flights Sarig flew four additional daytime sorties (two for napalm and strafing practice, one low altitude bombing training, and one low altitude navigation exercise) on 21 June 1961!

Shafrir Against Rockets

The first series of Shafrir test launches against an aerial target were staged from 20 June to 14 July 1961. One Vautour was tasked to launch the specially configured T-10 rocket and then to break away in order to avoid the slight chance that the Shafrir would home onto the Vautour instead of chasing the rocket. The other Vautour was tasked to launch the Shafrir fitted with a telemetry package instead of a warhead. This series of tests was plagued with technical problems and only two Shafrir missiles were launched; both missed.

Exercise SPUR

IAF Year 1961 end of Term 1 combat day at Ramat David was Exercise SPUR from 25 to 26 July 1961. Pilots flew up to four missions (including one nighttime mission) during the first day of the exercise and two missions on the second day.

Among the emergency posting crews were Flying School Commander Ezekiel Somech and Flying School Preparatory Squadron Commander Ephraim Ronen. Somech and Ronen were tasked to fly a BDA reconnaissance mission over the ranges that were used during the combat day, as an element of the exercise. Ephraim Ronen recalled:

> From my incident with [Maoz] Poraz on 4 March 1960 [until 26 July 1961] I was involved in a string of

The squadron staged two fly-pasts during Independence Day on 20 April 1961; Reuben Harel flew Vautour 03 in both fly-pasts. (PC)

Yossi Sarig flew four daytime sorties on 21 June 1961 – including low altitude bombing in the depicted Vautour 03 – after flying two nighttime reconnaissance test flights in Vautour 35; a pen with a roof is visible behind the runway, just below the taking-off Vautour 03. (PC)

Damage to Vautour 31 after the landing on 26 July 1961 was minimal, but still took time to repair, the next known flight of Vautour 31 was recorded on 4 December 1961, when it was flown to IAI at Lod. (PC)

five incidents/accidents. A few pilots even stated that they were not willing to fly with Ronen while other pilots said that Ronen is a talisman; after all I was a navigator and I was not responsible to any of the accidents but in all accidents when I was onboard the aircraft, the crew survived.

We flew our mission and returned to land. The procedure was to report to the control tower 'greens' thus indicating that all four landing gears – two main tandem units and the two balance wheels – were down. This time one bulb was not on. We aborted landing and checked switches but nothing happened so we flew over the control tower for visual inspection. The control tower reported that the rear main wheel was not down. The emergency procedure in case the front main wheel was not down was for the navigator to eject because in the Vautour B the navigator compartment was in the nose section. This was not the case so I was relieved, I would not have to eject. I was trained in parachute jumps and as Flying School Preparatory Squadron Commander I used to jump with every class but still I resented the idea of ejection.

We then tried everything. Emergency manual activation, high G maneuvers; nothing worked so we decided to prepare for an emergency landing. We still had a lot of fuel left – in an emergency landing minimum fuel is desirable in order to minimize the risk of fire – so we flew to the Mediterranean Sea, consumed fuel and returned to Ramat David. Somech suggested that our first approach would simulate landing so that we would be better prepared to land on our second approach. The idea was to fly final at a higher speed than usual so that the tail section would remain in the air for as long as possible after the front wheel touched down. We turned into final for our training simulation when the hydraulic pressure warning light came on. Instead of flying a simulation we landed immediately, the runway was cleared and all the rescue services were ready for us. Then Somech showed himself as an excellent pilot, he placed the front wheel on the ground right at the start of the runway and held the aircraft firmly until we sufficiently slowed down and the tail fin hit the surface; we yawed, the balance wheel in the direction of the yaw collapsed, we veered off the runway, the wingtip hit the ground and we stopped.

We exited the cockpits, I jumped to the ground, approached Somech, shook his hand and congratulated him for an extremely well done landing; Somech retorted, 'Ronen, I do not believe in superstitions but next time find someone else to fly with!'

Green Light for Landing at Damascus

As already mentioned, senior Squadron 110 pilots flew over enemy air bases during full moon nights to amass intimate knowledge of the enemy. Only the Vautour was authorized to fly such missions because of its performance and safety. Reuben Harel recalled:

> There was a lot of trust in the two engines safety of the Vautour. It was like in a contemporary F-4 Phantom advertisement, in the magazine Interavia, with an image of a Phantom flying on one engine and the title of the advertisement was IF YOU LOSE ONE, YOU STILL HAVE ANOTHER ONE TO GET YOU HOME.

Then it was decided to broaden the base of pilots who would be exposed to such an experience through the utilization of the Vautour B. Commanders and deputy commanders of IAF combat aircraft squadrons joined Vautour pilots, taking a seat in the nose section cockpit of the Vautour B for a grand moonlit view of enemy air bases. Squadron 101 Commander Joe Alon flew

Joe Alon (left) first flew the Vautour on 1 April 1959 and was Squadron 101 Commander from August 1960; on 31 January 1961 he flew a nighttime mission in the navigator's compartment of Vautour 37 with pilot Simon Ash. (PC)

Yossi Sarig (right) was authorized to land at Elmaza airfield, Damascus, in July 1961; the other Vautour pilots in this photo are Eli Bahat (center) and Oded Marom (left). (PC)

such a mission on the night of 31 January 1961 in the nose of Vautour 37, flown by Simon Ash in his last days as a Squadron 110 emergency posting pilot prior to taking over command of Squadron 105 Super Mystere from 12 February 1961.

The turn of Squadron 109 Deputy Commander Josef Zuk came on the night of 27 July 1961. The pilot of Vautour 35 was Yossi Sarig, and their mission was "watch and learn" while flying over Elmaza, on the outskirts of Damascus, Dmer, to the north-east of Damascus and Rayak in the Lebanon Valley. Yossi Sarig recalled:

> We flew all over the Middle East, mostly doing reconnaissance missions. We were well acquainted with the region and occasionally we took along with us deputy commanders of other combat aircraft squadrons. They would occupy the navigator seat, as 'passengers', and therefore were free to concentrate on observing details in enemy air bases that we flew overhead.
>
> One night, I took for a ride Josef Zuk who was Deputy Commander of a sister squadron in our base. Our mission was to fly at very low altitude, 100 to 200 feet [above ground level], on a full moon night, all the way to Elmaza air base, an airfield outside Damascus that was also the airport of the Syrian capital, then further to the east to Dmer air base and from there to Rayak air base in Lebanon. We were briefed to fly over each target a single round circle at an altitude of 2,000 feet, count the resident combat aircraft and get acquainted with the layout of the base and surrounding terrain compartments.
>
> When we arrived at our first target, we noticed an airplane with a rotating beacon over the airfield. I kept an eye on it and pulled up to 2,000 feet to start reconnoiter over the air base. I turned left so that my 'passenger' would have a good view of the airfield layout and would be able to count the MiGs that were parked in a line parallel to the main runway.
>
> The counting of the MiGs started an amiable argument; we each counted a different number. So we needed another pass to agree on the correct count. The second pass did not solve the issue. So, still keeping an eye on the unidentified aircraft, I reduced speed, lowered flaps and gear with an intention to simulate a landing pattern. I hoped that once flying slowly at an altitude of 50 feet along the runway, and parallel to the line of MiGs, we would be able to agree on the MiG count.
>
> Flaps and gear down, I turned the Vautour into final approach. To our surprise we received a green light followed by a green rocket, fired from the Damascus control tower, clearing us to land. The control tower probably mistook us for the other aircraft circling above and possible suffering from a communication failure.
>
> Zuk and me started to laugh, but I was still occupied flying the Vautour at 50 feet along the main runway with flaps and gear down, hopping to agree on the number of MiG. It all took seconds; flying very, very low and counting the MiG-17s. When we passed beside the control tower, they realized their mistake and no doubt had a very good view, and from very close, of a full size Israeli Vautour. In seconds all hell broke loose.

They immediately switched off all lights in the airfield. Seconds later, the lights in Damascus city were switched off in four segments; one after the other as though someone, somewhere was switching off four switches one after the other. Simultaneously, I retracted flaps and undercarriage, opened the throttles and when the anti-aircraft artillery opened fire we were already zooming upwards to 20,000 feet or so, at which point I ran out speed and rolled over. When we were upside down, a shell from radar controlled anti aircraft gun exploded maybe 300 feet below us. The explosion rocked the aircraft and the flash was extremely bright; Zuk yelled 'we were hit'. I rolled the aircraft, checked the systems, everything was in order so we dropped back to low altitude and completed our mission with passes over Dmer and Rayak.

K-37 Camera Qualification

The 18 June 1961 test on the nighttime K-37 verified that the modification of the camera mount alleviated vibrations. The next test flight – flown on the evening of 30 July 1961 – was aimed at figuring out the optimum parameters for best output. Various combinations of speed and altitude were evaluated from 390 to 470 knots and from 4,000 to 8,000 feet. It was concluded that the power of the flash was sufficient for photography at an altitude of up to 5,000 feet, but that for photography from 8,000 feet a full moon night was a must. The optimum parameters were thus an altitude of 4,000 to 5,000 feet and a speed of 470 knots, while photography at full moon night was recommended for best results.

On 8 August 1961 MAT4 recommended conducting one final test, at the recommended parameters of 5,000 feet and 470 knots, prior to qualification of the system. This test was flown on the evening of 13 August; pilot Reuben Harel and navigator Giora Bar-Nir photographed Hatzor from 5,000 and 8,000 feet at 470 knots. The night was moonless, but the quality of the photos was superb; it was possible to identify the exact types of aircraft that were photographed at Hatzor. This was partially related to a new Unit 121 procedure that was learned from their American colleagues, to use a fresh film that was previously stored in a refrigerator. The better photos were taken from 5,000 feet, so MAT4 recommendation was to photograph from 5,000 feet and 470 knots and to qualify the K-37 camera for operations.

Stick Bombing

At the time, the IAF viewed air base attack as its most important mission; the key to success in any future air superiority offensive. As such, air base attack practice was a core activity in IAF training. Hitting the runway in those days of ballistic bombing was a matter of aiming, wind and statistics. An attack along the runway axis was the simplest pattern, but it was prone to all bombs missing the runway in case of crosswind. To ensure a hit, the IAF practice was that all aircraft in a 4-ship formation would attack the same runway. If a leader missed due to crosswind then the wingman was expected to correct his aiming accordingly.

The Vautour was the only IAF combat aircraft capable of delivering more than two bombs in realistic air base attack mission profiles. While the attack pattern of all other IAF combat aircraft was a single bombing run along the runway axis – followed by strafing passes against parked aircraft – it was suggested that the Vautours would approach the runway at a certain angle and drop all their bombs in a single salvo. Because of the intervalometer, the bombs were released in intervals and created a stick of explosions, thus enhancing the statistics of hitting the runway.

Is this Vautour 11 or Vautour 15? Squadron 110 deployed two Vautours to Tel Nof on 1 August 1961. Reuben Harel flew Vautour 17 from Ramat David to Tel Nof, flew an aerobatic display over Tel Nof in Vautour 15, and returned from Tel Nof to Ramat David in Vautour 17. IAF practice was to add a digit to an aircraft number when on some sort of public display, like this visit to Tel Nof on 1 August 1961; the Vautour was also painted with a red nose cone, the first known application of this feature. (PC)

A Vautour head-on view with two 1,250 liter drop tanks under the wing's inner stations. The Vautour's air base attack configurations in 1961 included ten bombs (six in fuselage and four under wings) for very short range missions, eight bombs (six in fuselage and two under wings) and two drop tanks for short range missions, four bombs under the wings and fuel in the bomb-bay for longer range, or two bombs under the wings plus two drop tanks and fuel in the bomb-bay for the longest range missions. (PC)

Apparently not all of the Vautours were equipped with the original French intervalometer with adjustable time intervals. Most Vautours were fitted with an Israeli made intervalometer with fixed time intervals. Primarily due to this reason, plus a few tactical excuses, the stick bombing pattern was not adopted. Still, the Vautour could lift up to eight bombs in an air base attack mission configuration, so unlike other combat squadrons that flew a single bombing run along the runway axis followed by strafing passes, the Vautours were to fly two bombing runs against the same runway, releasing four bombs in each attack, and then fly the strafing passes. Squadron 110 Commander Jacob Agassi reported on 27 August 1961:

> We abandoned the stick [bombing] pattern and we bomb twice in the same manner as all other squadrons.

Vautour B Re-Engining Proposal

Another short term solution suggested to enhance the performance of the Vautour B was re-engining. The performance of the Vautour B was sufficient for short range high altitude reconnaissance missions over Egypt, as well as other Arab nations less well defended, but inadequate for long range high altitude operations against Egyptian defenses. The K-37 camera was accepted for nighttime operations, and this enabled the Vautour B to resume long range missions over Egypt, but the output of daytime photography was still superior. It was decided that an improvement in performance might offer the Vautour the required level of immunity while flying long range high altitude operations over Egypt.

Sud Aviation presented a proposal to re-engine the Vautour B with an Atar 9B in a Material Department meeting on 14 September 1961. The manufacturer was not fully aware of the exact configuration of the IAF Vautour B reconnaissance aircraft. The presented configuration was therefore different: convex leading edge wing, two 1,250 liter external fuel tanks, one 1,500 liter fuel tank in the rear suite of the bomb-bay, one K-38 camera in the front suite of the bomb-bay in an installation partially protruding underneath the fuselage, and 3.5 tons thrust Atar 101E-3 or 6 tons thrust Atar 9B with full reheat.

The compared mission profile was take-off, climb to maximum service ceiling, cruise and rapid descend back to landing. Take-off run was cut from 1,000 to 600 meters. The Vautour with Atar 101E-3 engines climbed to 43,000 feet in 12 minutes 50 seconds, while it was calculated that the Atar 9B aircraft would climb to 43,000 feet in four minutes 15 seconds; the respective service ceilings were 44,000 and 55,000 feet. The calculated service ceiling of the Atar 9B Vautour was not practical, because flutter was expected at 47,000 feet and because it was impossible to fly with external fuel tanks higher than 53,500 feet. Thus, the practical ceilings were 43,000 to 46,000 feet for the Atar 101E-3 Vautour and 46,000 to 49,300 feet for the Atar 9B Vautour if climbing through flutter would be certified. Cruise comparison was enlightening; at 46,000 feet the Atar 101E-3 Vautour starting speed was Mach 0.82, and it reached Mach 0.87 at the end of the cruise after an endurance of 2:29 hours and a range of 2,250 km, while the Atar 9B Vautour speed was Mach 0.905 to 0.92 with an endurance of 0:40 hours and a range of 650 km!

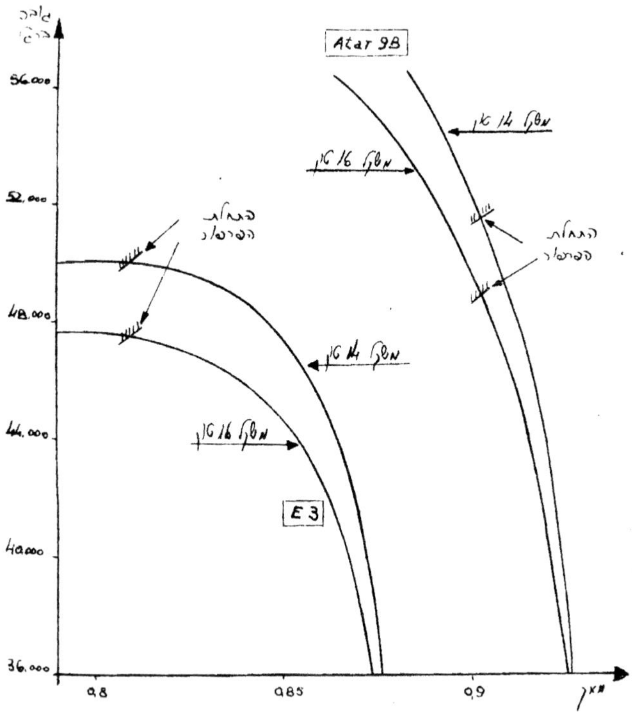

Mach number (X axis) comparison at various altitudes (Y axis) between an Atar 9B engined Vautour B (weighing 14 or 16 tons) and an Atar 101E-3 engined Vautour B (also weighing 14 or 16 tons); the limits are where flutter started. (IDFA)

Obviously, re-engining the Vautour B with the Atar 9B was not a solution for the problem that faced the IAF. Even if re-engining could have been accomplished within the short time frame remaining until the start of Egyptian MiG-21 and SA-2 introduction, the calculated gains in speed and ceiling were too small to matter, while the cost in range was unacceptable. The major improvements in take-off run and climb were irrelevant for the Vautour B reconnaissance mission. AIR2 summed up this issue on 24 September 1961:

> The improvement in performance critical for reconnaissance ... is too small and, considering the cost and time of the modification, not worthwhile.

Shafrir Against Flare

The second series of Shafrir test launches against an aerial target was staged from 17 September 1961. Technical problems persisted, and it was impossible to launch the Shafrir during the first test flight because the flares fitted to the T-10 rocket did not ignite and the Shafrir's seeker was unable to acquire the target. The test was suspended until the evening of 8 October, when the Shafrir was launched against a flare that Reuben Harel dropped from Vautour 05. The Shafrir flew right through the halo of the descending flare. In light of this success the test was resumed from 10 October, but all six subsequent Shafrir launches missed the T-10 rocket target. The conclusion of RAFAEL and the IAF was that the T-10 rocket was not a fair target for the Shafrir because the rocket was too fast.

Nighttime Close Air Support

Squadron 110 experimented with nocturnal close air support in a test of an operational scenario on the evening of 21 September 1961. The target was a simulation of a typical enemy company sized fortification, measuring 200 by 200 meters. The target was on a small hill and was slightly lighter in shade than the surrounding sand dunes. It was a full moon night, and a Forward Air Controller directed the Vautours to the target. When the Vautours were nearby the target two projectors were activated. The projectors were some five kilometers from the target, and their rays of light converged over the target. When the illuminating Vautour started to drop flares, the projectors were switched off and the attacking Vautours arrived to bomb, with 250 kg bombs, and to strafe.

The pilots reported that the crossing of the projectors' rays was clearly visible from 20 miles. The flares had to be dropped right over the target for best visual results. It was possible to see the target with one flare, but the view was much clearer under a cluster of three flares; either way it was impossible to identify details inside the target perimeter. It was impossible to assess battle damage, and smoke emanating from the target after bombing and strafing made it more difficult to acquire the target in the following attack.

It was also attempted to attack in the light of the projectors, but in this case it was impossible to bomb because the target was not visible – despite the full moon – at altitudes higher than 1,500 feet. Only strafing was possible with the support of the projectors.

A total of seven bombs were dropped, and four of these hit within the perimeter of the target. Half of the rounds fired during strafing hit the target. Overall, the test was considered successful. AIR3 indicated on 3 October 1961 that it was impossible to draw conclusions from a single test performed in ideal conditions and that more tests should follow. The prospects of flying more missions at night also prompted Jacob Agassi to notify AHQ on 27 September 1961 as follows:

> In light of the squadron's missions, and developments in all-weather fighting in our region, it is suggested to install rear warning (passive) in all Vautour aircraft that will undoubtedly fly many special nighttime operations for many more years.

Still Leaking

The IAF objective to cure all Vautours from fuel leaks by July 1961 was not accomplished. Leaks from fuselage tank 3 also started. When IAI forwarded to the IAF a survey of Vautour 20 on 14 August 1961, the estimated work was 5,700 working hours to seal the wings tank 7, 6,500 working hours to seal fuselage tank 3, and 7,500 working hours to repair a damaged wing; the estimated time frame to repair the wing was six months. Eventually the damaged wing of Vautour 20 would be repaired with parts from one of the four convex leading edge wing kits ordered for the Vautour B.

By 24 December 1961, only seven Vautour A aircraft had returned to Ramat David from IAI upon completion of the fuel tank sealing work. Vautour 17 was in periodic maintenance, so it was impossible to assess its status. As for the other six Vau-

Yossi Sarig flew Vautour 07 – photographed with 1,250 liter drop tanks – on the evening of 21 September 1961, when nighttime attack was tested; Reuben Harel flew Vautour 11. (PC)

IAI reported on 14 August 1961 that a convex leading edge wing kit was utilized to repair Vautour 20. This rather poor image of Vautour 20 – landing at Ramat David in 1960 – is extremely interesting because of the gunnery camera installation in the nose cone. (PC)

tours, only four of these were subjected to sealing of fuselage tank 3: two of these (4311 and 4314) were leaking again, while 4318 and 4320 were cured for the time being. As for wing tank 7 work, three aircraft (4314, 4318, and 4320) were cured; 4307 was leaking from right wing tank 7 and "sweating" from left wing tank 7, 4311 was leaking from the right wing tank 7, and 4312 was leaking from tank 7 left and right.

Return of the Convex Wing

The IAF decision to fit the Vautour B with the convex leading edge wing was not initially adhered to, even though the four kits were indeed purchased. In line with the IAF Commander's directive, the plan to fit the Vautour B with the convex leading edge wing was shelved on 6 March 1961 due to "time and expenses".

By January 1962, Vautour 31 was at IAI and was expected to remain there for at least three weeks. Material Department officers seized the opportunity to suggest revival of the plan to fit the Vautour B with the convex leading edge wing. The suggestion was to first fit Vautour 31, as it was already grounded at IAI, while the installation of the convex leading edge wing was estimated to consume only 600 working hours. Thus, the arguments of "time and expenses" were waved and the Material Department Chief added his approval on 16 January 1962. AIR2 followed with a formal IAF directive on 21 January.

A Night Over Cairo West

The IAF's top priority mission was accomplishment of air superiority. The chosen course of action was a campaign of con-

The Suez War Day was a major event in the IAF calendar, and mostly involved a turnaround competition between the Technical Flight teams (and sometime along the time line of these events an aircrew team was also added to this competition) and a fly-past. Vautour 05 was photographed at Ramat David during Suez War Day on 24 October 1961 with pilot Baruch Haviv in the cockpit; participants in the fly-past included Reuben Harel in Vautour 17 and Yossi Sarig in 11. (PC)

Vautour 21 readied for some sort of an open day at Ramat David; the displayed ordnance was obviously alternative options, including ten bombs (in the foreground) or six bombs in the belly (placed on bomb trolleys), with (under the wings) four bomb or four rocket pods or four napalm bombs or two drop tanks and two bombs (the latter loaded to the wing stations). (PC)

When Vautours "went public", a third digit – sometimes a prefix and sometimes a suffix – was temporarily added; in the case of Vautour 08 – on display at Tel Nof during a visit of an African personality – it was a suffix "4" that turned the Vautour into "084". IAF Commander Ezer Weizman is right behind the Vautour's drop tank (facing the camera with sunglasses), while Tel Nof Deputy Commander Yoash Tsiddon (facing the guests) is in between Weizman and the cameraman. (PC)

centrated attacks against enemy air bases. The objective of air base attacks was to neutralize the enemy force through damage to the runways and destruction on the ground of as many enemy aircraft as possible. Bombs were to be used to attack runways and guns against parked aircraft. The IAF actually structured its combat aircraft force around this crucial mission; IAF combat squadrons repeatedly practiced air base attack missions. An essential element of IAF air superiority plans – in addition to order of battle, planning and training – was precise intelligence.

Up to date photos of enemy air bases, especially Egyptian, were crucial for planning of the IAF's air superiority offensive. Squadron 110 photographed the Suez Canal air bases during the RETAMA mission on 23 February 1960, and again during a nocturnal mission on the night of 22 to 23 October 1961. The former was a daytime high altitude mission flown in Vautour 33 fitted with K-38 cameras; the latter was a nighttime low altitude mission flown in Vautour 35 fitted with a K-37 camera. This reflected changes in the perceived threat, as the MiG-19 entered Egyptian service from September 1960.

The nocturnal mission along the Suez Canal was a prelude to a more daring operation. This would be the Vautour's longest range, deepest penetration reconnaissance mission until that time. The objective was to photograph Cairo West air base and Abu Sueir air base, west of Ismailia. To achieve tactical surprise, the reconnaissance Vautour was to fly alone at very low altitude over the Mediterranean Sea, cross the Egyptian coastline well to the west of the Nile, fly south, turn left to photograph Cairo West from west to east, head east to photograph Abu Sueir, and only then return to Ramat David in high altitude flight over the Mediterranean Sea.

For such a challenging mission, the IAF Commander expected that Squadron 110 Commander would assign himself as pilot. But Jacob Agassi had already flown the nocturnal mission over the Suez Canal, and it was the turn of Deputy A Reuben Harel. Deputy B Yossi Sarig was also anxious to fly this unique mission, so it was agreed that Sarig would be assigned as Harel's navigator. The two deputies flew three preparatory flights together: a nighttime low altitude navigation training in Vautour 33 on 15 January 1962, a nighttime low altitude navigation training in Vautour 35 on 16 January, and a short nighttime Mission Order in Vautour 35 on 17 January. Sarig flew another short Mission Order, with Agassi in Vautour 35, on the evening of 22 January. They departed Ramat David in Vautour 35, bound for Egypt, at 01:00 on 23 January. Yossi Sarig recalled:

> With hindsight this sounds insane to assign two pilots for such a complex mission, but the two of us were so eager to fly this mission, and it was not clear whose turn it was, so we agreed to fly together. I volunteered to fly as the navigator.
>
> We selected a full moon night and planned a simple route to Cairo West, hoping that once in the Nile Delta area we would have no difficulty to get oriented between large towns and small villages. So we planned to fly along a straight line route from Ramat David to Cairo West. Flying at 450 knots, at an altitude of 100 feet; our only navigation aids were map and compass.
>
> At this speed, and with almost no wind factor, we calculated that we would cross the Egyptian coastline at a certain time. When the time came, all we saw were sporadic small yellow lights, which actually turned out to be fishing boats. As the navigator of this flight I started to wonder what was actually going on but I did not have a clue. Time now passed quickly, and no coast was in sight. We started to debate between ourselves what was going on but Harel had to concentrate in flying the Vautour as low as possible which, at that speed and altitude, was quite a demanding task. I on the other hand started to get goose skin pimples as I did not have a clue to what was really happening.
>
> Finally, a couple of minutes late, we crossed the coastline, but what we have expected to see, was not there. Harel was worried as I was, and asked 'what do we do next?' Since I was responsible for navigation, I really started to get nervous and was fearful of a 'disaster', a total navigation mishap. To fail in such a mission was not an option but Harel kept pushing me, asking me what to do next? Turn left? Turn right? This was not a good time for such questions as I did not have a clue where we were. I therefore asked Harel to give me some time to sort it all out. Full responsibility rested on my shoulders and I had to put aside my gloom thoughts and focus on what to do next.
>
> I looked at the map again and realized that crossing the Egyptian coast a few minutes late could only mean one thing; we are farther west to our planned track, a lot farther west. I made up my mind and said to Harel 'head due south, whenever we will cross the only road from Libya to Alexandria, turn left and follow it; this road will bring us to Cairo'. Harel slowed down from 450 knots, and some time later we indeed crossed the road from Benghazi, Libya, to Alexandria, Egypt; as I thought it was a desert road that could not have been missed.
>
> We turned left to fly along the road, accelerated and after 'eternity', actually a couple of minutes, we saw the lights of Cairo. We zoomed up to 8,000 feet and I had to set up the night camera and sync the flash bombs to burst exactly when the shutter opened. We flew two passes over Cairo West with our flashes probably seen all over the city of Cairo. Then, MISSION ACCOMPLISHED, we allowed ourselves to BUZZ the city at a very, very high speed, low altitude fly over.

The two deputies had to fly two photographic passes over Cairo West, one along the axis of each runway, in order to capture the whole target on film. They then completed their mission and returned to Ramat David after 1:35 flight hours. The film was processed and the printed photos were sharp and revealing. The AHQ officers preparing the IAF air superiority master plan now had the crucial intelligence data they so desperately needed.

Air Combat Training Accident

The qualification of new pilots developed into a routine process. At the start of every term, a small group of young pilots – mostly two, but sometimes three and rarely four – were assigned to Squadron 110. The young pilots were two terms after graduation from the IAF Flying School. In the first term, after winning their wings, they attended a Squadron 113 Ouragan Operational Training Unit course to qualify as fighter pilots. In their second term as IAF pilots they flew the Mystere, a preparatory phase prior to assignment to fly the Vautour.

Thus, a Vautour Conversion Course started in the squadron every term. The syllabus of the Vautour Conversion Course covered some 14 flights to enable the new pilots to fly the Vautour. Next they had to learn how to fight in the Vautour. The first term in the squadron was devoted to air-to-air training, and at the end of his first term in the squadron a new pilot was qualified as operational for air-to-air missions. Obviously the Vautour was not an air-to-air platform, so during his second term in the squadron, a new pilot mastered attack patterns to qualify as a truly operational pilot. The following phases in the qualification process of the regular squadron pilots was pair leader and then 4-ship leader; these two qualifications were not subjected to strict time frames, and every pilot was qualified as a pair leader and then as a 4-ship leader at his own personal pace. After about two years in the squadron the regular pilots were assigned to the Flying School as instructors and continued to fly the Vautour as emergency posting pilots. Having completed their roughly two years cadence as Flying School instructors, some pilots completed their IAF service contract and continued to fly with Squadron 110 as reservists, while others returned to the squadron and joined the ranks of the executives with individual pace promotion to Deputy B, Deputy A, and ultimately Squadron Commander.

The group assigned to Squadron 110 at the start of IAF Year 1961 Term 3 included four pilots: Amos Avni, Eitan Karmi, Mario Shaked, and Menahem Yarden graduated from Flying School Class 33 on 14 March 1961. They trained in the Ouragan during IAF Year 1961 Term 1, flew the Mystere during IAF Year 1961 Term 2, and soloed in the Vautour during December 1961. They were still first term pilots in Squadron 110 when a two versus

A Vautour with extended air brakes either side of the rear fuselage. (PC)

Vautour 11 (the suffix digit "6" is clearly added) in Pen 6 at Ramat David with 1,250 liter drop tanks; this Vautour was lost in an accident on 25 January 1962. (PC)

two air combat was practiced on 25 January 1962. Jonathan Shahar recalled:

> Simi [Saar] flew with Amos Avni and I was the leader of the other pair. The engagement started in a head on cross. After the pairs passed each other, we started to turn. Simi selected to climb in his turn and I dived. The engagement started at 25,000 feet so Simi climbed to 40,000 feet or so and I dived to 12,000 to 10,000 feet. When I reached the bottom of my dive I pulled up and saw them high above, some 20,000 feet higher. I saw Simi pulling some kind of maneuver and that Avni dropped down his nose and leveled his wings, starting a dive from about 40,000 feet. So I aimed at Avni and I caught him at an altitude of some 30,000 feet. We then dived steeply; I followed him at a range of about 1,000 meters, right behind him. I did not get any closer in order to avoid his slipstream, and I waited that he would turn so that I would be able to close and confirm a 'kill', 1,000 meters was too far for a kill. But he did not turn; he kept on diving straight down. When we passed through 15,000 feet I started to feel that handling was abnormal. The Vautour A had a conventional elevator [not a monoblock as in the Vautour B] and because of this it suffered from inversion of handling in some regions of transonic flight; in such cases the effectiveness of handling was considerably degraded. The proper action for recovery was to throttle back the engines and to deploy the air brakes. I realized that I was experiencing this phenomenon so I looked at the altimeter but I could not read it, the indicators were rolling that fast, and then I saw that the ground was getting closer and closer at a terrific pace. I throttled back, deployed the air brakes and pulled up. While pulling up I saw that the minimum altitude I reached during recovery was 8,000 feet. Then I rolled my wings to see what was Avni doing and I saw his Vautour flying into the ground, he did not recover from his dive.

PILLAR DAWN 110

The next series of Shafrir test flights – PILLAR DAWN 110 – started on 25 February 1962. The target was again a flare. Apart from the fact that the two launched missiles missed, the arrangement that one Vautour dropped flares and another launched the Shafrir proved complex. Therefore, a new concept was tested on the evening of 29 March when Reuben Harel in Vautour 23 dropped a flare and then locked the Shafrir's seeker on the flare. This procedure was accepted, and was used from 4 April to conclude PILLAR DAWN 110 with a successful hit.

Viscount Interception

Regular pilot Simi Saar and emergency posting navigator Oded Erez flew a reconnaissance training sortie over the Negev desert in Vautour 33 on 9 March 1962. It was during the course of this flight that IAF ground control vectored Vautour 33, an unarmed reconnaissance aircraft, to intercept an unidentified target. Oded Erez recalled:

> We were over the Negev when ground control asked us if we have enough fuel left. We were then ordered to fly at a certain altitude and heading, an interception! I said to Simi [Saar] 'tell ground control that we are a reconnaissance air-

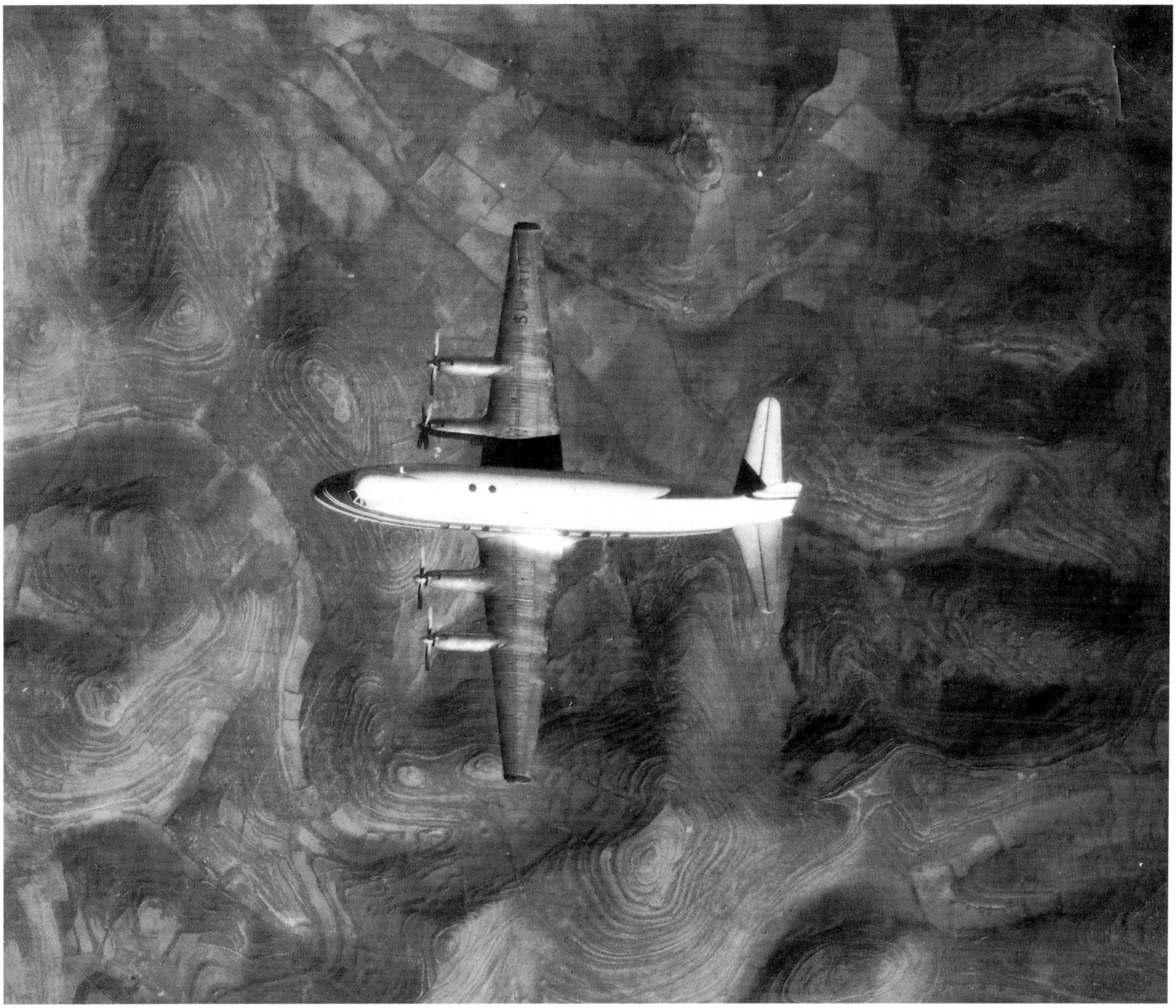

The Egyptian Viscount SU-AID that Oded Erez photographed from Vautour 33 on 9 March 1962; exactly a week later, on 16 March 1962, this very Viscount was written off in an accident when, on a flight from Jerusalem, Jordan, to Cairo, Egypt, it was diverted, due to bad weather, to Wadi Halfa, Sudan, ran out of fuel and crash landed in a field without fatalities among the 16 persons onboard. (PC)

craft, tell him that we are PICTURE' [in order not to reveal too much over the radio]. Simi transmitted to ground control 'I am PICTURE' and the ground controller retorted 'I know'. I was not sure what was this all about but I set the cameras on standby.

We crossed the Israeli-Jordanian border south of the Dead Sea and flew into Jordan. Ground control then gave us details of the target [as seen on the radar screen]. Previously I was a [Squadron 119 Vautour] night-fighter navigator so I asked for altitude data. At the time, ground control was unable to fix the exact altitude of a target but it was possible to produce a relative altitude so I asked ground control to place us 2,000 feet higher than the target. Ground control did as I asked and then, right at the corner of my eye, I saw, down below, a four engines airliner crossing our heading from left to right. At the time my proficiency in aircraft identification was poor, I referred to all four engines airliners as 'Skymaster'. I reported 'Skymaster' but Simi did not see so I leaded him to the target, like in my good old days as a night-fighter navigator, until we flew right above him at which point I activated the cameras and photographed what turned out to be an Egyptian Viscount. Apparently, they flew from Cairo to Amman and attempted a short-cut over the Negev Desert instead of the usual circumnavigation down the Gulf of Aqaba and up to Amman.

We landed and I was immediately summoned to the phone. AIR4 Chief was on the line and he questioned me what did I see. I answered that I saw a Skymaster and rushed to the photo lab. I took three photos, this was a mistake; I should have taken more. There was nothing in the first and third photos but right in the middle of the second frame was this airliner. The photos were delivered to AIR4 and I was reprimanded for my poor practice of aircraft identification, I should have known that it was a Viscount, not a Skymaster!

More Fuel

The Vautour was the Israeli premier long range heavy attack aircraft and reconnaissance aircraft. But even the long range performance of the Vautour was somewhat wanting. There was always a requirement for a little bit more fuel, more endurance, and more range for pure longer range operations, for safer margins in shorter range missions, and to enable the planning of innovative indirect approach routes to achieve tactical surprise over targets waiting for the Vautours to approach from a certain direction.

As indicated, AIR2 recommended on 11 May 1958 that 2,000 liter fuel tanks, to fit into each of the two suites inside the bomb-bay of the Vautour, should be purchased. Such tanks were not acquired, and on 14 March 1962, in a memo to AHQ, Squadron 110 Deputy A Reuben Harel resurfaced the issue of more fuel for the Vautour:

> An empty space exists in the Vautour A and B right behind the rear suite fuel tank. The B aircraft normally fly with cameras in the front suite and fuel tank in the rear suite so their range is shorter. The capacity of the rear suite tank is 1,500 liters and an inspection evaluated that this capacity can be increased to 2,000 liters.
>
> Since the B aircraft are currently flying a lot of operational missions it will be possible to fly farther at low altitude or to fly a high altitude indirect approach to a target.
>
> If the capacity of the tank is increased then the center of gravity might move backwards; a crucial issue, especially in the B aircraft, that should be carefully examined.

More fuel was always an issue in the Vautour; the fuel gauge was on the left lower part of the instruments panel, as illustrated in this view of the cockpit of Vautour 33. (PC)

Combat Debut

By 1962, the dispute concerning regional resources of water evolved into the Water War, a series of occasional armed clashes, mostly between Israel and Syria, that escalated towards activation of the Israeli National Pipeline, an Israeli national enterprise to pipe water from the Sea of Galilee to the Negev Desert launched in 1953 and scheduled for activation in late 1962, but was actually inaugurated on 10 June 1964. As work progressed towards activation, tension between Israel and Syria soared.

IDF forces attacked Syrian fortifications on the night of 16 to 17 March 1962. Squadron 110 was set on readiness and prepared to launch the standard nighttime attack formation: a 3-ship with an illumination leader loaded with flares and an attack pair armed with bombs. Weather was not favorable for air operations, with low clouds over the Golan Heights. When the battle evolved into an artillery duel, the IAF was ordered at 04:05 on 17 March to launch attack aircraft to bomb Syrian artillery batteries. The Squadron 110 readiness 3-ship was the first to take-off at 04:10, with Reuben Harel in the lead and Jonathan Shahar with Wing 1 Commander Menahem Bar as the pair. Cloud base was down to 1,500 feet, so the Vautours had to operate lower than usual. Harel dropped the flares over a Syrian artillery battery that was identified by the flash of firing. Shahar became the first pilot to attack enemy targets on the ground during the Vautour age of Squadron 110. Shahar's bombs were evaluated to have exploded some 200 meters from the artillery battery. Bar experienced difficulties and released his bombs much later; it is possible that his bombs landed on Jordanian soil.

Ouragans, Mysteres, and Super Mysteres were also scrambled but were not authorized to attack; 110 was the only IAF squadron to attack that night.

PILLAR DAWN 111

Another series of Shafrir test flights (PILLAR DAWN 111) started on the evening of 10 June 1962. The targets were still flares and missile functioning was still far from perfect, as

Reuben Harel flew Vautour 26 in the lead of the formation that attacked a Syrian artillery battery on 17 March 1962. (PC)

The Shafrir test aircraft Vautour 23 was lost in a fire on 22 July 1962; in this view that aircraft is in the HAS taxiway with T-10 rockets loaded onto the outer wing station and, under the inner wing station, a fuel tank that was modified with a camera to photograph testing. (BIAF)

only a single missile out of four launches hit the fairly static flare on 10 June. The second test was only slightly more successful; one of two missiles hit the flare on 14 July. A major setback was the total loss of the Shafrir test aircraft, Vautour 23, in a ground accident on 22 July. Jonathan Shahar recalled:

> It was circa 17:00, at the end of another working day. We exited the squadron building and we saw a small column of smoke emanating from the nearest HAS opposite the runway. I immediately said '23 has gone' and we rushed across the runway to the HAS. It was terrible; the fire was so intense that we could not get close to the HAS. By then a huge pillar of black smoke erupted from the HAS. Apparently, the aircraft was prepared for flight on the next morning and the electrical circuits were checked when a T-10 rocket was launched. Luckily, everybody escaped from the HAS. The rocket hit a fuel pipe and pressurized fuel splashed all over the place and burnt out the whole HAS including Vautour 23, vehicles, ammunition and everything. It was the most furious fire I have ever seen, a lot larger than the Six Day War fires that I subsequently witnessed over Fayid.

Vautour 09 was selected as the new Shafrir test aircraft and PILLAR DAWN 111 was resumed on 10 September 1962; the objective was to acquire a target Vautour during daylight. A second such flight was flown on 14 September and also included a daylight lock onto a flare. The first Shafrir launch from Vautour 09 was accomplished on 17 September, and was also the first launch against a flare during daylight; one missile hit the target and one missile failed to launch. An attempt to repeat this test the following day ended with disappointment, as the two launched missiles missed and thus ended PILLAR DAWN 111.

Mirage Impact 1

The first Israeli Mirage III Mach 2 fighters landed at Hatzor on 7 April 1962. The IAF planned to equip three squadrons with Mirages: the first (Squadron 101) to be based at Haztor, the second (Squadron 117) to be based at Ramat David, and the third (Squadron 119) to be based at Tel Nof. The first Mirages for Squadron 117 arrived at Ramat David on 7 July 1962.

The IAF was still evaluating acquisition of the Mirage IV as successor to the Vautour B reconnaissance aircraft, but such a purchase never materialized. Eventually, the IAF would equip with a pair of modified Mirages as reconnaissance aircraft, but these would only enter IAF service during 1964. Meanwhile, the impact of IAF Mirage introduction on Squadron 110 was still minor. A Mirage repercussion was a change of communication systems; the new Mach 2 fighter was equipped with UHF transceivers, while older IAF combat aircraft were fitted with VHF radios. Deputy A Reuben Harel suggested on 29 July 1962:

> The reconnaissance aircraft are currently escorted in their operational missions by Super Mysteres. It is reasonable that in the near future the task of escort will be taken over by the Mirage. The Vautour B is equipped with two VHF radio sets that are not compatible with the Mirage. It is therefore recommended to evaluate installation of UHF sets in all three

The Vautour was originally equipped with two VHF radio transceivers, but shortly after IAF introduction of the Mirage – which was equipped with two UHF sets – the Vautours had one VHF radio replaced by a UHF set, as illustrated in this cockpit view of a Vautour fitted with VHF radio-1 and UHF radio-2. (PC)

Compared to other IAF attack aircraft – actually former fighters (such as the depicted Mystere) relegated to attackers after service introduction of newer fighters – the offensive load of the Vautour was much more powerful, but the number of bombs was down as range was up; for this reason the IAF launched an effort to arm the Vautour with smaller bombs. (PC)

Vautour B aircraft as an urgent requirement rather than wait for an IAF general retrofit.

More Bombs

Neutralizing enemy runways was a key element of the IAF's air superiority offensive master plan. Hitting a runway with free fall bombs was a challenge, and the product, neutralization of a runway, was very much dependent on statistics. A proposed path to improve probability was to drop more bombs. However, the offensive load of the Mystere and Super Mystere in an air base attack configuration was limited to two 250 kg bombs, while the Ouragan could lift two 250 kg bombs plus rockets and the Meteor was armed with rockets only. The exception was the Vautour, which could lift up to eight bombs. However, the Vautour was also planned to attack the most distant enemy air bases, and in order to fly such missions the bomb-bay had to be fitted with fuel tanks and the effective offensive load was down to four or two bombs.

The IAF started to test the impact of 50 kg bombs, and in the wake of these tests Squadron 110 Commander Jacob Agassi signed the following memo on 27 August 1962:

> In light with recent tests with 50 kg bombs, it is suggested to explore the possibility to arm the Vautour with such bombs. The Vautour manual states that it is possible to load 36 50 kg bombs in the bomb-bay. It is necessary to fit special bomb racks to implement such a configuration. It seems that an investment in such bomb racks is worthwhile and should be considered.

AIR3 officer Adam Zivoni endorsed the Squadron 110 proposal on 6 September 1962. However, Zivoni recommended that the Vautour would lift 70 kg bombs, rather than 50 kg bombs, due to IAF inventory issues. The stumbling block of this proposal was not the exact type of small bomb, but the intention to load these bombs into the bomb-bay. The trade off in such a case was less fuel and an inability to reach distant enemy air bases. An alternative solution to lift more bombs was to develop special racks to be fitted under the wing stations. It was during 1962 that development of the IMI Light Bomb Rack and Quadruple Bomb Rack started. The LBR was designed to lift two small bombs and the QBR was designed to lift four bombs. Development of the LBR and QBR was expected to be completed within a year, but this objective was not achieved.

Electricity System Malfunction

Deputy B Jonathan Shahar and emergency posting navigator Oded Erez departed Ramat David on 2 October 1962 in Vautour 35 for a routine reconnaissance training sortie. It was on their way back to base that an emergency situation quickly developed. Oded Erez recalled:

> It was Friday and we flew back to Ramat David at high altitude. It is very cold at high altitude but the aircraft was equipped with a heating system that warmed up the cockpits and the cameras. We were south of Tel Nof when Jonathan told me that a generator warning light was on. One of the deficiencies of the Vautour was that there were no warning lights in the navigator compartment apart for warning lights related to specific systems. Jonathan's voice faded while he was speaking, an indication that the intercom would soon cease to function. I realized that the malfunction was total so I yelled 'wheels' because without electricity it was impossible to detract the landing gear in

Jonathan Shahar was Squadron 110 Deputy B until December 1962 and Deputy A from December 1962 until September 1964. (PC)

רשום - לזכות

השם והמשפחה: שחר יהונתן מס' צבאי: 324829/ק

דרגה: סרן יחידה: ט. 110

פרטי המקרה: ביום שלישי, 2 אוק 62 שעה 1345,

תוך בצוע גיחת אמון צלום, הופיעה אזעקת גנרטור. הטייס

בצע הפעולות הנכונות, כעבור דקות ספורות הופיעה

אזעקת מצבר. הטייס הוריד בזריזות גלגלים ומדפים

2/..

עזר ויצמן אלוף

מפקד חיל האויר

תאריך:

The IAF Commander "Green Note" awarded to Deputy B Jonathan Shahar after landing Vautour 35 on 2 October 1962; a "Green Note" was awarded after a feat and a "Red Note" after a mishap, and both were attached to the relevant page in the pilot's logbook. (PC)

> the Vautour. Immediately afterwards we lost all electrical systems.

As already indicated, the Vautour depended heavily on electrical power; an electricity system failure at high altitude was therefore a very serious, and dangerous, malfunction. Jonathan Shahar recalled:

> The Vautour had two generators so a malfunction in one generator was not serious trouble. I notified Oded [Erez], disconnected the malfunctioning generator and we continued to cruise at 40,000 feet. Then all the lights inside the cockpit started to fade out. I said over the intercom 'Oded, can you hear me?' and he replied something but I could no longer hear anything. Instinctively, I dropped down the undercarriage. I knew the Vautour very well and I was aware that there was no manual emergency pump to extend the landing gear; no electricity, no wheels! I saw a green light that indicated lock down of the rear main wheels. Next to lock down was the front main gear but by then none of the lights functioned. However, I sensed the click that signaled to me that the front main gear was also locked down. I saw that the left balancing wheel was down but the right balancing wheel was not down.
>
> Without electricity there was no heating and the cockpit was filled with vapors. I did not see the instruments panel. Over the instruments panel was an air inlet that was used to ventilate the cockpit and refresh the pilot. I pointed this air vent at the windshield. The airflow defrosted a small circle on the windshield. My only view outside the aircraft was through this small 'peep-hole'. I descended and decided to land at Tel Nof.
>
> We arrived at Tel Nof, the busiest IAF air base at the time, without electricity, with no communication but I decided to ignore everything and to land. I flew a circuit and landed, applying our special technique to minimize damage to the right wing tip. At the last moment I swerved the aircraft to the right so that the wing would not hit the ground, applied full brakes and stopped on the runway. The wing dropped gently to the ground but the aircraft was not damaged even a bit.

PILLAR DAWN 112

The malfunctions that surfaced during the PILLAR DAWN 111 series of test flights were addressed and the Shafrir was deemed ready for PILLAR DAWN 112, which started on 4 October 1962. The first test flight involved two Vautours, with the Shafrir Vautour acquiring the target Vautour in order to verify the functionality of the missile. The second PILLAR DAWN 112 test flight was flown on 24 October. For the first time a Shafrir fitted with a proximity fuse was launched from an aircraft. The target was a flare, and the two launched missiles hit the target one after the other. This success was not repeated on 27 October

Vautour 14 on display at Hatzor during a 28 October 1961 visit of donors from the USA; a suffix/prefix digit was not added to the Vautour, which has its nose cone painted red. (PC)

when, for the first time, the objective was to launch Shafrir missiles fitted with warheads and proximity fuses from an aircraft. One missile failed to acquire the target and was not launched; the launched missile hit the target, but the proximity fuse did not detonate the warhead.

The fourth and final PILLAR DAWN 112 test flight was only flown on 12 January 1963. The Shafrir Vautour was tasked to acquire a Fouga trainer. This was actually a preparatory phase prior to the next series of Shafrir tests against a CT-20 target in France. The same engine type powered the Fouga and the CT-20, so for this reason the Shafrir's seeker was tested against a Fouga. The conclusion was that the heat signature of the Fouga was insufficient to ensure proper acquisition by the Shafrir's seeker; as a lesson, it was suggested to augment the heat signature of the CT-20 with a flare during the planned Shafrir test against this realistic target.

Mirage Impact 2

The Mirage was designed as an all-weather interceptor to operate effectively during day and night. This new concept rendered obsolete the night-fighter mission. The IAF decided that Squadron 119 – the IAF night-fighter unit – would close down after the first two Mirage squadrons achieved all-weather operational capability. The redundant Vautour N night-fighters would then boost the depleted ranks of Squadron 110. The set schedule was that the Vautour N aircraft would join Squadron 110 by 1 May 1963.

It was time to discuss how Squadron 110 would operate the Vautour N. AIR2 officer Adam Zivoni launched this effort on 21 October 1962. At the time, the main mission of the Vautour N was all-weather interception and the secondary mission was attack, including radar bombing against maritime targets. It was expected that the Mirage would take over the all-weather interception mission, but there were indications that the Vautour N's DRAC 25 radar was superior to the Mirage's Cyrano. Therefore, there was a school of thought among the ranks of the IAF that if this was the case, then the all-weather interception qualification of the Vautour N should be preserved as a backup. As for attack, because of its radar system the Vautour N was unable to lift bombs in the bomb-bay and also had less fuel than the Vautour A. If the Mirage took over the mission of radar bombing of maritime targets and the all-weather interception backup idea was rejected, then it was suggested to remove the DRAC 25 radars from the Vautour N aircraft so that the bomb-bay would be capable of lifting bombs. In this case the Vautour N would operate as an attack aircraft alongside the Vautour A aircraft of Squadron 110.

Another implication of this planned assignment was repainting. This issue had already been discussed, but the Squadron 119 Vautour N night-fighters were still silver, so on 24 October 1963, AIR3 issued a directive to repaint the Vautour N aircraft in the standard IAF camouflage scheme a-la the Vautour A scheme. The IAF technical echelons challenged this directive, claiming that from a maintenance perspective it would be better to retain the Vautour N aircraft in silver. AIR2 ended the argument on 28 October:

> We are facing two conflicting requirements: the operational requirement for a camouflage scheme and the maintenance requirement for silver finish. In the future, most of the Vautour N missions will be flown at low altitude; therefore the operational requirement is overwhelming and our direc-

Photomontage was the Photoshop® of the 1960s, and in this photomontage the photos of Squadron 110 Vautour 21 (flipped horizontally) and Squadron 117 Mirage 60 were superimposed on a Ramat David typical scene. The introduction of the Mirage impacted Squadron 110 in many ways – including assignment of Vautour pilots to Mirage squadrons and changes in IAF reconnaissance priorities – but the first impact was positive: the reassignment of the Vautour N aircraft from Squadron 119 to 110. (PC)

The Squadron 119 line at Tel Nof. This photo was taken in October 1958. By 1963 the all-weather fighter rendered the night-fighter obsolete and the IAF planned to convert Squadron 119 to the Mirage, reassigning its Vautours to Squadron 110 and its Meteors to Squadron 107. (PC)

Vautour 15 was lost in an accident on 5 November 1962; pilot Simon Zimon ejected. (PC)

tive, to repaint the aircraft in a camouflage scheme, remains unchanged.

MAT3 then addressed – on 31 October 1962 – the planned transfer from a maintenance point of view. At the time, the inventory of Squadron 110 included 13 Vautour A and three Vautour B aircraft that operated from five HAS complexes and a single pen. Wing 1 Maintenance Squadron was in charge of B-Level maintenance and D-Level work was contracted to IAI. The transfer of six Vautour N aircraft from Squadron 119 to 110 would result in a total inventory of 22 aircraft. Considering that, at any given time, three aircraft were undergoing B-Level periodic maintenance and three aircraft were undergoing D-Level work at IAI (including fuel tank sealing), then a total of eight HAS complexes would be needed (each HAS housed two aircraft). Technical manpower would have to be boosted to cater for an increase from two to three aircraft in B-Level work. The A-Level increase from 12 to 16 aircraft posed no problem because the squadron standard already included eight technical teams, with each technical team tending to two aircraft in a HAS. It was suggested to prepare a list of Squadron 119 equipment that would be transferred with the aircraft from Tel Nof to Ramat David; this depended upon the decision concerning the future of the Vautour N's radar; retained or removed, and if retained, would it be maintained as an operational system?

Overall, MAT3 evaluated that the transfer would not pose any insurmountable problems, but it was recommended to prepare in advance two more pens for Squadron 110 at Ramat David.

Elevator Malfunction

The internal IAF potential to increase the number of aircraft assigned to Squadron 110 from 16 to 22 suffered a setback on 5 November 1962, when Vautour 15 was lost in an accident. Jonathan Shahar flew in the lead of a formation tasked to practice "operational range"; an attack training at the range without the presence of a range controller. When a range controller was present, the attack training was leveled "restricted range" and mostly involved basic attack patterns, while an "operational range" practice simulated an actual attack mission.

On their way back to base pilot Simon Zimon lost control of Vautour 15 due to an elevator malfunction. Jonathan Shahar and sub-leader Beni Zohar watched Vautour 15 enter a spin and ordered Zimon to eject. The pilot landed safely, but the Vautour was wrecked.

Radar Reflections

Squadron 119 Commander Ovadia Nachman was a staunch supporter of the school of thought to retain radar in the Vautour N aircraft to be transferred to Squadron 110 and to preserve the operational qualification of the Vautour N as a night-fighter. Nachman presented his case on 23 November 1962. Apparently, the Vautour's DRAC 25 was superior to the Mirage's Cyrano in low altitude interception scenarios. Additional reasons for retaining the DRAC 25 were the firepower of the Vautour N, the teething troubles of the Mirage's R-530 air-to-air missile, the inability of the Mirage to fly supersonic all-weather interception due to lack of an interception computer, the employment of the Vautour N as backup to the Mirage in the all-weather interception mission during wartime, and the Vautour's longer endurance and better suitability to operate against slow targets.

Ovadia Nachman also cited the DRAC 25 as an effective aid for all-weather attack missions, easing navigation and enabling identification of high value targets, such as air bases, ports, power stations and bridges. Once such a target was identified, the Vautour N could launch a radar bombing attack or drop flares to mark the target for follow up attackers. The DRAC 25 was especially efficient against maritime targets and was able to detect a ship at a range of up to 80 miles.

Retaining radar would limit the bomb-bay to installation of fuel tanks only, but Ovadia Nachman figured, correctly, that this issue should not be overemphasized in light of the ongoing development of the Light Bombs Rack and Quadruple Bombs Rack. With LBR or QBR under the wings, the Vautour N was expected to lift the same multi-bomb load as a Vautour A.

The most problematic issues of the proposal to retain the DRAC 25 in the Vautour N were logistics. In order to overcome the logistics issues – manpower, spares and training for so few systems – Ovadia Nachman suggested to preserved the radar

Simi Saar (left) took over from Jonathan Shahar as Deputy B in December 1962 and was photographed with, from left to right: Yigal Yanay, Oded Einav, Eli Mor and Gideon Dror. (PC)

operational qualification of the Vautour N, but to replace the DRAC 25 with Cyrano, as fitted to the Mirage, or even with the improved Cyrano 2; the latter was optimized for attack in addition to interception. Nachman summed up:

> The Vautour all-weather interception qualification should be preserved at least until it would be proved that the Mirage is superior, and enhanced by the addition of missiles. The Vautour's radar should be retained as a weapon system that enables development of attack patterns. It is desirable to replace DRAC 25 with Cyrano 2 to ease logistics, enhance capabilities and enable training and development of Cyrano [for Mirage]. All changes should be gradual so current proven weapon systems and methods must be retained until more modern weapon systems and methods will be proven.

VULTURE

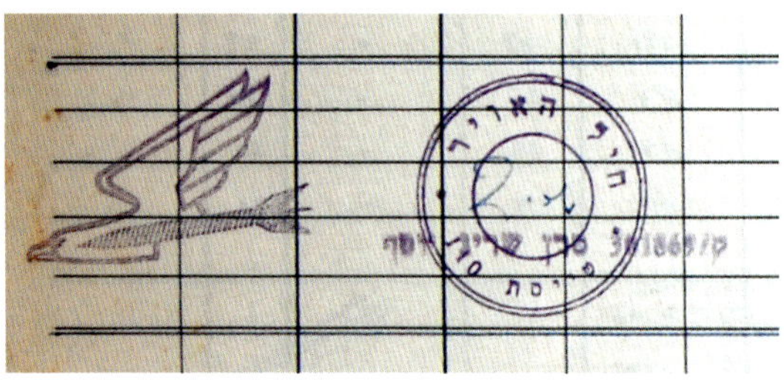

Squadron 110 Acting Commander Yossi Sarig signed logbook pages of Squadron 110 pilots for October and November 1962. (PC)

Water War clashes continued, but air power was rarely used. The IAF was the Israeli counterbalance to the superiority of the Syrian artillery, but from a diplomatic perspective, air power involvement was viewed as an escalation and was thus authorized only under exceptional circumstances. Another round of clashes along the Israeli-Syrian border started on 4 December 1962. The exchange of fire intensified along the time line to the point that the Israeli Government authorized air strikes on 9 December, a decision that was made public on 11 December.

AIR3 issued an Operation Order on 12 December 1962 (Operation VULTURE) to cover possible involvement of air power. Operation VULTURE alerted IAF units for the possibility of gradual involvement: Plan A was strikes against Syrian fortifications along the border, Plan B was attacks against Syrian artillery batteries in the Golan Heights, and Plan C was an air superiority offensive against Syrian air bases. Squadron 110 was to set six Vautours on readiness. Strafing was banned, and minimum altitudes for bombing runs were 1,500 feet above ground level for bombs with impact detonators and 2,000 feet for aircraft attacking with proximity fused bombs set to explode above the surface.

VULTURE D-Day was 13 December 1962 and H-Hour was 10:00. The IAF switched to Readiness C, so emergency posting crews reported to their units and reserve air crews were mobilized. The crews were briefed and the aircraft were armed, but actual activation of VULTURE was postponed several times due to bad weather; the last VULTURE D-Day and H-hour was 23:10 on 22 December. Then the situation calmed down and another round of Water War clashes ended without IAF involvement.

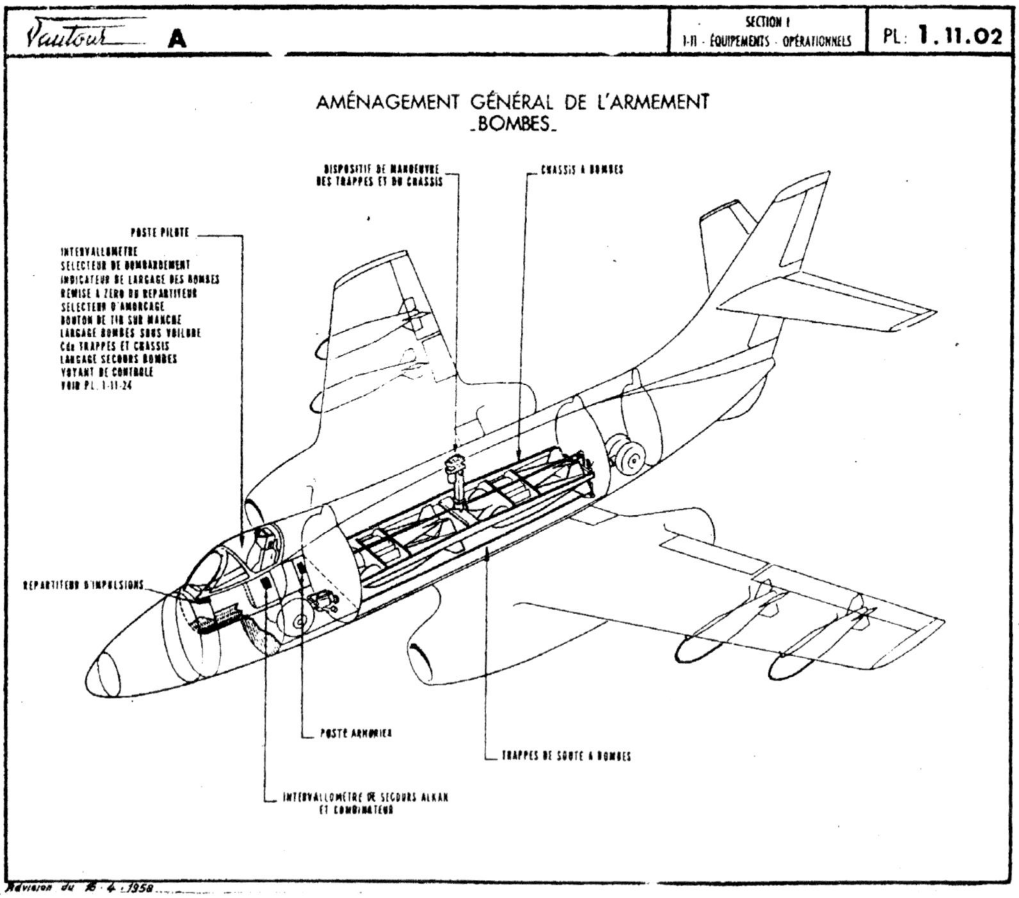

The Vautour A was equipped with a rather complex bomb-bay installation that included doors that lowered downwards and slid sideways to the fuselage sides, while the bomb racks were attached to an elevator that was lowered downwards. When the bomb-bay doors opened they exposed the bombs to the airstream prior to actual drop. (IDFA)

Amalgamation of Squadrons 110 and 119

The target date of Vautour N transfer to Squadron 110 was fast approaching, and AIR2 issued a paper to discuss the implications of this action on 31 December 1962. The baseline assumptions were that the Squadron 110 Vautours were to remain in IAF service at least until 1968, and that by 1968 the combined fleet of 21 aircraft would be depleted to 17 due to attrition.

Modifications explored, in order to fit the Vautour N for service with Squadron 110, were: removal of radar, reconfiguration of bomb-bay for bombs, and replacement of the navigator's cockpit with a fuel tank. Removal of radar was essential if the bomb-bay was to be fitted for bombs, because a number of radar electronic components – including the computer, an oxygen bottle, and an oil reservoir – were inside the bomb-bay. The modification of the bomb-bay to be loaded with bombs also necessitated structural changes, including the installation of bomb-bay doors instead of the covers used in the Vautour N, installation of elevators for carriage of bombs, and extensive re-wiring, including re-design of the pilot's cockpit to remove radar control boxes and to install bombing control boxes. The installation of two fuel tanks in place of the rear cockpit would also require removal of radar, re-wiring, and the transfer of essential control boxes from the rear cockpit to the front cockpit, and would only add 938 liters of fuel.

AIR2 evaluated the proposed changes, except perhaps removal of radar, as unjustified in light of the massive work required to modify the bomb-bay (it was estimated that bomb-bay modification of the six aircraft would last three years) and the rather negligible improvement in total fuel capacity if the rear cockpit would be replaced by fuel tanks (an estimated gain of up to 15 minutes flying time at low altitude).

The remaining options were to transfer the Vautour N aircraft from 119 to 110 "as is" or to remove the radar. The advantages of transferring the aircraft "as is" included an option to fly unique missions, such as all-weather lead-ship, nighttime target markers, radar bombing, maritime strike, all-weather interception, and electronic warfare, as well as development platforms for new radars and weapons. The down sides were mostly logistics: the assignment of additional resources to maintain the Vautour N's radars in working order. Additionally, the utilization of the Vautour N aircraft for any of the aforementioned missions would be on top of utilization as attack aircraft, and would erode the original motivation to transfer the Vautour N aircraft to 110. Squadron 110 suffered from the assignment of a smaller than standard number of aircraft (15, including only 12 attack aircraft) and low availability, with an average of only seven serviceable Vautour A aircraft. The reassignment of the Vautour N aircraft from 119 to 110 was aimed to bolster the attack force of 110, but if new missions would be added then the basic problem would not be solved.

The recommendation of AIR2 was to transfer the aircraft "as is". To support the "as is" transfer, AIR2 suggested to assign five navigators to Squadron 110 and to allocate the resources required to ensure the future serviceability of the Vautour N's radar systems.

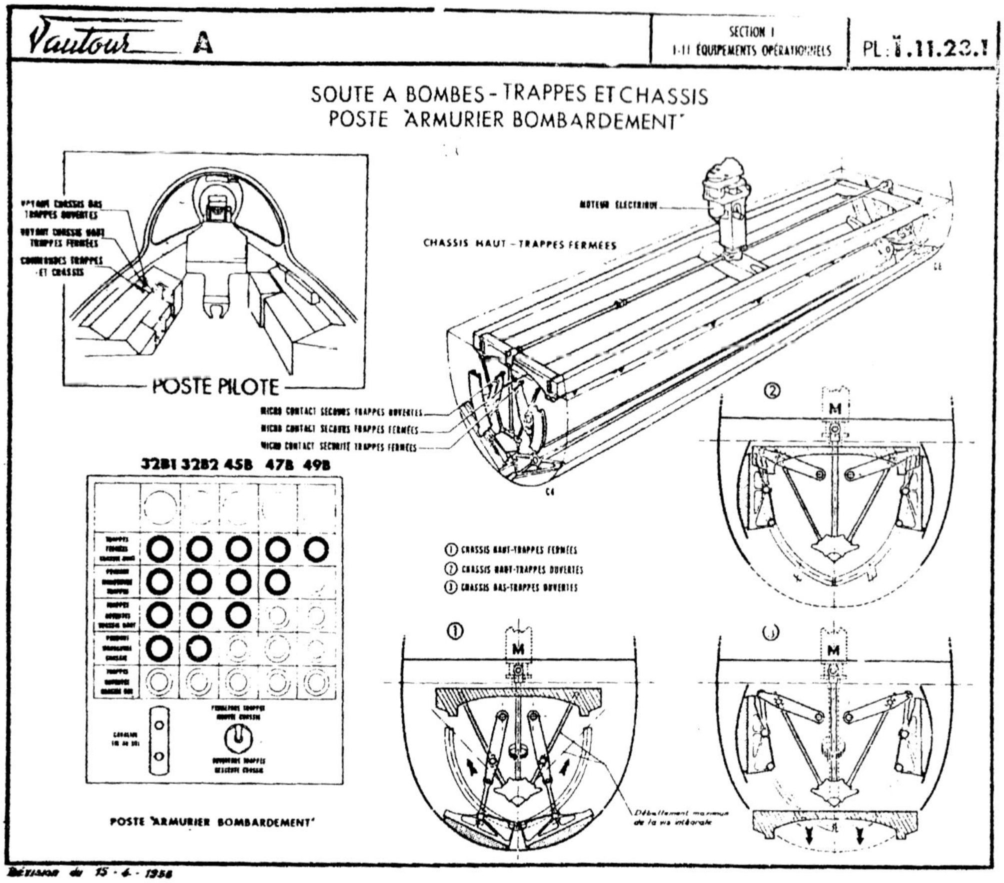

The Vautour N was not equipped with the Vautour A bomb-bay, as illustrated, and AIR2 evaluated, on 31 December 1962, that modification of the Vautour N with the Vautour A bomb-bay would be an extremely complex engineering undertaking. (IDFA)

Larger External Fuel Tanks

While AIR2 evaluated that an addition of 938 liters of fuel, for a gain of up to 15 minutes flying time at low altitude, was negligible, Squadron 110 continued to forward proposals to increase the fuel capacity of the Vautour. Deputy A Jonathan Shahar suggested on 1 February 1963 to replace the standard 1,300 liter external fuel tanks with larger 1,700 liter tanks. Shahar wrote:

> This extra capacity of fuel will enable the aircraft to fly low-low-low mission profiles at 450 to 500 knots to all air bases in the region, including Cairo West. The squadron views such a modification as very important for major improvement in operational efficiency and attack power.

Radar Removal Directive

Regardless of the prior AIR2 recommendation to transfer the Vautour N aircraft from Squadron 110 to Squadron 119 in "as is" condition, AIR2 issued an order on 10 February 1963 to remove the DRAC 25 radars from the Vautour N aircraft by 25 April 1963, five days short of the planned assignment to Squadron 110. However, the DRAC 25 was not to be completely withdrawn from IAF service, as four radars out of 11 in the inventory were to be preserved in serviceable status for future refit if such a requirement would emerge.

Radar removal work was far more complicated than simply taking out the antenna in the nose section. There were radar components in the fuselage from nose to tail, including three generators and three electronic boxes in the empennage, as well as associated wiring and control panels in the cockpits. Moreover, radar removal was set to surface center of gravity issues. Due to three reasons – center of gravity, complexity of removal, and operational aspects – the AIR2 order remained on paper and the Vautour N would be reassigned to Squadron 110 in an "as is" configuration.

PILLAR DAWN 113

The IAF decided in late 1962 to acquire 200 Shafrir missiles; these were planned to enter service with the Mirage squadrons. While IMI was gearing up to manufacture the RAFAEL designed missile, development and tests continued. Until then the Shafrir was tested against T-10 rockets (with little success) and against flares (with mixed results). The next phase was to test the missile against a proper target over a proper missile range, and such a facility was only available to Israel in France. AIR3 issued a request on 24 January 1963 to arrange a ferry flight from Israel to France for Vautour 09. The proposed route was from Israel to Istres via Brindisi. The proposed date was 6 February, and the cover story was that Jacob Agassi was flying the Vautour (radio call sign 4X-FBE) for repairs in France.

On the eve of the ferry flight, Vautour 09 was damaged in an accident and had to be replaced in haste. Vautour 03 was the second Shafrir replacement test aircraft, and Jacob Agassi flew to France on 7 February 1963. The series of test flights lasted from 18 February to 7 March, with the Vautour operating from Istres over the Ile du Levant range. A dozen missiles were allocated: four without warhead and/or proximity fuse, four with warhead, and four with

Vautour 09 was damaged during preparations for flight to France for Shafrir tests; the aircraft is fitted with 1,300 liter drop tanks (that replaced the original 1,250 liter tanks) and has its nose cone painted red; it would return to fly in December 1964. (PC)

warhead and proximity fuse. Seven of these missiles were launched against a CT-20 target. Five launches were against a CT-20 flying straight and level; three missiles hit the target, including one missile that succeeded in shooting down the CT-20. Two launches were against a maneuvering CT-20 and both missed.

PILLAR DAWN 113 proved that the Shafrir was capable of intercepting a target that was not maneuvering, but the warhead's impact was inadequate and the missile was unable to intercept maneuvering targets.

Ras Banas

Emergency posting pilot Yossi Sarig, with regular navigator Alex Meltzer in Vautour 31, and Deputy A Jonathan Shahar in Vautour 18 departed Tel Nof on 20 March 1963 for a long range mission. The objective of this mission was to pinpoint and photograph a new Egyptian air base recently constructed near Ras Banas, on the coast of the Red Sea north of the Egyptian-Sudanese border, to support Egypt's involvement in Yemen and as an alternative base for the EAF force of Tu-16 heavy bombers, supposedly safely beyond the reach of Israeli air power.

The straight line distance from Ramat David to Ras Banas is nearly 1,000 km. To fly such a mission in the Vautour B was a challenge, because the aircraft was fitted with only one additional internal fuel tank, in the rear suite of the bomb-bay (the cameras occupied the front suite), while the escorting Vautour A flew with a maximum fuel load, including two additional tanks in the bomb-bay. Yossi Sarig recalled:

> [Squadron 110 Commander Jacob] Agassi was abroad and I was acting commander of the squadron. During this time frame I met [Ezekiel] Somech, who was at the squadron for some reason. Somech said, by the way, that the Egyptians constructed a new air base for their Tu-16 bombers at Ras Banas and that the IAF cancelled a plan to fly a reconnaissance mission to Ras Banas based on his evaluation that the Vautour's range was insufficient for such a mission. This enraged me because I was acting commander and I should have been consulted if such a mission was possible or not. I phoned IAF Commander Ezer Weizman and asked him to investigate the feasibility of a reconnaissance mission to Ras Banas in a Vautour. I had a reputation of someone who flew a lot of reconnaissance missions so he said 'OK, go ahead'.
>
> I started all sorts of experiments how to extend the range of the Vautour, an effort that lasted over a time frame of several months. There were all sorts of crazy ideas, even to cruise on one engine, but none offered a solution. I then sought consultation from Josef Gideoni, an IAF reserve pilot and an engineer who was an expert in propulsion, and asked him for methods to preserve fuel but still, no solution. Then someone suggested that I should consult Daniel Arber, another IAF reserve pilot who was an El Al Boeing 720 Captain. Arber presented to me the optimal cruise that was used for El Al long range flights: starting at optimal altitude and slowing down as fuel was consumed and weight was lost. This opposed the French optimal cruise solution to fly at maximum RPM and constantly climb as fuel was consumed and weight was lost. The problem with climbing in a reconnaissance mission was that at a certain altitude you started to 'paint', to produce contrails that revealed your presence. Slowing down seemed to be a good idea so I started to test this solution and it worked!
>
> I reported that a solution has been found, that it was possible for a Vautour to fly to Ras Banas: a deployment [south] to Tel Nof, parking the aircraft at the start of the runway, jacking up the tail section so that it would be possible to fully fuel the aircraft, take-off in the direction of Ras Banas, fly the mission and return to land at Eilat [a short civil airport] with braking chute. So it was. We deployed to Tel Nof, took-off and climbed to 45,000 feet. This was the planned cruise altitude but south of Eilat we noted that we were starting to 'paint' so we descended to 33,000 feet or so. We arrived over Ras Banas but did not find the air base where we were told it was. We circled overhead until we saw it, flew one photographic run at 30,000 feet and another at 20,000 feet. The navigator suggested to return, he was worried about our fuel state, but I did not know when next would someone fly over this distant place so I descended to 10,000 feet for a third photographic run and only then climbed back to high altitude for the return flight and where did I land? Not at Eilat and not at Tel Nof but at Ramat David. The next time that IAF aircraft would fly over Ras Banas was during the Six Day War when our Vautours raided Ras Banas in a mission that was planned based on the photos that we took in 1963!

Vautour 18 over the Gulf of Aqaba; Jonathan Shahar flew 18 to Ras Banas on 20 March 1963 as an escort to reconnaissance Vautour 31. (PC)

Convex Wing for the Vautour A

The IAF decided to refit only the Vautour B aircraft with leading edge convex wings. When this decision was made the IAF inventory included four Vautour B aircraft, so four kits of leading edge convex wings were initially purchased. One kit was used to repair a damaged Vautour A (4320), so a replacement kit was ordered. By the time the installation of the convex leading edge wings was authorized, there were only three Vautour B left. AIR2 Chief Isaac Sela summed up this situation on 28 June 1962:

> After the installation of convex wings to a Vautour A and to a Vautour B (still at IAI), we still have kits for two more aircraft. Another kit is due to be delivered within a short while. Air Department Chief ordered on 25 June 1962 that Vautour A aircraft will not be fitted with convex wings. The available kits will only be used to refit Vautour B aircraft.

The availability of a redundant kit of the convex leading edge wing resulted in a change. AIR3 officer Adam Zivoni reported on 28 March 1963 an urgent operational requirement to fit convex leading edge wings on a Vautour A, and suggested to remove the wings from the damaged Vautour 09. AIR2 officer Josef Afik elaborated on 31 March:

> An urgent operational requirement to fit convex leading edge wings in a Vautour A has emerged... Vautour 08 was transferred to IAI for this purpose. The issue is top priority. Target date for completion of installation is 5 April 1963.

Radar Removal Decision

The discussion concerning modifications to Vautour N aircraft due to be transferred to Squadron 110 was concluded on 26 April 1963, when Air Department Chief notified Material Department Chief as follows:

> Two Vautour N aircraft will be transferred to Wing 1 with functional DRAC [25] systems that are to be maintained in operational status. A decision concerning schedule for removal of radars from the other aircraft is postponed until AIR2 evaluation will be completed. The aircraft will be transferred without special equipment except for radar ground test bed table. AIR1 will examine

Vautour 08 with the convex leading edge wing; the leading edge of the wing is not continuous inner and outer of the wing fence. The aircraft in the background (Vautour 18) has an original wing. (PC)

A Vautour A with original wing; the leading edge is continuous inner and outer of the wing fence. (PC)

Gideon Magen flew Vautour 17 during the Independence Day fly-past on 29 April 1963. (PC)

if there is a requirement to add manpower to the Wing 1 standard for DRAC maintenance.

The schedule slipped, and Vautour N aircraft were not transferred during April 1963. Minor modifications had to be implemented, while other issues were to surface only after Squadron 110 had accumulated experience with mixed operations of Vautour A and N aircraft. One such modification that was planned for implementation prior to transfer of aircraft was replacement of the Vautour N fixed sight with a Mark IVE gyro sight as in the Vautour A. Squadron 119 Commander Ovadia Nachman reported on 24 June 1963:

The decision to replace the Vautour N's fixed sight with a gyroscopic sight was taken a year ago. Since then this issue has been the subject of planning and implementation has not been started yet. As the transfer of aircraft is imminent, this issue has to be assigned with higher priority. The gyro sight is important for operations (without it there is no reasonable chance to hit enemy aircraft with gunfire in case of an aerial engagement) and for training (without it, it will be impossible to practice air-to-air tracking and ranging or air-to-air gunnery). I therefore ask to speed up implementation of this issue.

IAF airpower after April 1962 included, from left to right: Mirage, Mystere, Vautour, Super Mystere, and Ouragan. (PC)

Ovadia Nachman also suggested installing an additional radio compass control box in the front cockpit of the Vautour N, so that in Squadron 110 service it would be possible to operate the radio compass from either cockpit, while in Squadron 119 it was only possible to operate the radio compass from the navigator's cockpit. This modification was recommended, because it was planned that in Squadron 110 service the Vautour N aircraft would fly with pilot only, or with a crew of pilot and navigator (depending on mission requirements), but in Squadron 119 a crew of pilot and navigator was the standard.

Trrr..., Boom, Puff, You are Blind

Reconnaissance missions over enemy territory continued. The short range incursions into Jordan, Lebanon, Syria, and northeast Sinai, Egypt, were treated almost as routine. Only the longer range missions, which amplified the prospects of enemy interception, still stimulated the adrenaline of the experienced "photographers", as were nicknamed the members of a small group of pilots and navigators qualified to fly reconnaissance missions. Such a longer range mission was flown on 18 June 1963, when Jonathan Shahar with Arie Raviv (in Vautour 33) and Gideon Magen with Achikar Eyal (in 31) flew a mission of 1:15 duration to photograph in the sector of the Straits of Aqaba, at the southern tip of Sinai.

Meanwhile, the nocturnal reconnaissance qualification of the Vautour was refined in testing and training; the latter included low altitude navigation flights. Jonathan Shahar and Alex Meltzer flew a nighttime reconnaissance test in Vautour 33 on the evening of 20 June 1963, as Jonathan Shahar recalled:

> We practiced low altitude navigation at night, flying over areas in the Negev Desert, were no human made obstacles existed, at an altitude of 100 feet [above ground level] in the darkness, without lights, without instruments. We adjusted ourselves to the night and our ability to fly at low altitude during the night was exceptional. Then we started to practice nighttime photography; the most shocking kind of flight that I ever flew. The training profile was an out leg over the Mediterranean Sea at low altitude, 100 feet above the surface of the sea. The radio altimeter display was only in the navigator's compartment. There were only three general indicators in the pilot's cockpit: a green light OK, orange light too high and red light too low. The navigator constantly reported, '90 feet', '95 feet', '100 feet', '90 feet', '80 feet' and so on. We flew five minutes over the sea, turned and flew back towards land. Our target was En Shemer [airfield] so as we crossed the coastline, we climbed abruptly to 5,000

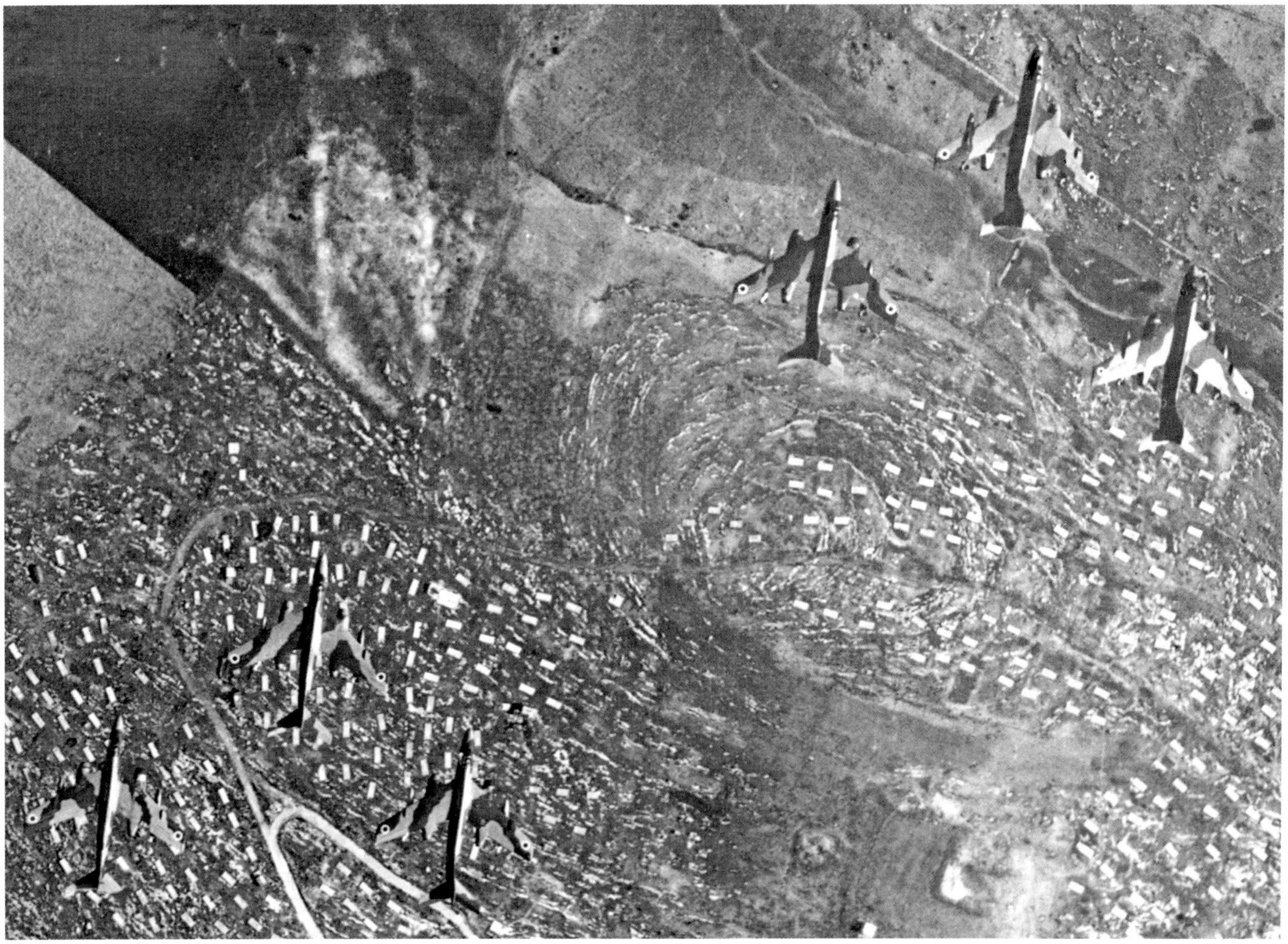

A Vautour B photographed six Vautour A aircraft. (PC)

feet, set speed to 470 knots, flew straight and level and started the photographic run. The camera had a feature that was known as Image Motion Compensation to prevent relative motion between the camera and the surface down below. The camera compartment was right behind and below the pilot's cockpit so you clearly heard the unique sound – trrrr... – of the Image Motion Compensation engine moving the camera. Then the system dropped a flash, actually launching the flash from the bottom of the aircraft with the sound of a boom. Finally, the flash ignited in a puff right underneath the aircraft and blinded the pilot. So it was trrrr..., boom, puff, you are blind and then again trrrr..., boom, puff, you are blind and again. When the system finished taking three images, we immediately dived back to low altitude and flew back to the sea at low altitude only to repeat this pattern again and again. We flew some five such patterns in a single flight. I did not fly such an operational mission but definitely, this was the most shocking kind of flying I ever experienced.

4

SQUADRON 110 VAUTOUR

1963 - 1967

From the summer of 1963, the IAF Vautour force of two squadrons (110 and 119) was scaled down to a single squadron. At the time, IAF planning expected a gradual diminishing in the inventory of the Vautour fleet due to normal attrition, until retirement sometime after 1968. The Vautour was still the most capable IAF long range attack aircraft, but enemy air defenses – including MiG-21 fighters and SA-2 missiles – were much more capable from 1963 than they were during the initial time frame of IAF Vautour service from 1957 to 1963. For this reason, the IAF started to evaluate an operational requirement for a new "heavy attack aircraft" (to be ultimately met with the introduction of the F-4 Phantom from 1969). More urgent was a requirement to introduce Mach 2 reconnaissance aircraft to replace the Vautour B as the IAF's main type of reconnaissance aircraft, and this effort was also initiated in 1963 when modified Mirage III aircraft started to fly reconnaissance missions, eventually taking over the longer range, deeper penetration operations from 1964. Thus, from 1963 started a Squadron 110 era that was dominated by three processes: the aging of the Vautour and IAF evaluation of potential successors, the loss of the Vautour B's status as the IAF's premier long range reconnaissance aircraft, and the fact that – for the time being and regardless of the improved enemy defenses – Squadron 110 Vautour was the "heavy cavalry" of Israeli air power.

Standardization of Aircraft Configurations

Squadron 119 closed as a night-fighter unit on 7 July 1963. The Vautour N aircraft were reassigned to Squadron 110, and former Squadron 119 Commander Ovadia Nachman took over from Jacob Agassi as Squadron 110 Commander. Prior to this, Agassi signed a memo concerning Vautour standardization issues on 2 July:

> Several months ago, in a discussion at the office of Air Department Chief, it was agreed that the Vautour N aircraft would be brought to standard configuration, except for bomb-bay, before assignment to Squadron 110. During that same discussion I mentioned the issue of Vautour N sight and Material Department Chief reassured me that the aircraft will be reassigned with Mark IVE sights. Now, the Vautour N aircraft are transferred to Squadron 110 without such sights. The aircraft are indeed equipped with a sight capable of coping, somehow, with attack patterns but certainly not suitable for air-to-air exercises. This will adversely impact the planning of flights schedule in the squadron. The change of sight (as well as a change of sight for the Vautour B aircraft) should be assigned with higher priority because the squadron already faces many other problems in the planning of its flights schedule that result form lack of standardization in the configuration of the aircraft (aircraft without certain fuel tanks, an aircraft with a convex wing).

T-4, Syria

Most of the more interesting and challenging Squadron 110 Vautour reconnaissance missions were flown over Egypt. The distances to potential targets were longer and the Egyptian defenses were vastly superior to Jordanian, Lebanese, or Syrian defenses.

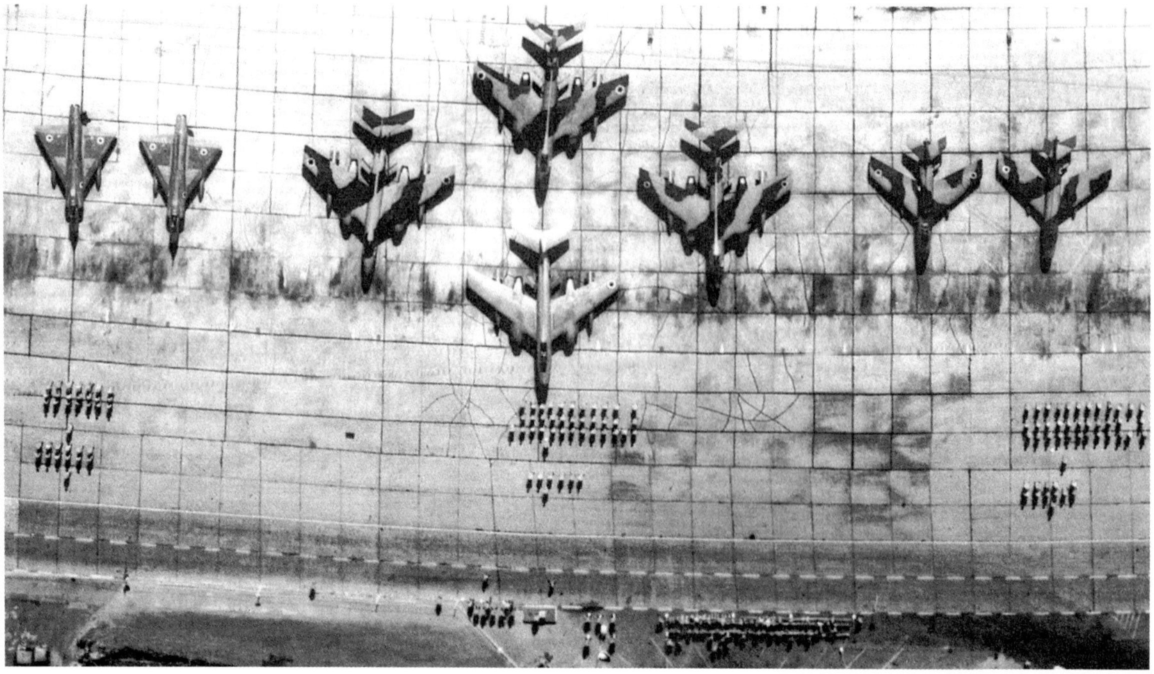

The new look of Squadron 110 from July 1963 included those three camouflaged Vautour A aircraft and a silver Vautour N aircraft as depicted in this view (along with Squadron 117 Mirages on the left and Squadron 109 Mysteres on the right) during the August 1963 Ramat David command hand-over parade, when Ezekiel Somech took over command of the base from Menahem Bar. (PC)

A few reconnaissance missions flown over nations other than Egypt were engraved in the annals of Squadron 110 Vautour history, and the 29 July 1963 flight to photograph T-4 in Syria was one such memorable mission.

Deputy A Jonathan Shahar flew lead with emergency posting navigator Oded Erez in Vautour 33; Gideon Magen with Alex Meltzer flew as wingman in Vautour 31. The objective of the mission was to photograph a new Syrian air base. Israeli Intelligence reported that T-4 was constructed somewhere along the old Iraq Petroleum Company pipeline from Iraq to Tripoli. Pumping stations along the pipeline were numbered, and the new Syrian air base was reportedly constructed near T-4. The plan was to fly over the Mediterranean Sea to Tripoli, climb to 40,000 feet and fly along the IPC pipeline until the new air base could be pinpointed and photographed. Oded Erez recalled:

> We knew that the Syrians were constructing a new air base somewhere nearby Palmyra but we did not know the exact location. It was in the afternoon, the sun was behind us and we were quite safe at high altitude; there were no SA-2 missiles yet and in case of interception we had the advantage of altitude and were able to escape before the MiGs would climb so high. Our maps were awful, the scale was too large and it was impossible to discern details, but suddenly I saw something glittering. Seemingly a Fata Morgana it could have been the surface of a road or a water source. I told Jonathan and as we got closer we saw that this was the air base, the glittering object was only part of a runway that had just been coated; days earlier it would have been much more difficult to spot it. We photographed as planned and then, because we were already there, we descended to low altitude to have a look at the antiquities of Palmyra.

The plan was after the two aircraft had photographed T-4 they would split. The wingman was to return to Israel via the Golan Heights, photographing additional targets along this route. The leader was planned to head north and photograph Aleppo air base. Jonathan Shahar recalled:

> We flew to Aleppo and I was amazed at what we saw. We were accustomed to live in a desert and to fly over a desert but north Syria looked from above exactly like England or France, a lot of rivers and magnificent agricultural fields all over the place; wow, it was a delight to fly over such a landscape.

Vautour N's old look. The Vautour N aircraft were transferred from Squadron 119 to 110 in silver finish; Vautour 62, with 1,250 liter drop tanks, was photographed departing Ramat David with only a pilot in the front cockpit. (PC)

From DRAC 25 to Cyrano

The Vautour N aircraft were indeed delivered to Squadron 110 with DRAC 25 radars fitted, but preserving two such systems in operational status proved more difficult than originally envisioned. By 12 August 1963, AIR3 noted that Squadron 110's readiness reports did not mention two DRAC 25 systems as operational. Squadron 110 Commander Ovadia Nachman responded to the AIR3 enquiry on 27 August:

> It is impossible to make serviceable two DRAC 25 systems due to lack of experienced radar technicians and essential ground equipment.

The following day, the Material Department also issued a negative view concerning the feasibility to maintain in working order only two DRAC 25 systems. Material Department cited DRAC 25 low serviceability (only 35% probability to complete a mission without a malfunction), manpower shortage, and the logistics burden as reasons why the DRAC 25 should not remain in IAF service. If operational considerations were overwhelming, then Material Department suggested DRAC 25 should be replaced with Cyrano, thus solving the issues of manpower shortage and the logistics burden (because Cyrano was already installed in the IAF Mirage fleet).

Air Department discussed this issue on 16 September 1963 and accepted the suggestion of Material Department to install Cyrano in the Vautour N for air-to-ground missions, maritime strike, and as a test bed for development of air-to-air and air-to-ground radar techniques and tactics. In line with this decision, on 27 September AIR2 Chief Yoash Tsiddon forwarded to Material Department a request for information concerning the required investment in such a modification. For proper reply, on 14 October 1963 IMOD representatives inspected at Bretigny a French Flight Test Center Vautour N that was thus modified.

The IMOD representatives reported that Vautour N 301 was modified for test purposes. The modification was quite extensive and included a new nose section, removal of the guns pack, installation of a new electricity system, changes in the cockpits, and rewiring. It took eight months to thus modify Vautour N 301. The most surprising aspect of this evaluation was the fact that the French Flight Test Center undertook to return Vautour N 301 to the FAF upon completion of the flight test program in its original configuration. Therefore, the French Flight Test Center suggested to swap aircraft with the IAF; the latter would save the money and time to modify a Vautour N with a Cyrano, while the former would save money and time to return Vautour N 301 to its original configuration. The FAF would receive a Vautour N night-fighter as agreed with the French Flight Test Center, and the IAF would have a Vautour N fitted with Cyrano without the troubles of modification.

Buzz over Merchavya

Squadron 110 Intelligence Officer Reuben Nir was a member of Kibbutz Merchavya, some 15 km southeast of Ramat David; his younger brother Isaac Nir would be assigned to the squadron as a Vautour pilot from March 1964. Reuben Nir joined Jonathan Shahar for a joyride in a Vautour B; a flight – possibly flown on 30 September 1963 in Vautour 35 – that almost ended in personal disappointment for both of them. Jonathan Shahar recalled:

> A certain gasket inside the engine used to fail at high speed. The engines went through a modification and each aircraft that had its engines modified was tested in a high speed, low altitude flight. One day a Vautour B had to be tested so I took Intelligence Officer Reuben Nir for a joyride. We speeded to 500 knots at low altitude for a certain time frame and when we completed the required profile I told him that we completed the flight and that we would soon land. He then asked me to fly over Kibbutz Merchavya; 'sure' I retorted 'just tell me where exactly to fly over' so he, from his grand view in the nose section, pointed me exactly where to fly over Merchavya.
>
> We returned to landing, I reported that the aircraft was OK, Nir thanked me for the ride and we went our separate ways. The next morning he arrived at the squadron and said 'Jonathan, we are in trouble'. We flew right over the dining room at 13:00 when it was full of Kibbutz members [including Squadron 113 Ouragan Operation Training Unit student Isaac Nir]. We really shocked the building and this was way too much for some of the Kibbutz members. When Reuben Nir returned home that evening and told everyone that it was him who flew over the dining room, those Kibbutz members who were unhappy with this feat considered an approach to IAF Commander so that the pilot would be grounded. If this story would have reached the IAF Commander then I assume that this would have been the end of my career in the IAF. Somehow he managed to calm them down but it was a close call indeed.

Vautour N's new look; AIR2 reported, on 13 August 1963, a decision to repaint the N aircraft in the same scheme as the A aircraft. Vautour 67, with 1,300 liter drop tanks, was photographed departing Ramat David. (PC)

A camouflaged Vautour N "buzzing" an IDF truck along a road in the Negev Desert; Vautour pilots mastered low flying while training to penetrate enemy defenses underneath radar coverage. (PC)

Vautour N Gyro Sight Tests

MAT5 issued a requirement on 1 October 1963 to test the Vautour N gyro sight modification as fitted to Vautour 69. Ten test flights were planned in three phases: first phase (one flight) to test the installation in flight (including firing without a target) at an envelope of up to 500 knots, Mach 0.95, 30,000 feet and 4 G; second phase (three flights) to test strafing a target on the ground; and third phase (six flights) to test air-to-air gunnery.

Jonathan Shahar and Alex Meltzer departed Ramat David in Vautour 69 at 10:44 on 6 October 1963. They climbed to 8,000 feet and accelerated to 500 knots, then another climb to 20,000 feet at 0.8 Mach and acceleration to Mach 0.9 (420 knots). Then followed a climb to 30,000 feet and acceleration to 0.95 Mach. Data from the new sight was recorded during this climb and throughout the flight. Next were tight turns: the first was a 3.5 G turn at 20,000 feet and the second was a 4 G turn at 15,000 feet. The sight functioned flawlessly up to 5 G, when it started to vibrate. They then flew to a range to strafe: two bursts with the two lower guns, two bursts with the two upper guns, and two bursts with all four guns. Vibration levels were normal, and the only unusual issue was a jam in the upper right gun.

Next phase in the qualification process of the Vautour N gyro sight was air-to-air gunnery, and AIR3 issued, on 22 October 1963, a Mission Order for such a test to be flown on 23 October.

On One Engine to Cairo and Back

Egypt introduced the SA-2 from 1963, the first Middle East nation to deploy a surface-to-air missile system. In the wake of this development, the IAF was authorized to fly a reconnaissance mission to capture on film the first Egyptian SA-2 batteries.

A Vautour strafing a target at a range. (PC)

AIR3 planned Operation MOMENTUM to enable two Vautours to penetrate deep into Egypt at low altitude, photograph south of Cairo, and return safely. D-Day was 18 October 1963, and the assigned crews were emergency posting pilot Yossi Sarig (at the time Flying School Instructors Squadron Commander) with emergency posting navigator Achikar Eyal (an AIR3 staff officer) in Vautour 35 and Squadron 110 Commander Ovadia Nachman with regular navigator Alex Meltzer in Vautour 33. Yossi Sarig recalled:

> This mission was carefully planned, but even so, took many weeks to be approved mainly because it actually meant to fly as far as Cairo, in broad daylight regardless of added threat from the new SA-2 missiles. When it was finally approved, all the decision makers and members of various operational circles involved in this mission knew that it was to be a daring important mission that had to succeed.
>
> The route was planned to be flown from Ramat David south, descend to low level near Eilat and continue south, then, half way to the Straits of Aqaba, turn west across the Sinai Desert up to Lake Bitter in the south sector of the Suez Canal, then to Helwan that was known to be the center of the Egyptian SA-2 force, including a depot and a missile base, north-west to Cairo West air base and back over the Nile Delta. This route actually passed right over SA-2 and MiG bases.
>
> When we started our westerly track into the Sinai Desert, I realized it was going to be a tough mission. It was a hazy day with very low clouds from horizon to horizon that resulted in poor visibility; occasionally we had to fly as low as 50 feet [above ground level] in order to remain below clouds.
>
> Roughly halfway across the Sinai Desert I lost oil pressure in one engine. Flying in radio silence, I knew that I would have to sort things out all by myself. I had to slow down from 450 knots to a slower, one engine speed. The weather, as bad as it was – low level clouds mixed with haze and partial fog – at least guaranteed protection against enemy interception. So I decided to press ahead. I did not switch off the troubled engine, but throttled back to idle power and reduced speed to 360 knots.
>
> The other Vautour, flying beside me in wide open line abreast, overtook me but [Ovadia] Nachman realized immediately that something was wrong. He slowed down, joined

A picture of a Pyramid taken during the 18 October 1963 mission from the cockpit of a Vautour, a cockpit frame is visible in the lower left corner of this photo. (PC)

Ramat David Wing Commander Ezekiel Somech (right) presents the 30 mm guns pack of Vautour 05 to President Shazar during the latter's visit to Ramat David on 30 October 1963. (PC)

me in close formation and, with hand signals, asked if something was wrong. I gave him the thumbs up OK signal and we continued on our track.

Not wanting to worry my navigator, I did not say anything about the engine. He saw the fuel flow gauge dropping, but I dismissed the problem as a 'non issue', and explained that the slower speed was necessary to ensure our accurate navigation in such poor visibility.

The mission was flown in radio silence; we just had to report, to a communication relay aircraft, codes that represented our position along the route. Because we flew slower, we also reported those codes a couple of minutes late [compared to the original plan]. Nobody said or asked anything.

So I flew the whole mission on one engine with the other engine in idle power. From my personal point of view the most moving aspect of this mission was not the engine issue but a picture of the Pyramids that I took, with a Leica camera that was strapped around my neck, while piloting the aircraft as very low altitude. The commander of the IAF Photography Technical Services Unit handed to me that Leica camera prior to the flight and asked me to take photos along the flight; 'I know that you will fly at low altitude, I know that you will fly fast but whenever you can, take photos, maybe you will capture something interesting' and so I did. With this camera I brought back home to Israel a marvelous photo of the Pyramids; the first time that an Israeli photographed the Pyramids from the air!

Our egress point from Egyptian territory was nearby Port Said at which point I climbed, slowly, to a safe higher altitude, broke radio silence and only then reported an engine failure.

It was a crazy mission and nobody asked any questions for years. The only one who guessed that something was left to be told was [IAF Commander] Ezer [Weizman]. He was puzzled by the late transmission of the position codes and figured that there was more to it than poor weather. Years after I left the IAF, at a social party, he forced me to have a couple of drinks, in order to loosen up, and then asked me to tell him the whole story again. Step by step we went [through the mission] until he came with the big one, 'I need to know, when and where did you actually lose the engine?' He said he had figured it all by himself a long time ago, but needed my confirmation.

Even now I cannot explain my decision to press ahead on one engine, but I guess it was a period in our IAF life that one could not imagine to abort a mission. I also probably did not want to destroy the safety and reliability reputation of our two engines, long range Vautour.

Vautour N Cyrano Operational Requirement

In November 1963 AIR2 issued an operational requirement for a Vautour N fitted with Cyrano radar. The planned missions for the Vautour N with Cyrano were long range attack, maritime strike, marking of targets for other attack aircraft, and test flying future developments of Cyrano for the benefit of the Vautour N and the similarly equipped Mirage III.

To suit the Vautour to Cyrano the cockpits had to be modified. The navigator's cockpit was to be modified for operation of Cyrano, and the pilot's cockpit was to be optimized for low altitude operations. Radar indicators (including a screen) were to be fitted in the pilot's cockpit, while basic flight instruments and a warning panel were to be installed in the navigator's cockpit. The requirement for installation of Cyrano was that it would not negatively impact the guns pack and the bomb-bay.

The Cyrano for the Vautour N was to be a modified version of the Cyrano Ibis, optimized for air-to-ground operations without a deterioration of the original air-to-air qualities. The radar was required to present on the crew screens an undistorted image of a 120 degree segment in front of the aircraft. Sufficient resolution was required to display railways, highways, and power lines. The displays were to offer three screen options: range up to 45 km, range up to 110 km, and range up to 220 km. For low flying, the radar was required to display obstacles to enable safe flying down to 200 feet above ground level. Bombing was to be possible in two modes: blind radar bombing from an altitude of 400 to 3,000 feet against a target on which the radar could lock, or an optical bombing with the regular sight, but with accurate range feed from the radar. Finally, this improved Cyrano was to be capable of operating up to three hours at low altitude without malfunctioning.

The squadrons of the Ramat David wing flew two fly-pasts during Suez War Day on 30 October 1963, including one over visiting President Shazar. Squadron 110 Commander Ovadia Nachman flew the first in Vautour 62 and the second in 61; Deputy A Jonathan Shahar flew both in Vautour 21. The lead Vautour in this photo is 61; the second is also a Vautour N, while the third and fourth are Vautour A aircraft. The two silver Vautour N aircraft are flying with 1,250 liter drop tanks, the first Vautour A is not loaded with drop tanks, and the second Vautour A has 1,300 liter drop tanks. (PC)

Gun Deflectors

The Cyrano to Vautour N was an ambitious requirement that would consume a lot of resources. A much more minor Vautour N issue was sensitivity of engine cut during high altitude gunnery. The French solved this issue through installation of gun deflectors, but this appendage had a negative impact on performance. Only Vautour 69 was fitted with such deflectors. By 1963 the Vautour was no longer expected to intercept enemy aircraft at high altitude, and gunnery at high altitude, for whatever reason, was a remote option. Therefore, in an AIR2 meeting on 25 November 1963, it was decided to abandon plans for installation of gun deflectors in IAF Vautours. Instead, the gunnery envelope of the Vautour N was restricted to speeds higher than 210 knots and altitudes lower than 35,000 feet.

Vautour 69 was the only IAF Vautour to be fitted with gun deflectors, the metal plates covering the nozzles. The IAF decided in November 1963 not to fit gun deflectors to its Vautours; actually, the gun deflectors were removed from 69. (PC)

Strategic Situation

Israel is a small nation; the distance from the Egyptian border to Tel Aviv is a mere 40 miles, and from the Jordanian border to Tel Aviv only ten miles. The IDF was significantly smaller than regular Arab armies, and Israel relied on reserves to counter this imbalance. Mobilization of reserves took at least 24 hours, and Israel's major concern was an Arab surprise attack in a scenario that would repeat the RETAMA crisis of February 1960. If Israeli Intelligence failed to give ample warning of an impending enemy attack, the IAF was to fill in the reserves' mission as a counterbalance until reserves mobilization could be accomplished. Therefore, IAF organization enabled a switch from

A Vautour on display with ordnance, including five optional under wing configurations: bombs, napalm bombs, rocket pods, T-10 rockets and (as attached to the wing) two 1,300 liter drop tanks and two bombs. (PC)

peacetime to wartime in a very short time frame, and was aimed to enable nearly full scale operations without mobilization of the bulk of IAF reserves.

IDF Deputy Chief of Staff Isaac Rabin presented the Israeli view of the Middle East strategic situation in a meeting at the USA State Department on 12 November 1963. The air power aspects of this presentation focused on the perceived Egyptian threat. IDF evaluation was that the EAF was capable of attacking Israel with 330 tons of bombs in a single wave, or with 870 tons over a 24 hour time frame. The IDF expected that this capability would evolve into 550 and 1,400 tons, respectively, by 1965. The evaluation was that an EAF offensive would focus on seven targets in Israel: the three IAF air bases (Hatzor, Tel Nof, and Ramat David) and four densely populated areas. The capability of other Arab air forces to add hitting power to the EAF offensive was marginal; for this reason Israel viewed air superiority over Egypt as the key to victory in a future war. In a worst case scenario of a surprise Arab attack, the IAF was expected to be capable of air defense, recovery from enemy air offensives, and counterattacking the enemy air bases. The weaknesses of the IAF were the small number of air bases, the short distances that challenged detection and interception, and the small number of IAF combat aircraft, "especially attack aircraft such as Vautour". Thus, the impact of an IAF counteroffensive against the EAF – which operated from 14 major air bases – was expected to be less than adequate.

Rabin stated that "if a first strike against Israel would destroy the IAF, it will be possible for the enemy to occupy Israel". Such was the importance of air superiority that the IDF's chosen course of action was a preemptive strike, and the IAF focused its planning on a surprise attack against Egyptian air power.

The Four Playing Card Suits

For quite some time, from IAF Year 1962 Term 3 when Flying School Class 36 graduate Josef Arazi was assigned to Squadron 110, no new pilots were assigned to the squadron until the start of IAF Year 1963 Term 3. Then, four Flying School Class 40 graduates reported to 110: Herzle Bodinger, Micha Gil, Ran Goren, and top student of Class 40 Zvi Shlomis. Micha Gil recalled:

> We attended an Ouragan OTU and towards the end of this course we were already notified of our planned promotion paths. The plan was to train one Term in the Mystere and then progress to our next type. The options were quite limited, Super Mystere or Vautour. I figured that the Vautour suited me well; it was a special aircraft with special missions operated by a special squadron, different in many ways from other IAF combat aircraft squadrons. When I was asked about my personal preference, I answered 'Vautour' and I was very pleased when I learnt that I was indeed to be assigned to the Vautour.
>
> Four of us were assigned to the Vautour and Herzle [Bodinger] – a very talented illustrator – suggested that each one of us four would have his helmet painted with one of the four playing cards suits and so it was. We were the playing cards suit foursome.
>
> Conversion courses started at the start of a Term when everyone was on vacation so we reported to an almost deserted squadron, the squadron building was older than the more modern facilities that we were used to at Hatzor [during the Ouragan OTU] and at Tel Nof [during the Term of flying Mysteres]. Vautour conversion started with a ground school and then Exercise 0 which was a high speed run along the runway; Exercise 1 was the solo flight. When we practiced Exercise 0, the squadron commander was in the control tower and he told us when to throttle back. One of us was told to throttle back too late so he took-off and landed straight ahead; a sort of a solo flight during Exercise 0. Serviceability was an issue and

The "playing cards suit foursome" on the right – left to right: Ran Goren, Zvi Shlomis, Herzle Bodinger and Micha Gil – facing a "photographer" during the annual squadron day party, January 1964. (PC)

Helmets of the "playing cards suit foursome". (PC)

Vautour 69 in flight when it was still a Squadron 119 aircraft; the Squadron 119 badge (originally the Squadron 110 Mosquito badge) is barely visible on the vertical stabilizer. The gun deflectors were removed by the time this photo was taken. (PC)

> my solo was flown in the late afternoon hours because I waited for an aircraft to be made serviceable and as a consequence my first landing in the Vautour was almost in darkness.
>
> Overall the Vautour was a challenging aircraft. The pilot had to exert a lot of physical power and there were many systems in the aircraft that the pilot had to activate correctly but the aircraft was superb; as a bomber it was very stable and very accurate.

Vautour N Gyro Sight Air-to-Air Gunnery

Tests of the gyro sight in Vautour 69 were resumed in December 1963 with a series of air-to-air gunnery flights. On 13 December AIR3 issued a Mission Order for six flights with a Squadron 107 Meteor as the target tow aircraft. A Squadron 110 pilot flew the Squadron 107 Meteor, and test flights started with two sorties on 16 December, when Deputy A Jonathan Shahar flew Meteor 31 and Deputy B Simi Saar flew Vautour 69; the former remarked in the debrief form "it is very cold in the cockpit of the Meteor at 15,000 feet in minus 12 degrees Celsius".

Squadron 110 Training Officer Beni Zohar flew two sorties in Vautour 69 on 22 December 1963, with Oded Einav piloting the Meteor. The rounds of each gun were smeared with different colors to enable post mission examination of hits on the drogue target, but during the first test flight that day the target fell to the sea. The final two sorties were flown on 23 December, again with Zohar as pilot. Zohar fired from all four guns during the first flight, but the upper left gun jammed after five rounds, while firing was stopped after some 60 rounds per gun in order to avoid shooting down the target. During the second flight, Zohar fired at the target all rounds from the two lower guns.

Vautour Replacement Study

The new IDF Chief of Staff, Isaac Rabin, entered office on 1 January 1964 and presented his views to senior commanders on 6 January. Rabin expressed his evaluation that the top priority concerning the IAF was to improve its attack force. The following day, 7 January, AIR2 issued a study of possible re-equipment options for two classes of attack aircraft: a light/medium attack aircraft (to replace the Ouragan, the Mystere, and the Super Mystere) and a heavy attack aircraft to succeed the Vautour. The envisioned operational requirement was for an attack aircraft capable of operating at low altitude over long ranges while penetrating enemy defenses and lifting a significant offensive load.

Four Vautours – two N aircraft and two A aircraft, all with 1,300 liter drop tanks – lead similar formations of Mysteres, Super Mysteres and Mirages. Mirage deliveries to Israel were completed during 1964, and the IAF started to evaluate candidates to replace the Vautour "heavy attack aircraft" and the Super Mystere/Mystere/Ouragan "light attack aircraft". (PC)

Evaluated options for such a new heavy attack aircraft were the F-4 Phantom, the Mirage IV, the Buccaneer, the Mirage IIIF2, and the Super Vautour. AIR2 recommended the Phantom as the IAF future heavy attack aircraft, with the Mirage IV as the second choice, the Buccaneer as the third choice, and the Super Vautour as the least favorable option.

The Super Vautour was actually the personal preference of AIR2 Chief Yoash Tsiddon. A paper project, the Super Vautour was to be a re-engined and re-winged Vautour with modern systems. Shorter wings and Rolls Royce Spey engines were expected to boost performance and revive an old design. Tsiddon's vision was that the Super Vautour would be developed in France and produced in Israel. Eventually, the Super Vautour remained a paper project, while Squadron 110 would never operate any of the types that AIR2 listed in this study as options for a Vautour replacement.

Vautour 70 Inspection

A team of IAF technical officers inspected the FAF Flight Test Center Vautour N 301 at Bretigny from 8 to 13 January 1964; Material Department Chief Arie Hillel circulated their report on 19 January. The report revealed that Vautour N 301 had flown only 441 flight hours since delivery to the FAF on 21 September 1956. The aircraft was damaged in a belly landing on 2 March 1959, but was repaired and was fitted with the convex leading edge wing. Many of the systems in this aircraft were non-standard in comparison to IAF Vautours; as an early production aircraft, it was equipped with an E-95 ejection seat, while IAF Vautours were fitted with the E-96 ejection seat. The French fitted Vautour N 301 to test the weapon system of the Mirage III. During this modification the aircraft became even more non-standard in configuration, including replacement of the guns with Cyrano units, installation of test equipment in the front suite of the bomb-bay, modifications to the cockpits, a re-engineered communication system, a smoke generator, and a winch for target towing.

Previously, on 12 November 1963, AIR2 Chief Yoash Tsiddon recommended to reject the proposed swap with Vautour N 301 "based on personal acquaintance with this aircraft". Tsiddon evaluated that Vautour N 301 was a worn out, non-standard aircraft, and re-installation of the guns pack and fuel tanks in the front suite of the bomb-bay would not be practical.

Regardless of the AIR2 Chief's recommendation, the IAF Commander approved the exchange of Vautour N 301 (to become IAF Vautour 70 in due course) with an IAF standard Vautour N (that the French requested to be delivered by 27 January 1964),

Flying in formation in early 1964, left to right: Vautours 05, 07, 08, 67, 12 and 20. (PC)

Superb view of a 3-ship, with Vautour 20 leading 03 (left) and 18 (right). (PC)

because this aircraft was expected to enable development of the Cyrano as an air-to-ground optimized radar in a much shorter time frame than a test program that would involve the modification of an IAF standard Vautour N to Vautour N Cyrano configuration.

IAF Readiness Order

The unique status of the IAF – as Israel's shield and the only IDF arm capable of operating almost immediately at maximum power – was re-validated in a new AIR3 Readiness Order that was issued on 13 February 1964. The AIR3 order defined readiness level per every IAF unit based on two criterions: daytime or nighttime, working hours or other hours.

Squadron 110 was tasked to maintain four Vautours on readiness during working hours. These Vautours were to take-off in attack configuration within an hour from scramble; the attack configuration was six 250 kg bombs in the bomb-bay, two 500 kg bombs under the outer wing stations, two external fuel tanks under the inner wing stations, and full fuel and full gun ammunition. In other hours Wing 1 was to rotate readiness for four aircraft between the three squadrons, but when other hours overlapped nighttime a single 3-ship formation was to be on readiness; one aircraft (always a Vautour) in illumination configuration and two aircraft (rotated between the wing's squadrons) in attack configuration.

Additionally, Squadron 110 had to maintain a Vautour B on reconnaissance readiness, ready to take-off within two hours during working hours or within four hours during other hours. To support this readiness a selection of cameras had to be maintained in serviceable status, including three high altitude K-38 cameras with 36" lens, one K-38 with 24" lens, one Zeiss BDA/mapping camera, and one K-37 nighttime camera.

Panoramic Camera

Operation MOMENTUM was the final penetration deep into Egypt of Vautour reconnaissance aircraft. Egyptian defenses were bolstered with MiG-21 interceptors and SA-2 missiles, so the mission of long range reconnaissance over enemy territory was switched from the Vautour to the Mirage. The Vautour was not deprived of the reconnaissance qualification, but instead relegated to second tier. In order to improve the operational capability of the Vautour as a second tier reconnaissance aircraft a lesson from MOMENTUM had to be implemented. Ovadia Nachman recalled:

> Yossi Sarig, with navigator Achikar Eyal, flew lead and I, with navigator Alex Meltzer, flew as wingman. The objective was to photograph [SA-2] missile bases that had just been constructed. We flew very low over Sinai; it was very difficult to fly and navigate because of low clouds. We may have deviated a little from our planned route and suddenly I saw it [the SA-2 base] on the side so I rolled my aircraft, pointed at it the camera that was at the underside of the aircraft and we photographed in such an oblique position. It was a SA-2 base still under construction and this was the first photo that we took of a SA-2 base. Then we flew onwards to Cairo West and Yossi Sarig lost an engine, he returned safely on one engine. En route to Cairo West we flew past the Pyramids and my navigator Alex Meltzer photographed the Pyramids with his private camera; he later presented to me a few very nice photos of the Pyramids as a souvenir from this flight.

A proper answer to such a shortcoming was a panoramic camera, and Israel acquired the American Fairchild F-415A camera, IAF code GOVERNOR. The new camera was planned for both the Vautour and Mirage; on 23 February 1964 MAT4 issued a requirement for testing of GOVERNOR in a Vautour B. The prototype installation was in the front camera compartment of Vautour 31 with a mount that was generally similar to the mounts of the Zeiss BDA camera or the K-37 night camera; the major difference was a new window to enable panoramic photography from horizon to horizon. IAI modified Vautour 31, so the first two test flights were to be flown from Lod to measure vibration levels. The second phase of the test program was to include four flights from Ramat David with actual photography from 250 to 40,000 feet at various conditions.

The Squadron 110 introduction of the Vautour N was inevitable in line with the diminishing number of IAF Vautours, but certainly complicated mission planning, because the A and N aircraft differed in configurations; the N aircraft – with their distinctive black nose cones – could only haul fuel in their bomb-bay. (PC)

Swap 69 for 70

Vautour 69, IAF Number 4469 and originally French Vautour N 330 (the 30th serial production Vautour night-fighter), was selected to be swapped for the Cyrano equipped Vautour N 301 (the first serial production Vautour N). Vautour 69 was the newest of the IAF Vautour N aircraft and had only recently been fitted with the prototype Vautour N gyro sight retrofit.

In line with IAF request a French crew flew Vautour N 301 to Israel, and the aircraft arrived at IAI on 20 March 1964. Jonathan Shahar flew Vautour 69 from Ramat David to IAI on 22 March, and the French crew returned to France in Vautour N 330. The following day, 23 March, Material Department Chief Arie Hillel inspected Vautour N 301, IAF

Yossi Sarig flew Vautour 05 – first from the left in this photo – on the evening of 22 March 1964, practicing nighttime attack with marking by IDF projectors.

Number 4470, at IAI. Hillel reported that Vautour 70 arrived in good condition, and that the option to refit the guns pack, previously rejected, should be reconsidered, as such an installation was, in his view, possible. Further inspection of this issue on 9 April concluded that 2,300 working hours would be needed to install guns, and the Material Department Chief withdrew his original evaluation on 19 April when he recommended that Vautour 70 should not be armed with guns.

Shafrir 2

Shafrir 1 disappointed the IAF, but was an essential stepping stone towards development of a family of extremely successful Israeli air-to-air missiles; the first of these was the Shafrir 2. RAFAEL started development of the improved missile well before AIR2 issued an operational requirement on 23 March 1964. The new missile was to be capable of intercepting MiG-17, MiG-19, MiG-21, Il-28, and Tu-16 aircraft flying at an altitude of up to 30,000 feet and maneuvering at up to 4 G.

The AIR2 operational requirement earmarked the Shafrir 2 to arm the IAF Mirages, but for the longer term the new missile was also planned to be fitted to the Vautour (priority A), Super Mystere (priority B), Ouragan (priority C), and Mystere (priority D).

Serviceability Nadir

The IAF Vautours accumulated 2,101 flight hours during IAF Year 1963; an average of 700 flight hours per term and an average annual utilization of some 100 flight hours per airframe. The plan for IAF Year 1964 Term 1 called for 885 flight hours, but MAT3 evaluated, on 22 April 1964, that such an objective was unrealistic. At any given time there were seven A and N aircraft undergoing A-Level to D-Level maintenance, so there were only ten A and N aircraft, at best, serviceable in the squadron. Almost half of these had to be in readiness during working hours in line with the aforementioned AIR3 Readiness Order. Actual experience indicated that even an objective of ten A and N serviceable aircraft was optimistic, MAT3 re-evaluated on 22 June 1964 that only eight Vautour A and N aircraft could be expected to be in serviceable status in the squadron.

SARA Shortage

Almost a year after Squadron 110 absorbed the former Squadron 119 Vautour N aircraft, issues relating to this event still surfaced. One such issue was the standard of SARA personal radio transceivers. The SARA transceivers were an element of the emergency kit attached to an ejection seat and were aimed to enable an air crew to communicate with rescue forces after ejection. The addition of six Vautour N aircraft added 12 ejection seats to a

The 16 April 1964 Independence Day fly-past over Beer Sheba included this formation of 16 Vautours - quite an achievement considering serviceability. Ovadia Nachman with Alex Meltzer flew lead in Vautour 31, while Jonathan Shahar with Menahem Eini flew Vautour 33. In addition to the two B aircraft, the 16-ship formation included four N aircraft and ten A aircraft, with Micha Gil flying in Vautour 21. (PC)

Transfer of the N aircraft – including 62, which was photographed in June 1964 prior to a joyride with Intelligence Officer Reuben Nir in the back seat with pilot Beni Zohar – complicated logistics for Squadron 110. (PC)

maximum Squadron 110 inventory of 30 seats. The standard was not changed, however, and remained 21 transceivers, with 18 for attachment to ejection seats and three as reserve units. Squadron 110 Commander Ovadia Nachman reported on 28 April 1964:

> There are 22 ejection seats [in Vautour aircraft] at Ramat David and in time of wartime it is reasonable to estimate a requirement for 25 to 28 transceivers apart from units in reserve. There are 16 SARA transceivers in the squadron; in all two seats aircraft the SARA is only fitted to the pilot's ejection seat.
>
> The standard should be amended to an adequate number, as per wartime. For the meanwhile, the squadron should be issued with at least 22 transceivers.

GOVERNOR Qualification

In line with MAT4's request, AIR2 issued a GOVERNOR test order on 25 February 1964. Three tests were flown from Lod from 10 to 14 March to verify the installation. Then Vautour 31 was ferried from Lod to Ramat David on 19 March. Five tests were flown from Ramat David from 23 March to 19 April to explore the flight envelope and to photograph from 250 to 40,000 feet; Yossi Sarig with Alex Meltzer flew the 25 March test, while Jonathan Shahar with Achikar Eyal flew the 14 April test. Resulting photographs were sharp, and the resolution was adequate for identification of aircraft on the ground from altitudes up to 20,000 feet. Photography from 40,000 feet enabled pinpointing large targets, such as air bases, at long ranges.

The problem with GOVERNOR was high temperatures. Previously, a heating system was installed in the reconnaissance Vautours to warm up the front suite of the bomb-bay so that the cameras would function during high altitude photography. At low altitude, high speed photography an opposite problem surfaced; when a temperature of 65 degrees Celsius was measured in the front camera compartment, the temperature on the ground was 20 degrees Celsius. Thus, MAT4 recommended on 24 April 1964 to qualify GOVERNOR for operations, but to restrict low level photography speed to 500 knots at a ground temperature of up to 23 degrees Celsius, to 450 knots at 23 to 27 degrees, and to 400 knots at 27 to 30 degrees. This recommendation was accepted and AIR2 qualified GOVERNOR on 29 April.

Regardless of this qualification, AIR2 ordered, on 30 April 1964, additional tests of low altitude, high speed GOVERNOR photography in order to evaluate the impact of high temperatures in the camera compartment. Two test flights were to be flown at 500 knots at 250 feet in order to achieve 60 degrees Celsius in the camera compartment, and two subsequent flights were to be flown at 550 knots so that 65 degrees would be measured in the camera compartment. This program was slightly amended on 3 May, when AIR3 added a fifth test to be flown at 500 knots at 250 feet, but with a take-off time at 05:15 in order to evaluate effectiveness at low sun angles. Additionally, AIR3 defined GOVERNOR missions as photography of air bases (to discover deployment, to pinpoint potential targets, and for BDA during wartime), "point" targets (such as radar stations, SAM batteries, ports, etc.), along roads, along planned routes for discovery of new targets, and in combination with visual reconnaissance to detect enemy deployment of forces. In light of these missions, and the limited scope of the test flights, it was suggested to conduct an operational evaluation of GOVERNOR to test the system in realistic mission profiles.

The second series of GOVERNOR tests was flown from 6 to 19 May 1964; Ovadia Nachman with Alex Meltzer flew the first (this was actually the low sun angle test with take-off at 05:03), and Jonathan Shahar with Meltzer flew the last. Photography was proved possible from 15 minutes after sunrise, but the high temperature issue was not solved; as a result of these tests the speed was limited to 450 knots at an outside temperature of 35 degrees Celsius and to 410 knots at 40 degrees. MAT4 therefore suggested, on 11 June, to install an air conditioning system in aircraft fitted for photography with GOVERNOR.

A GOVERNOR panoramic photo, taken on 26 March 1964 from 250 feet, of the Squadron 119 Mirage complex at Tel Nof. (IDFA)

The Squadron 107 Meteor line in a GOVERNOR panoramic photo taken from Vautour 31 on 26 March 1964. (IDFA)

Prior to this, Vautour 31 took part – alongside the two new dedicated Mirage III reconnaissance aircraft – in a special display at a Squadron 119 HAS on 20 May 1964. Guests were potential users of the new IAF reconnaissance assets, and in addition to briefings beside the aircraft, Vautour 31 flew a demonstration over Tel Nof with subsequent processing of the film and printing of photos. The importance of GOVERNOR was such that AIR3 issued an operational requirement on 8 June for three Vautours (and three Mirages) capable of operating the panoramic camera. To fully exploit GOVERNOR the issue of high temperatures had to be ironed out, and AIR2 ordered the installation of an air conditioning system on 16 June that would enable a GOVERNOR equipped Vautour to fly as fast as 550 knots at low level at an external temperature of up to 45 degrees Celsius and without any time limitations.

New Vautour Proposal

The Vautour's manufacturer forwarded a price proposal for 25 New Vautour aircraft on 20 May 1964. The unit cost for such an aircraft powered by refurbished British Rolls Royce Avon engines was 7,072,000 French Francs (down to 6,307,000 FF for purchase of 50 aircraft). The unit cost for similar aircraft powered by new French SNECMA Atar 8C engines was 8,305,000 FF in a production batch of 25 or 7,370,000 FF for production of 50 aircraft. Since the Vautour was no longer in production, deliveries of New Vautour aircraft could start two years after contract signature and span a time frame of 14 months for 25 aircraft (at a maximum production rate of two aircraft per month), or 19 months per 50 aircraft (at a maximum production rate of three aircraft per month).

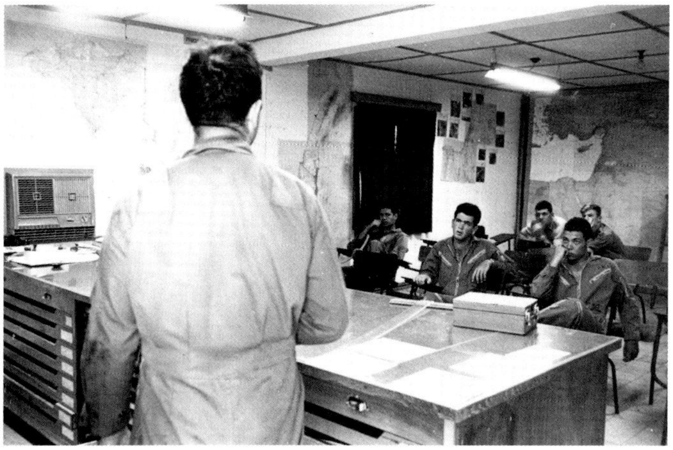

Beni Zohar (back to camera) briefs, left to right: Micha Gil, Zvi Shlomis and Isaac Nir. Sitting in the back, in front of the wall map, from left to right: Jonathan Shahar and Gideon Dror. (PC)

Herzle Bodinger (left, lower) and Isaac Nir (right, higher) on the ladder in front of Vautour 20 a short while before the accident. (PC)

Crash Landing in a Vineyard

Atar engines powered the IAF Mirages (Atar 9B), Super Mysteres (Atar 101G), and the Vautours (Atar 101E). From March 1963 until June 1964 the IAF lost four Mirages and two Super Mysteres due to engine related malfunctions; the Vautours also suffered from engine issues, but the two engine safety proved itself very well, and not a single Vautour was lost due to in flight engine cuts until 28 June 1964. On that date four Vautours prepared for a flight from Ramat David to the Negev Desert; as was often the case, one of the four Vautours malfunctioned and the formation departed as a 3-ship, with Squadron 110 Commander Ovadia Nachman leading in Vautour 08 and Herzle Bodinger as his wingman in Vautour 20. Herzle Bodinger recalled:

> By June 1964 we were relatively experienced pilots and we were just about to return to the Flying School for service as instructors while continuing to fly Vautours as emergency posting pilots; this was the normal track of service. The French aircraft were excellent from the perspectives of human engineering, ease of handling, aerodynamics and the concept of the onboard systems. The weakness of the French aircraft was in propulsion and there were three types of aircraft in IAF service that were powered by various versions of the Atar: the Super Mystere, the Mirage and the Vautour. All of these Atar versions suffered from malfunctions, in the engine control unit, during the summer of 1964. The engine control unit monitored fuel flow to the engine; when a pilot pushed the throttle forward to maximum RPM, the engine control unit managed optimal acceleration of the engine to maximum RPM per input of parameters such as altitude, pressure and temperature. Likewise, when a pilot set the throttle at a certain fixed setting, the engine control unit maintained constant RPM even if the aircraft climbed or dived.
>
> The malfunction in the engine control box was related to a certain diaphragm. When this diaphragm failed, the engine was no longer controllable and would be stuck at maximum RPM until falling apart. This Atar malfunction resulted in a string of accidents that ended in ejections from Mirages and Super Mysteres; the Super Mystere squadron even had a display of the ejected seats. The Mirages and Super Mysteres were grounded but the Vautour, with two engines, continued to fly.

Isaac Nir examines the right engine of Vautour 08 during a pre-flight inspection in early June 1964. (PC)

Vautour 20, fitted with 1,300 liter fuel tanks, departs Ramat David sometime prior to the 28 June 1964 accident. (PC)

Vautour 08 – which Ovadia Nachman flew on 28 June 1964 – as photographed earlier that month. (IDFA)

It was pure coincidence that I flew two test flights in Vautour 20 prior to the accident. Vautour 20 underwent engine changes and after an engine change a certain pattern of test flight had to verify that the aircraft was serviceable. This included a climb to high altitude, a check of several parameters, a dive to maximum speed and another series of tests at low altitude. Both engines in this aircraft were unable to maintain RPM at 25,000 feet. I landed, reported, the engines were repaired and when this work was completed, I flew the second test flight. The same phenomenon repeated so the aircraft was declared serviceable for low flying only.

The day of the accident was very hot and we were planned to fly at noon an operational test to evaluate a certain mission profile that included very accurate navigation, bombing of a target hidden in the Negev Desert with a single bomb and then one strafing burst. It was very hot and when we taxied to take-off, one of the four Vautours reported a malfunction and aborted the flight. We took-off as a 3-ship formation with Squadron 110 Commander as leader.

We followed briefing precisely; navigation, bombing run, strafing pass and when we re-grouped I sensed severe shuddering from the left engine and warning light of fire in the right engine. I reported to the leader and he got closer to inspect my aircraft. When he finished this inspection he ordered me to switch off the right engine. He did not tell me anything but he saw that the engine lost some parts and was in poor shape. The left engine vibrated wildly above 70% or 80% RPM so I had to set it to a lower RPM that produced less thrust. Obviously, I would have ejected right there if both engines were lost but one engine still somehow functioned and I decided to remain in the aircraft for as long as the aircraft would be able to fly straight and level. The aircraft managed to fly straight and level at 10,000 feet, rather slow at slightly more than 200 knots but it kept going. There were only three air bases; southernmost was Hatzor, then Tel Nof and Ramat David in the north. Squadron 119 Vautour was based at Tel Nof so the maintenance echelons at Tel Nof were well acquainted with the Vautour and the squadron commander previously commanded 119; he therefore led me to Tel Nof.

We approached runway 36 and everything was still OK. I lowered flaps and landing gear, still everything was OK. Then the engine started to lose power and within seconds I realized that I would not make it to the runway. I did not even have the time to report, I had to react immediately. I placed my hand on the ejection handle but I was below the minimum altitude for safe ejection in this type of seat. What shall I do? I cannot land and I cannot eject. In a split second I decided to land in a vineyard, I noted a gravel track in the vineyard and turned right towards this track; I do not know if landing on this track changed anything but it was an instinctive action. I landed on that track and the aircraft rolled some 200 or 300 meters, swerved and stopped. During this landing the aircraft entrenched in the ground so there were very loud noises from this digging and, due to electrical short cuts, the guns fired all ammunition. When the aircraft stopped it was well entrenched so that the edge of the cockpit, normally high above the surface, was leveled with the ground. The lower part of the cockpit was below the surface of the ground and when I looked around I saw that the vines were on the same level as I was.

I tried to open the canopy and it opened. I then wanted to report over the radio that I was OK but the communication system did not function any more so I figured that I should get out of the aircraft but I could not move. I hardly managed to breath and I suffered severe pain. I assumed that I fractured my spine but I was able to move my legs so I was relieved that I was not paralyzed and decided to wait for help while avoiding movement of my body.

The first to reach me was the farmer. He ran to the aircraft, asked me what happened, which was a rather pointless question, but he was eager to help, and it was a very hot day, so he asked me if I would like to drink water. I tried to answer but was unable to speak, I was hardly breathing. He interpreted my silence differently and slapped me in the face, twice, probably thinking that I was in some sort of a shock and again asked 'do you want water?' so I quickly nodded my head for a yes. He went away to get me water and before he returned, the aircraft was already surrounded by a lot of IAF service persons who ran to the aircraft from Tel Nof. Then a fire engine arrived and a helicopter landed beside my aircraft. All of this happened very quickly, the farmer did not have the time to return with water. The trouble was that everyone tried to pull me out of the cockpit but I was worried about two issues: the ejection seat was not safe and my injury. I was quiet sure that I was wounded in my back and if this was the case then every incautious movement might worsen my injury. Then I saw, in the crowd around me, the Squadron 116 Deputy A Arie Ben-Or. He was a big person so I signaled to him with my finger to get closer, he got closer and I whispered 'Arie, I think that I fractured my spine, try to pull me out as I am seated and carry me to the helicopter'. He put his big arms around me and carried me like a baby straight to the helicopter.

The helicopter flew me to hospital. My suspicion of a back injury was confirmed but I was not operated, I was plastered and hospitalized for several months until my back injury healed. When still in plaster, I was offered to continue my service as a ground controller 'for the time being, while you are grounded' but I refused, I feared that such a temporary assignment would lead to a permanent assignment and I wanted to return to fly. My proposition was to continue my service at the squadron as a sort of an assistant operations officer. The squadron commander, by then there was a new squadron commander, accepted this proposition and I returned to the squadron. When my recovery process was completed, I asked to fly. There were no Vautours with two pilot cockpits so I flew in a Meteor with a safety pilot to explore how much G load I would be able to tolerate. Then I attended a medical committee and was cleared to return to fly in the squadron. Some years later and many years later, I faced two interesting repercussions.

The first, some years later, was after the Six Day War in 1967. The IAF started to computerize medical files and someone noted that my medical fitness category was 37; well below the minimum for pilots. Prior to the accident my medical fitness category was 97, the maximum, but it was changed to 37 after the accident. I was summoned to the IAF Medical Unit and was told that I cannot fly with such a low medical fitness category. This amazed me as I just flew in a war but I asked what can be done. The IAF Medical Unit suggested that I would apply to the IDF Medical Service so that they would change my medical fitness category. I did but the IDF Medical Ser-

A Vautour in a vineyard; 4320 resting beside the track on which Herzle Bodinger landed on 28 June 1964. (PC)

Debris, including a complete balancing wheel assembly, that Vautour 20 departed from prior to coming to rest. (PC)

vice refused to change my medical fitness category, claiming that this was the responsibility of the IAF Medical Unit. I was caught in a 'Catch 22' situation and only after a lot of effort the IDF Medical Service and the IAF Medical Unit assembled a special committee that changed my medical fitness category to 45, enough for a pilot who would become IAF Commander.

The second, many years later, happened after I completed my cadence as IAF Commander in 1996. Two years later I was asked if I was willing to serve as a Flying School instructor in reserve service. I could not be happier; I served in the Flying School from instructor to commander and I loved to instruct flying. So I was a Flying School instructor in reserve service until 2006. That same year I received a letter. The writer of this letter was the farmer who came to my aid in the vineyard back in 1964. He described the incident from his point of view; when he returned with water the aircraft was already surrounded by IAF service persons he was not allowed to get any closer. He saw that I was evacuated in a helicopter but he did not know anything, not the name of the pilot and not even if the pilot survived. Ever since that time he tried to learn what happened to the pilot that 'he helped to rescue' – in his words and I accept his view – but he never received an answer. Then he met someone, again told his story and that someone told the former farmer that the pilot in question was me, a former IAF Commander. The former farmer was stunned and immediately wrote a letter to me, asking me if I can confirm that I was indeed the pilot that 'he helped to rescue'. I immediately called him on the phone, confirmed that I was the pilot and invited him to the ceremony that the IAF arranged for my farewell flight upon completion of my reserve service as a Flying School instructor; an invitation that he gladly accepted.

Meteors from 107 to 110

None of the Vautours had duplicate flying controls to enable training of new pilots. Therefore, a first flight in the Vautour was always a solo flight for the pilot in command. To prepare pilots for a first flight in the Vautour, the Vautour Conversion Course included a couple of preparatory flights in a Meteor training aircraft with an instructor in the back seat and a Vautour Conversion Course student in the front cockpit. The objectives of these training sessions were to practice flying a relatively fast two engine jet fighter and to experience single engine handling.

Squadron 110 relied on Ramat David based, Meteor equipped Squadron 117 (until 1962) and Squadron 107 (from 1962) for availability of Meteor trainers for the Vautour Conversion Course flights. Squadron 117 operated the Meteor as a frontline fighter until 1962, and Squadron 107 was organized as a reserve squadron, with a skeleton of regular service persons, to operate the Meteor during wartime. This change reflected the decay of the Meteor as an IAF combat aircraft, and on 30 June 1964 AIR1 issued an order to deactivate Squadron 107. All Meteors were to be withdrawn from IAF service except for the Meteor trainers.

Squadron 110 was one of the principle customers of Meteor training services, as well as the only IAF combat aircraft squadron to operate two engine fast jets. Therefore, the aircraft inventory of the squadron was boosted, as 110 became the caretaker of six Meteor trainers.

Vautour 70 Operational Requirement

Inspections of Vautour 70, still stored at IAI, revealed that the aircraft was less standard than originally estimated. As a result, AIR2 issued an operational requirement on 5 July 1964 for installations to modify Vautour 70 as a test aircraft.

Vautour 70 would be utilized to test weapon systems in general, and particularly the improved Cyrano. The modifications that AIR2 requested were a radar screen in the front cockpit (for viewing only; operating the radar would continue to be from the rear cockpit), a Horizontal Situation Indicator as in the Mirage, a radio altimeter, and rewiring of the wing racks with matching control boxes in the relevant cockpits. The latter included the wiring of one station for the R530 air-to-air missile, wiring two stations for Shafrir 1 or 2 missiles, the wiring of one station for launch of generic air-to-ground missiles, and the wiring of all

The Squadron 107 line at Ramat David prior to deactivation; the Meteor trainers were reassigned to Squadron 110 and the Meteor fighters were retired. (PC)

four station for bombs. Missile launch control boxes were to be fitted in the front cockpit, bombs drop control boxes were to be fitted in both cockpits (if possible), and a CSF-95 sight – as in the Mirage – was to be fitted in the front cockpit.

Systems to be retained "as is" were the fuel system, the extensive communication system (two UHF sets and one VHF transceiver), and the radio compass. IAF modifications relating to flight safety issues (such as an EJECT warning light in the rear cockpit) were to be implemented. Overall, Vautour 70 was to be prepared for testing the improved Cyrano as soon as possible, because the associated flight test equipment was loaned to the IAF for one year.

Shafrir Service Acceptance Test

The Vautour's role in the development of the Shafrir did not end after the missile entered IAF service from 1963. Prior to service introduction, the Shafrir tests were mostly sterile and involved launches against non-maneuvering targets. Testing of a more realistic scenario was made possible only after acceptance of the first Del Mar target in May 1964. The planned test was therefore viewed as a "Shafrir service acceptance test," and ten missiles were allocated for launch against a maneuvering Del Mar target (actually turning at up to 2 G). The target tow aircraft was to be a Squadron 110 Vautour.

PILLAR DAWN 115 was flown from 7 to 16 July 1964. The Vautour crew was pilot Jacob Rafaeli and navigator Alex Meltzer; the launch aircraft was a Squadron 117 Mirage. Eight missiles were launched and only one missile hit a target; three missiles "got close" to the Del Mar target and four missiles did not even get close to the towed target.

A Vautour A flying over Vautour dummies at Ramat David in June 1964. (PC)

Vautour Replacement Mirage

The IAF requirement for an influx of modern attack aircraft was redefined on 25 June 1964 to include 50 to 60 heavy at-

Jonathan Shahar flew Vautour 12 over the wings parade at Tel Nof of Flying School Class 44 on 16 July 1964; four Class 44 graduates – Michael Dvir, Giora Goren, Jacob Tal and Abraham Vilan – were assigned to Squadron 110 from April 1965. (PC)

tack aircraft (eventually to materialize with the introduction of the Phantom), 100 to 120 medium attack aircraft (to be fulfilled by acquisition of the Skyhawk), and eight to 12 Mirage IV aircraft. That same month, Israel asked France for the purchase of 15 Mirage IV aircraft. These were to be designated Mirage IVRJ and be capable of lifting and launching four AS-30 air-to-surface missiles. For all purposes, the Mirage IVRJ could have been the ultimate replacement for the IAF Vautours. Negotiation went ahead, up until a draft contract was drawn on 23 July 1964, but never progressed any farther.

Sud Aviation summed up a 14 August 1964 discussion with AIR2 Chief Yoash Tsiddon concerning the Super Vautour operational requirement. The 23 ton Super Vautour was to be designed in light of two baseline mission profiles: a short range, low-low-low mission against a target 250 miles away with an offensive load of five to six tons and a long range, low-low-low mission against a target at a distance of 500 miles with an offensive load of three tons. The Super Vautour was to be powered by Rolls Royce Spey engines, equipped with a comprehensive avionics suite (including Cyrano radar, navigation computer, and electronic countermeasures), and armed with powerful weapons (including the four 30 mm guns pack, but with 150 rounds per gun, plus one 3,000 kg bomb or six 1,000 kg bombs or 12 500 kg bombs or 32 120 kg bombs or four air-to-surface missiles, or various combinations of these weapons).

Test Vehicle or Operational Aircraft

Yossi Sarig took over from Ovadia Nachman as Squadron 110 Commander in late July 1964. At the same time, three of the four playing card suit pilots – Micha Gil, Ran Goren, and Zvi Shlomis (Herzle Bodinger was in hospital, recovering from his injury) – were assigned to the Flying School, in line with the then normal track of service, for a Flying Instructors Course and subsequent service as instructors while remaining in the squadron's roster of air crews as emergency posting pilots. Micha Gil recalled:

> The most significant issue during our service as Flying School instructors and Squadron 110 emergency posting pilots was the extraordinary high number of malfunctions in the Vautours. As a result of these serviceability issues, Vautour flights were very limited and every flight counted; we flew so few hours [in the Vautour] that I often wondered how we managed to preserve our operational qualifications.

Such a sorry state of affairs magnified the importance of every aircraft. The aforementioned AIR2 issue of an operational requirement to modify Vautour 70 as a test aircraft triggered Wing 1 Commander Ezekiel Somech to inquire, on 23 September 1964, if Vautour 70 would be a test vehicle or an operational aircraft:

> This aircraft arrived in Israel six months ago. It has not been added to the squadron's inventory and there is no schedule for its assignment to the squadron. Furthermore, I was informed that work on this aircraft has not started yet.
>
> Considering the number of Vautour aircraft – and their many missions – it is impossible to forgo this aircraft as an operational aircraft. The modification of this aircraft into a flying laboratory, at the expense of its capabilities as a combat aircraft, will result in a situation that this aircraft will not be capable of flying operationally at all.
>
> My opinion is that this aircraft should be assigned to the squadron as soon as possible as an operational aircraft. Any required modifications for test flying, that will take a long time to implement anyway, should be accomplished in the wing or by occasionally flying the aircraft away [to D-Level provider IAI] in the same manner as done with other aircraft.

Ejection from Vautour 35

The gloomy situation that Micha Gil described and Ezekiel Somech reported was exacerbated on 5 October 1964, when pilot Yigal Yanay and navigator Zvi Har-Nevo ejected from Vautour 35. The crew practiced a low altitude reconnaissance mission profile, including a photographic run over a simulated SA-2 defended target in the Negev Desert. When Vautour 35 was pulled up to fly the photographic run, the notorious diaphragm failed. The impacted engine exploded, and this explosion ripped hydraulic pipes. Control of the aircraft was lost and the crew ejected.

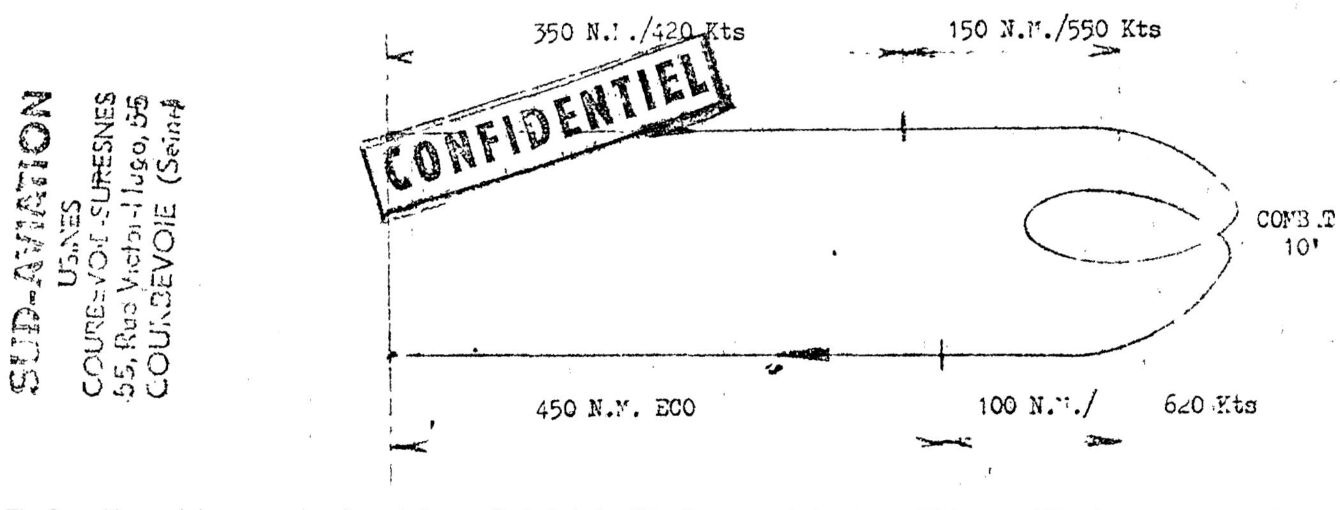

The Super Vautour's long range baseline mission profile included a 400 mile economical cruise at 450 knots, a 100 mile penetration at 620 knots, ten minutes combat, disengagement for 150 miles at 550 knots, and economical cruise for 350 miles at 420 knots. (IDFA)

IAF Chief Test Pilot Daniel Shapira (right) evaluated the Mirage IV in five flights from 23 February 1965. (PC)

Friday the 13th

The night of 13 to 14 November 1964 was set for an IDF GHQ Reconnaissance Unit special operation across the Egyptian-Israeli border, called Operation KINGFISHER. In support of this operation, the IAF issued on 12 November, at 14:50, an order to upgrade readiness from dusk on Friday, 13 November. Squadron 110's part was to maintain two standard nighttime attack 3-ship formations (a lead aircraft loaded with flares and a pair armed with bombs), plus a reconnaissance Vautour. Yossi Sarig recalled:

> We were to be on readiness with two nighttime attack 3-ship formations. The Vautours suffered from a lot of malfunctions, serviceability was low and on top of that there were engines that did not start without any apparent reason. I realized that in order to be on the safe side, I should have a 50% margin. The squadron was dispersed in HAS complexes and there were two aircraft in every HAS. It was impossible to gather at the line all the aircraft that I planned to prepare for that readiness. As a result, I realized that a lot of maintainers would have to remain on readiness with the aircraft. I consulted with the Technical Flight Commander and we decided that the night from Friday to Saturday would be closed; all squadron service persons would be confined to base. Most of our service persons resided in the north of Israel, not too far from Ramat David, so our idea was to give them a leave on Friday. Then they would all return in the afternoon and we would spend the evening and night together, to boost the squadron's morale and esprit de corps, while also ensuring that we would be able to launch the required number of aircraft on readiness.
>
> Ramat David Commander [Ezekiel] Somech was abroad and Squadron 117 Commander David Ivry was acting commander. Naturally, I reported my decision to close the squadron for the weekend because of this readiness. Ivry liked this idea and decided to close the base.
>
> Friday, circa 14:00, I was at home, having a shower, when I heard the base siren. There were no telephones at the base neighborhood but as squadron commander I was entitled for a point-to-point phone. I operated the point-to-point phone and heard a phrase that I never heard before or after: 'you are authorized to scramble and you are authorized to bomb'. I asked 'where?' and the answer was 'take-off, all details will be transmitted when you are in the air'. I called the squadron, ordered to scramble everybody and rushed to an aircraft.
>
> We took-off and [IAF Commander] Ezer [Weizman] contacted me directly, 'the Syrians are bombarding the Galilee, Mound Azaziat is yours, attack there' and indeed

Vautours 08, 18 and 26 as photographed during the 29 October 1964 Suez War Day turnaround; pilots participating in the Suez War Day fly-past included Squadron 110 Commander Yossi Sarig with Alex Meltzer in Vautour 62 and Deputy B Isaac Golan in 07. Micha Gil flew the depicted Vautour 26 on 13 November 1964. (PC)

A Vautour bombing Mound Azaziat on 13 November 1964. (PC)

when we were over Mount Tavor we already saw the Upper East Galilee covered with smoke. I planned my attack, asked permission to attack from the ground controller and was authorized to attack. I reported to my wingman '1 going in' and as I started my bombing run I noted a pair of Mysteres flying over Mound Azaziat, head on and passing right underneath me. Apparently, the Squadron 109 Mysteres managed to start their engines faster, they had only one engine to start, and they took-off ahead of us. By the time we bombed, they strafed. Anyway they cleared the area and our bombing was very accurate, the Mound was engulfed in smoke.

Operation KINGFISHER had not started yet, and the scramble of the aircraft was not southwest to Egypt, in support of KINGFISHER as expected, but northeast to Syria. Another Water War clash erupted and the IAF was ordered, at 14:12, to attack Mound Azaziat. The first formation to take-off was a Squadron 109 Mystere 4-ship that departed Ramat David at 14:47. Next was a Squadron 110 pair: Yossi Sarig with Alex Meltzer in Vautour 62 as leader and Micha Gil in Vautour 26 as wingman. Micha Gil recalled:

I was an emergency posting pilot qualified for daytime operations only. It was Friday and I was on readiness, this was not pre-planned but someone on readiness had to go somewhere and he asked me to take his place. A short while later we heard a siren, the squadron commander Yossi Sarig told me to join him and Deputy B Isaac Golan shouted 'we are authorized to attack'. We all ran to the aircraft and took-off.

Prior to this flight I only once – during one of the combat day exercises – dropped two 500 kg sand bombs from the wing stations and I never before dropped bombs from the bomb-bay. There was an environment of pressure to attack as fast as possible, we were afraid that we would be ordered to abort our attack, so we dropped all bombs in a single salvo of six [250 kg] bombs from the bomb-bay and two [500 kg] bombs from under the wings. I was totally unaware that it took some time for all the bombs to drop and my final two bombs overshot the target. We were eager to strafe but we were ordered not to strafe because many more aircraft were on their way to attack Mound Azaziat.

Those aircraft included a Squadron 110 3-ship formation that departed Ramat David at 14:55 with Isaac Golan leading Oded Einav and Boaz Gafni. They attacked right after the pair; one aircraft was in the same configuration as the pair, and the other two aircraft were armed with four 500 kg bombs. Thus, the five Vautours dropped a total of 14 500 kg bombs and 18 250 kg bombs. Then the 3-ship formation flew three strafing passes in the face of scattered Syrian anti-aircraft fire; a 0.5" round hit Golan's Vautour 65 in the cockpit, penetrated the cockpit via the "bullet proof windshield," and exited via the canopy, narrowly missing the pilot's head while glass splinters stabbed Golan in his face.

Overall, the IAF scrambled 49 combat aircraft, and 19 of these aircraft actually attacked: eight Mysteres with eight napalm bombs and eight 250 kg bombs, four Mirages with eight 500 kg bombs, two Super Mysteres with two napalm bombs and two 250 kg bombs, and the Vautours with their 32 bombs! This was the best demonstration of the value and impact of a heavy attack aircraft; 14 light attack aircraft dropped 28 bombs weighing significantly less than the 32 bombs that only five Vautours dropped.

More Vautours for 110?

Super Vautour discussions continued, and on 1 December 1964 Sud Aviation suggested to skip prototypes and to start production. Based on a production run of 50 "Vautour Spey" aircraft, Sud proposed a unit cost of 8,500,000 French Francs and progressive transfer of manufacture to IAI, with deliveries to be completed by 1967.

An Air Department Chief document dated 11 December 1964 presented a survey of available attack aircraft in England (Buccaneer, Canberra, Harrier and Hunter) and in France (Mirage IIIE, Mirage IIIF2, Mirage IV and Super Vautour). From these types, the preferred heavy attack aircraft was the Super Vautour. However, the Super Vautour was the only paper project among the evaluated eight types, and as such was still a concept, and a long way from realization. Meantime, until the Super Vautour became a reality, it was recommended to acquire additional Vautours to top off the depleted inventory of Squadron 110.

IAF Commander presented the 11 December 1964 survey in an IDF GHQ meeting on 1 February 1965. In the wake of this presentation, a budget was allocated for acquisition of two Vautours and 16 Super Mysteres. Negotiations with France for purchase of Vautours and Super Mysteres was launched, but both types were still in FAF service, and the French could not spare any of these FAF aircraft for sale to Israel.

The Return of Vautour 09

A Vautour that returned to fly during December 1964 was 4309. As indicated, Vautour 09 was damaged in an accident in February 1963 and arrived at IAI on 7 March 1963 for a survey to figure out repair options. While IAI was preparing this survey, the convex leading edge wings of Vautour 09 were removed and fitted to Vautour 08; the latter's wings were fitted to Vautour 09.

IAI completed the survey and forwarded to the IAF a proposition on 15 May 1963 that the IAF evaluated to be exaggerated. The IAF therefore launched its own survey to figure out how exaggerated the IAI's offer was. Discussions between the IAF and IAI (both were branches of the same organization, the IMOD) continued until terms were agreed upon in July 1963. Essential structural parts were then ordered from the manufacturer and delivered in February 1964, which is when actual repair work started.

Originally, repair was scheduled for completion by August 1964, but this slipped to December 1964 when IAI test pilot Zeev Tavor – a Squadron 110 Vautour reserve pilot – started test flying the repaired aircraft. By March 1965, Vautour 09 was again in Squadron 110 service.

First Shafrir 2 Launch

The first series of Shafrir 2 test flights was PILLAR DAWN 117, from 18 December 1964 to 7 January 1965. The test was planned to evaluate the functionality of the Shafrir 2 launcher as attached to a Vautour aircraft. If this functionality was proven then three missiles, without seeker and without warhead, would be launched.

Previously the Vautour Shafrir test aircraft (Vautour 03) underwent D-Level maintenance at IAI from 22 January 1964 and did not return to fly until November 1964. It was during this time frame that AIR2 ordered, on 7 April, to remove the Shafrir installation from Vautour 03, because the Mirage had already taken over as launch platform in the Shafrir 1 tests. However, the Shafrir 1 test pattern was repeated in the Shafrir 2 tests, with the Vautour as initial launch platform and the Mirage taking over along the development time line as the ultimate test launch aircraft.

The first phase of PILLAR DAWN 117 was completed successfully; the functionality of the Vautour's Shafrir 2 launcher

AIR2 ordered on 29 December 1964 to repaint the Vautour B aircraft in a camouflage scheme; the repainted Vautour 33 was actually photographed landing after modification with the GOVERNOR air conditioning system during 1965. (PC)

The IAI team that repaired Vautour 09; IAI test pilot and Squadron 110 reservist Zeev Tavor stands below the cockpit. (PC)

The Squadron 110 anniversary evening in January 1965; IAF Commander Ezer Weizman is standing, while facing the camera, from left to right, are: Squadron 110 Commander Yossi Sarig, Flying School Basic Squadron Commander Jonathan Shahar, and Flying School Commander Jacob Agassi. (PC)

was indeed proven. The test progressed to its second phase, and Squadron 110 Commander Yossi Sarig was honored as the first pilot to launch a Shafrir 2 missile, a feat that was achieved on 30 December 1964 from Vautour 03.

Engine Explosion

Two Vautours were lost to engine malfunctions in June and October 1964; then, on the evening of 11 February 1965 the right Atar engine of Vautour 21 exploded in flight. Serving at the time as Flying School Basic Squadron Commander and a Squadron 110 emergency posting pilot, Jonathan Shahar recalled:

> It was a nighttime navigation exercise. We took-off from Ramat David, headed south-west to Zichron Yaakov, set course south and short of Sharon Junction I suddenly heard a BOOM, an explosion in the aircraft, at night, and the aircraft started to shudder so violently that I could see nothing. I immediately realized that something happened to one of the engines but I did not know which engine and what happened to it. I throttled back the left engine and nothing changed so I throttled it back to full power and throttled back the right engine and shuddering continued but less violently.
>
> I switched off the problematic engine, reported to ground control, returned to Ramat David, landed and taxied to the HAS. It was dark so I could not see but I was sure that the right engine exploded to shreds. As I taxied into the HAS I looked at the faces of the maintainers who welcomed me and expected to see shocked faces looking at a ruined engine but they looked normal and greeted me as though I returned from an eventless regular flight. I opened the canopy and asked the maintainers what happened to the right engine? They saw nothing, the engine looked perfectly OK.
>
> They then started to inspect the right engine from the front but everything looked OK so they had a look in the left engine. At this point I was really worried. That I – I viewed myself as 'King of Vautour' – switched off the wrong engine?
>
> Exactly then arrived a vehicle with a team of [B-Level] maintainers from the [Wing Maintenance Squadron] Vautour Workshop. They did not have a look at the inlet; they simply crawled into the nozzle of the right engine and started, to my great relief, to pull out ruined metal parts. It was indeed the right engine that exploded, a turbine fell apart and all the damage was internal, in external inspection everything looked normal.

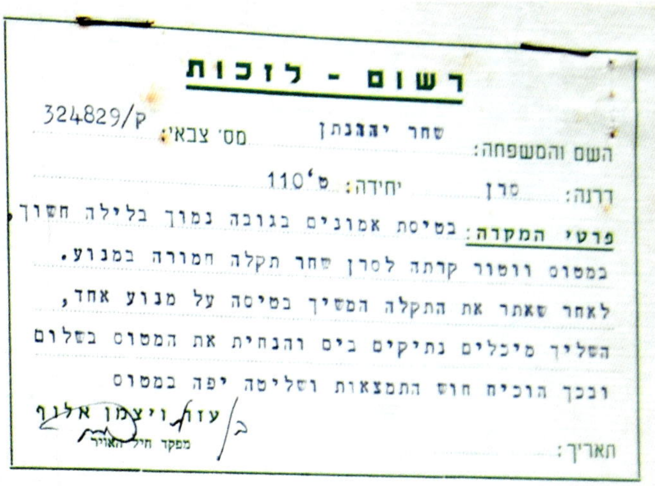

רשום - לזכות

השם והמשפחה: שחר יהונתן מס' צבאי: 324829/p

דרגה: סרן יחידה: ט' 110

פרטי המקרה: בטיסת אמונים בגובה נמוך בלילה חשוך במטוס ווטור קרתה לסרן שחר תקלה חמורה במנוע. לאחר שאתר את התקלה המשיך בטיסה על מנוע אחד, השליך מיכלים נתיקים בים והנחית את המטוס בשלום ובכך הוכיח חוש התמצאות ושליטה יפה במטוס

עזר ויצמן אלוף
מפקד חיל האויר

תאריך:

IAF Commander Ezer Weizman signed the "Green Note" awarded to Jonathan Shahar after the 11 February 1965 nighttime incident in Vautour 21. (PC)

Vautour Replacement MoU

The IAF aspired to acquire an American combat aircraft to fulfill the "heavy attack aircraft" Vautour replacement requirement. US aircraft were superior to French aircraft, primarily in propulsion and weapon systems, with corresponding superiority in range, load and hitting power. French offerings were deemed unsuitable for war in the Middle East as envisioned by the IAF.

The USA was reluctant to supply combat aircraft to Israel. Purposeful talks between Israel and the USA, concerning combat aircraft, started in 1964. USA policy was that combat aircraft would only be sold to Israel if Israel would not be able to acquire equivalent types in Europe. Such policy was productive for Israel, because the only European types available for Israel were French aircraft, and the small selection of French attack aircraft was unsuitable to IAF requirements.

Initial USA response to this state of affairs was an offer of surplus B-66 Destroyer bombers for the IAF "heavy attack aircraft" requirement. Israel rejected this offer and requested modern types: the A-6 Intruder or the F-4 Phantom. A series of discussions with USA representatives started in Israel on 12 February 1965. IAF Commander Ezer Weizman explained, on 28 February, that the Destroyer was unsuitable to Israeli requirements, and suggested that the USA could finance the Super Vautour for the IAF as an alternative to direct sale of advanced aircraft. The product of this round of discussions was a memo-

Some time prior to the 11 February 1965 engine explosion Vautour 21 was involved in this take-off accident,. The landing gear collapsed due to incorrect assembly of parts in B-Level maintenance, and pilot Yossi Sarig deployed the brake chute and stopped the Vautour on the runway, just past the old Ramat David control tower. (PC)

IAF maintainers surround Giora Goren after his Vautour solo flight; four Flying School Class 44 graduates – Dvir, Goren, Tal and Vilan – were assigned to the IAF Year 1965 Term 1 Squadron 110 Vautour Conversion Course. (PC)

randum of understanding that was signed on 10 March 1965. The USA undertook to sell combat aircraft to Israel if equivalent combat aircraft were not available from other suppliers.

PILLAR DAWN 117A

RAFAEL implemented lessons from the PILLAR DAWN 117 test of the Shafrir 2. A PILLAR DAWN 117A preparatory series of tests was flown from 18 to 25 March 1965. The objective of these four flights was to measure temperatures and vibrations in a Shafrir 2 missile attached to a Vautour. Deputy B Isaac Golan flew two of these flights, both in Vautour 03: the first on 22 March and the second on 24 March.

The actual PILLAR DAWN 117A test was flown from 25 to 30 April 1965. Squadron 110 Commander Yossi Sarig flew four sorties in Vautour 03 and launched five Shafrir 2 missiles.

Vautour N Reconnaissance

The demise of Vautour 35 reduced the number of Vautour B aircraft to only two, which was below the original requirement for three Vautour type reconnaissance aircraft. To bridge this operational gap AIR2 issued, in April 1965, an operational requirement to modify a Vautour N as a reconnaissance with a KA-52 panoramic camera.

In response to AIR2's requirement, MAT4 reported on 25 May 1965 that preliminary design of a Vautour N GOVERNOR modification was completed. The camera compartment replaced radar components (the DRAC 25 was not yet removed from the Vautour N aircraft due to center of gravity issues), and an air conditioner was an element of the design. Additionally, a window had to be installed in the underside of the aircraft. MAT4 anticipated that IAI would modify a Vautour N to this GOVERNOR modification within three months, with subsequent test flights over a time frame of a fortnight. As was often the case, this estimate was grossly overoptimistic and the Vautour N GOVERNOR would not be qualified for operations before August 1966.

Scramble to Syria

The largest ever IAF Vautour formation (18 aircraft) flew over Haifa on 6 May 1965 during the Israel Independence Day military parade. Israel Independence Day military parades were aimed to boost internal morale and to deter external enemies. Maximum participation in IAF fly-past over the parade served both objectives, and the squadron's ability to contribute 18 Vautours, out of a total IAF effective inventory of 21 Vautours and a squadron actual inventory of 20 Vautours (Vautour 70 was yet to be delivered to 110), was a truly magnificent effort.

A week later, on 13 May 1965, Squadron 110 Vautours were again scrambled against Syria. Israel initiated an exchange of fire, another Water War clash, in order to strike Syrian en-

Squadron 110 service persons in front of the squadron building in May 1965. (PC)

The 18-ship formation of Vautours flying over Haifa on Independence Day 6 May 1965: Yossi Sarig flew lead in Vautour 21, Isaac Golan flew 08, and Ovadia Nachman with Alex Meltzer flew 62. (PC)

Participants (not all) in the 18-ship, 6 May 1965 Independence Day fly-past, from left to right: Adam Zivoni, Gideon Magen, Uri Margalit, Oded Einav, Alex Meltzer, Daniel Ilan, Herzle Bodinger, Yigal Yanay, Squadron 110 Commander Yossi Sarig, Simon Zimon, Boaz Gafni, Jacob Rafaeli, Deputy A Simi Saar, Beni Zohar, Moshe Ofer, and Menahem Eyal. (PC)

gineering equipment. Four Vautours were scrambled in case the Syrians would respond with artillery fire; Gideon Magen, Herzle Bodinger and Daniel Ilan took-off first, and Squadron 110 Commander Yossi Sarig, flying Vautour 14, joined them. The Syrians restrained, so the Vautours were not authorized to attack. They flew a racetrack pattern on hold until the exchange of fire ended, having lasted for a mere 36 minutes from start to finish.

Combine Test Order

Vautour 70 was grounded at IAI, pending inspections and modifications, from arrival in March 1964 until early 1965. While the aircraft was grounded, AIR2 clarified, on 25 August 1964, the missions of the improved Cyrano as terrain avoidance and radar bombing. As work on Vautour 70 progressed, AIR2 finalized armament requirements on 16 October 1964: no guns would be fitted, the bomb-bay would not be configured for weapons, all four wing stations would be cleared for bombs, and only the two outer wing stations would be wired for missiles.

Meanwhile, local development of the ground attack optimized Cyrano radar – Israeli name Combine – duly progressed. MAT8 forecasted, on 23 December 1964, that manufacture of the prototype Combine would be completed, including post manufacture lab tests, by 1 March 1965.

AIR2 issued an order on 30 May 1965 to test Combine in Vautour 70. The primary objective of this first series of test flights was to validate the functionality of the system; a secondary objective was a preliminary assessment of the Combine's operational capability, including recognition of large, prominent targets, such as runways.

The series of test flights was expected to include some 30 sorties over a time frame of six months from 1 June 1965. The first phase of circa ten flights was to tune the radar's mapping mode; the second phase of some eight sorties was to evaluate the operational capability of this mapping mode; and the third phase was planned to tune and evaluate the radar's low altitude flight obstacle avoidance feature during the course of some 15 flights. The air-to-air mode of

A Squadron 110 parade in June 1965 with the unique large tankers flanking Vautour 66 and Vautour 03 with FAR vehicles in the center. (PC)

the radar was to be tested in separate flights over the same time frame.

Assigned test pilots were Daniel Shapira and AIR2 Chief Yoash Tsiddon; assigned test navigator was AIR2 electrical engineer Itche Erez. Combine test flights started on 20 June 1965, when Tsiddon, with Erez, flew Vautour 70 and continued until 27 February 1966; Itche Erez flew a total of 12 Combine flights during this time frame. The indigenous radar excelled in maritime strike scenarios (a torpedo boat was detected from a range of 80 km during the third test flight on 23 June when flown by Tsiddon with Erez), but the radar failed in its "raison d'etre", the production of proper mapping over land.

French Vautour 07 to IAF Vautour 28

The FAF could not spare Vautours for sale to the IAF, but Israel did purchase from France a single surplus Vautour. This was a pre-production aircraft in A configuration (Vautour 07); not to be confused with the Israeli Vautour 07, IAF Number 4307, while the French Vautour 07 would eventually become IAF Number 4328.

Technical Officer Gabriel Gidor reported from France, on 2 June 1965, the French offer to sell Vautour 07 to Israel for 2,000 dollars. The non-standard aircraft was grounded and lacked engines and weapons. Gidor suggested to use Vautour 07 as a Technical School instructional airframe, or to recondition the aircraft for flight either by IAI in Israel or by the manufacturer in France; Sud Aviation had already agreed to recondition the aircraft at a cost of 650,000 French Francs (135,000 dollars).

Material Department Deputy Chief Israel Rot asked the IMOD, on 13 June 1965, to accept the French offer and to purchase Vautour 07 for a mere 2,000 dollars, including spares specific to this non-standard airframe. Due to this non-standard configuration Vautour 07 was intended to serve as an IAF Technical School instructional airframe.

Israeli acquisition of this Vautour was indeed approved and, prior to eventual shipment to Israel, it was intended to use parts from French Vautour 07 for repair of Israeli Vautour 20, as related hereinafter.

Shafrir 1 of 8 Again

The failure of the first Shafrir 1 acceptance test prompted a second such test from 13 to 27 June 1965. During the first Shafrir 1 acceptance test – from 7 July 1964 – only one missile out of eight launched hit the Del Mar target. The second acceptance test did not fair any better. Again, a Squadron 117 Mirage launched eight missiles against a Del Mar target towed by a Vautour, and again only one missile hit the target. Four missiles exploded nearby the target but did not destroy the Del Mar, and three missiles missed.

Fifth Commander

Levy Zur took over from Yossi Sarig as Squadron 110 Commander from July 1965. Zur was the first commander of Squadron 110 Vautour without a background of being a 110 service person and without previously serving as a Vautour pilot; he first flew the Vautour on 16 July 1965.

OFRA

Electronic warfare grew in importance, and IAF efforts in this field emphasized the Vautour. Even before the establishment of MAT9 as an outgrowth of Material Department's Special Systems Section, the Squadron 119 Vautour N night-fighters were equipped with an internal electronic warfare system, IAF code MEIRA, during 1961. MEIRA was aimed against emissions from communication transmitters, radar systems and IFF emitters.

Another internal electronic warfare system to be fitted to Vautour was IAF code OFRA, a communications jammer. Material Department's Special Systems Section Chief Josef Naor reported on 15 October 1965 that the prototype OFRA installations in a Vautour A (4309), a Vautour B (4233), and a Vautour N (4461) satisfied requirements. It was therefore recommended that AIR2 would announce these installations as standard, so that a modification order would be issued to modify accordingly the whole Vautour fleet. AIR2 ordered on 31 March 1966 to fit OFRA to all Vautour aircraft.

PILLAR DAWN 118

The Shafrir 2 was developed to arm the Mirage fighters. Therefore, it was decided to cease test launches from the Vautour and to utilize a Squadron 117 Mirage as a Shafrir 2 test platform. During the next

Another view of the June 1965 parade showing Vautour 03, a large tanker, and four Meteor trainers. (PC)

Squadron 110 Commander Yossi Sarig handed over command of the squadron to Levy Zur in July 1965; from left to right: Giora Goren, Yossi Sarig, Levy Zur, Beni Zohar, Menahem Eyal, Deputy B Isaac Golan, Yigal Yanay, Deputy A Simi Saar, Boaz Gafni, Oded Einav, Abraham Vilan, and Jacob Tal. (PC)

Vautour 31 tested the Vautour B Shafrir 1 combination in November 1965; by then the Vautour B aircraft were camouflaged in line with an AIR2 directive issued on 29 December 1964. (IDFA)

The Shafrir 1 control box in the cockpit of Vautour 31. (IDFA)

series of test flights (PILLAR DAWN 118 from 1 to 15 November 1965) the Vautour's participation was limited to target towing.

Six missiles were allocated to PILLAR DAWN 118 and three were launched. The first scored a direct hit against a non-maneuvering target, the second narrowly missed a 2 G maneuvering target, and the third missed a 2.5 G maneuvering target. In light of this pattern, it was decided to discontinue PILLAR DAWN 118 for analysis and modifications prior to resumption of test flights.

Shafrir 1 for Vautour B

The Vautour B was not armed with guns, and was therefore defenseless during reconnaissance missions. In reconnaissance configurations the outer wing stations were free, unlike in Vautour attack configurations, so this coupling – lack of guns and availability of wing weapon stations – was the logic that motivated a requirement to arm the Vautour B with Shafrir 1.

The Vautour B Shafrir 1 test flights were flown within the time frame of PILLAR DAWN 118. Vautour 31 flew seven test flights from 2 to 18 November 1965, with a ninth sortie flown on 9 December 1965; Squadron 110 Deputy B Isaac Golan flew six of the test flights from 2 to 18 November, the final four of these with navigator Alex Meltzer.

The first test cleared the flight envelope of the Vautour B Shafrir 1 combination up to a speed of 560 knots, as fast as Mach 1.009, as high as 42,000 feet, and with maneuverability limitations of a roll rate of up to 80 degrees per second and a maximum 5 G.

Six Shafrir 1 missiles were launched from the second test flight on 4 November 1965. The missiles were launched at speeds

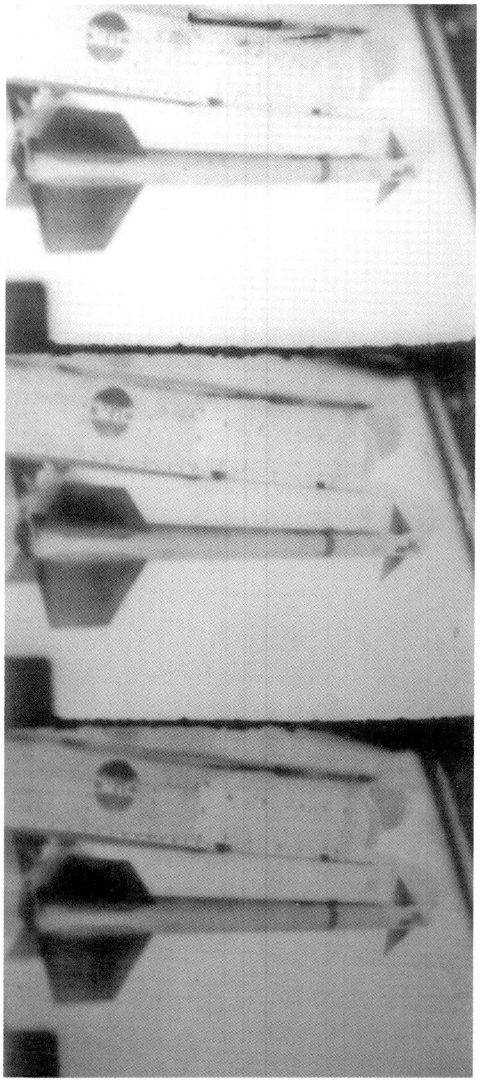

Launch of a Shafrir 1 missile from Vautour 31 during the Vautour B Shafrir 1 qualification test in November 1965. (IDFA)

from 253 to 555 knots, at altitudes from 1,771 to 34,645 feet, at Mach 0.52 to 0.89, and at 0.2 to 2.57 G. No adverse impacts surfaced during the launches. Additionally, the sixth flight, on 8 November, tested the effective envelope of maximum engagement against a Mystere target aircraft, and this was concluded to be up to 2.5 km range at an up to 20 degree angle off.

MAT5 compiled the Vautour B Shafrir 1 test report on 17 January 1966; an optimum aiming technique for the missile was recommended, and it was suggested that AIR2 would qualify for service the Vautour B Shafrir 1 combination. AIR2 qualified Vautour B Shafrir 1 on 22 August 1966.

Vautour 20 in France

The IAF inventory of 20 operational aircraft (Vautour 70 was not yet delivered to 110) was deemed too small. Attempts to purchase additional Vautours from France failed because production had ceased years before, and the FAF was not willing to dispose of any of its Vautours.

Every IAF Vautour counted. In line of this dictum, MAT2 ordered Air Maintenance Unit 22, on 15 July 1964, to transport the wreckage of Vautour 20 from Tel Nof to Lod so that IAI could evaluate if a rebuild was an option. The survey concluded that rebuild was indeed an option, but not in Israel. The manufacturer was approached and dispatched to Israel a team of engineers to survey the wreckage at IAI.

The wreckage of Vautour 20.

Eventually, Vautour 20 was transported to France for repair. The initial evaluation of the manufacturer was that rebuilding would cost 250,000 dollars. This evaluation inflated to 400,000 dollars, which the IMOD also approved, until the final blow came on 4 November 1965. It was then that the manufacturer forwarded to the IMOD its final offer: work 2,216,000 French Francs and parts 4,786,000 FF, excluding engines, armament, ejection seat, communication set, bomb-bay fuel tanks, external fuel tanks, convex leading edge wing (half a kit), and 23 other parts that the IMOD had already undertook to deliver!

The French final offer was valued at over one million dollars and was rejected. Material Department Chief Arie Hillel ordered MAT3, on 15 November 1965, to arrange the return of the wreckage from France to Israel.

Shafrir 1 Roadmap

The disappointment from the Shafrir 1 service acceptance test – and the promise that the Shafrir 2 displayed during the PILLAR DAWN 118 test – prompted an Air Department discussion, held on 6 December 1965, concerning IAF air-to-air missile policy. The basic facts were that the Shafrir 1's probability of a hit was less than 50% against a target maneuvering at up to 2 G (this presented probability can only be described as a fantasy in light of results from the two Shafrir 1 service acceptance tests), while the Shafrir 2 had already demonstrated equivalent or better performance during this early phase of its development program.

Air Department recommended terminating acquisition of Shafrir 1 beyond the 50 launchers and 100-120 missiles already ordered. The Shafrir 1 would continue to arm the Mirage squadrons pending service introduction of the Shafrir 2. Additionally, 12 Super Mysteres and the two Vautour B reconnaissance aircraft were to be armed with Shafrir 1.

Light Bomb Rack

The 13 November 1964 action demonstrated the Vautour's capability to attack with eight bombs a target not too far away from Ramat David. Equivalent long range strike, with eight

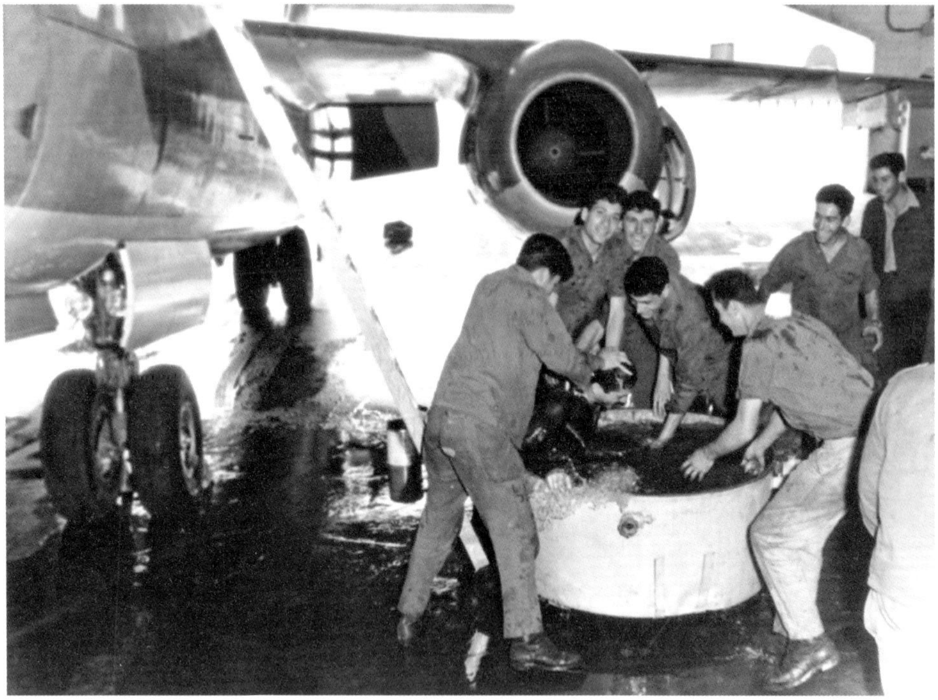

Cheerful maintainers baptize a pilot after a solo flight in December 1965; Flying School Class 46 graduates Absalom Friedman and Ran Zur were assigned to the IAF Year 1965 Term 3 Vautour Conversion Course. (PC)

lighter bombs, had not achieved operational status because the Light Bomb Rack and the Quadruple Bomb Rack were still in development. A step forward was the 4 January 1966 AIR2 order to test LBR and subsequent AIR3 issue, on 9 January, of an Mission Order to cover two LBR test flights that were flown on 10 January; the LBR test aircraft was Vautour 62. The objectives of the test were to qualify the LBR on the Vautour for flight (at up to 550 knots, Mach 0.9, 4.5 G and 100 degrees per second roll rate) and for bombing (at up to 500 knots, at 0.5 to 4.5 G, and at a dive angle of up to 45 degrees).

A third LBR test was flown on 5 March 1966; four 70 kg bombs – two from each LBR – were dropped during every test flight, and MAT5 recommended on 29 May to qualify the Vautour LBR combination as operational.

1,300 Liter Fuel Tanks

A payload/range improvement that reached operational status was replacement of the original 1,250 liter external fuel tanks with 1,300 liter tanks of an IAF standard type as fitted to Super Mysteres and Mirages. External fuel tanks were in short supply and pilots were not encouraged to jettison these tanks; certainly not in training. Squadron 110 Deputy A Isaac Golan surfaced this issue on 14 February 1966:

> The new squadron training plan includes an exercise that has to be flown with external fuel tanks but includes an element of air combat training. There is no knowledge in the squadron concerning limitations on air combat maneuvering with empty external fuel tanks. The aircraft manual does not ban air combat with empty external tanks so our lack of knowledge in this field may

IAF combat aircraft on display at Tel Nof, left to right: Vautour, Mirage, Super Mystere, Mystere, Ouragan and Fouga. The ordnance in front of Vautour 17 (prefix digit "5" was added specifically for this display) included bombs, napalm bombs, rockets and rocket pods, but not the small bombs associated with Light Bomb Rack and Quadruple Bomb Rack, as QBR was only qualified in 1966. (PC)

Painted with prefix digit "5", Vautour 17 lands at Ramat David with 1,300 liter drop tanks. (BIAF)

> have resulted from lack of previous requirement for air combat [training] in such a configuration. Our opinion is that two experienced pilots should test air combat with empty external tanks so that this can be approved if no problems will surface.

Only in the wake of this Squadron 110 inquiry was the question of safe jettison asked. Flight Safety Branch stated on 17 February 1964:

> Despite replacement of the Vautour's original 1,250 liters external tanks (with cruciform stabilizers) with standard 1,300 liters tanks (with parallel stabilizers) it has not been determined yet whether the new tanks behave like the old tanks during jettisoning. It is therefore suggested to check if the two tanks behave in similar fashion during jettison and if there are limitations concerning jettison of 1,300 liters tanks from Vautour.

PILLAR DAWN 118A

The three Shafrir 2 missiles remaining from PILLAR DAWN 118 were launched during PILLAR DAWN 118A, from 20 March to 3 April 1966. As usual, a Vautour B towed a Del Mar target and a Mirage launched the missiles. The first missile missed, the second passed two meters from the target, and the third hit the target.

No Replacement for Vautour

The IAF requirement for light attack aircraft and heavy attack aircraft was partially fulfilled during early 1966. Israel ordered the American Skyhawk and the French Mirage 5. The former was purchased as the future IAF light attack aircraft; the latter was a substitute for the IAF heavy attack aircraft requirement as a result of the US' refusal to sell to Israel aircraft such as the Phantom or the Intruder.

Neither type was a true Vautour successor. The IAF was well aware of this fact, and had already asked for stop gap purchase of additional Vautours, as IDF Chief of Staff Isaac Rabin reported on 2 May 1966:

> We suggested to the French that we would not buy American aircraft if they would let us purchase 20 Vautours and a few Super Mysteres.

The trouble was that both types – the Super Mystere and the Vautour – were no longer manufactured, the original production was quite limited, and the FAF was still operating these aircraft as frontline types, so none was actually available for export to Israel.

Cockpit Standardization

Squadron 110 operated three Vautour variants, while some of the aircraft, like the GOVERNOR equipped Vautour 66, were fit-

Participants of the 25 April 1966 Independence Day fly-past simulate the arrowhead formation on the ground, from left to right and front to rear: Micha Gil, Ran Zur, Alex Meltzer, Isaac Golan, Zeev Tavor, Ovadia Nachman, Boaz Gafni, Elhanan Raz, Levy Zur, Simi Saar, Michael Dvir, Shlomo Keren, Gideon Magen, Beni Zohar, Simon Zimon, Yigal Yanay, Absalom Friedman, and Adam Zivoni.

One of the three rehearsals flown prior to the 25 April 1966 Independence Day fly-past was in 16-ship formation; the aircraft trailing the leader is Vautour 70, piloted by Simi Saar. (PC)

The 14-ship arrowhead formation on Independence Day, 25 April 1966; Deputy A Isaac Golan with Alex Meltzer flew Vautour 33 in the lead of the left column of four Vautours; Ovadia Nachman flew Vautour 07 (third in the left column of four Vautours), Simi Saar flew 66 in the middle (right behind the leader), and Micha Gil flew 62. (PC)

ted with unique systems. As a result there were more than three cockpit variations, and on 8 May 1966 the issue of a standard Vautour cockpit was discussed in an AIR2 meeting attended by Squadron 110 Commander Levy Zur and representatives from MAT2, MAT4 and MAT5.

The objective to achieve a standard Vautour cockpit was never achieved; operational requirements always outpaced modifications schedules. However, this meeting did outline an ideal cockpit configuration. The right panel was to be allocated to communication systems only, from front to rear: VHF transceiver controls (to be replaced with UHF-2 in due course), UHF-1 transceiver controls, Radio Compass controls, communication system switches and IFF controls. Radar scope and control panel were to be removed from the Vautour N pilot cockpits and replaced with fuel flow-meters (previously fitted in the rear cockpit of the Vautour N) and GOVERNOR control panel, respectively.

The most challenging aspect of cockpit standardization was the various weapon systems, because each Vautour variant differed from the other two in this aspect. A cockpit standard of weapon systems controls was actually an impossible goal. The Vautour N was equipped with a rockets control system different than that of the Vautour A, the Vautour B was uniquely armed with Shafrir 1, and it was planned that the Vautour N would be the first Vautour variant to introduce into service the new anti-runway bombs.

The output of this meeting was not a standard Vautour cockpit, but a list of standardization feature priorities from installation of gun rounds counters as the highest priority, installation of an engine oil level light, armament modifications, communication panel and GOVERNOR control panel, down to Vautour N front cockpit fuel flow-meters.

The cockpit of Vautour 62 after removal of the radar scope and installation of gun rounds counters. (PC)

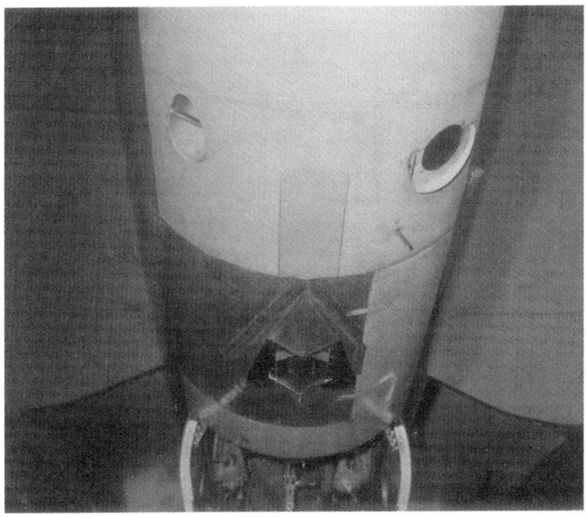

Configuration comparison between Vautour 31 (left) and Vautour 33 (right) after installation of the GOVERNOR air conditioning system in Vautour 33; the marked features on 33 are the additional air inlet (left) and outlet (right). (IDFA)

GOVERNOR Air Conditioning

Also on 8 May 1966, AIR2 qualified the GOVERNOR air conditioning system. This milestone was achieved almost two years after AIR2 ordered (on 16 June 1964) the installation of an air conditioning system that would enable a GOVERNOR equipped Vautour to fly as fast as 550 knots at low level at an external temperature of up to 45 degrees Celsius.

AIR2 ordered a test of the GOVERNOR air conditioning system, as fitted in Vautour 33, on 4 August 1965. The test report was issued on 27 October 1965 and further tests were ordered. An issue that surfaced during these tests was the vulnerability of Vautour 33 to engine malfunctions in such prolonged low level fast flights. Squadron 110 reported on 5 November 1965 that Vautour 33 managed to fly 20 minutes at low altitude while clocking 500 knots without engine malfunction or abnormal oil consumption. Then Vautour 33 was due for maintenance, so the squadron suggested that tests should be resumed with Vautour 31, but this suggestion was not endorsed.

Tests with Vautour 33 resumed on 9 and 10 February 1966. The first test included a 30 minute run at 500 knots at an altitude of 500 to 2,000 feet. The second flight included a similar profile, plus photography at the end of the low level run. External temperature was 18 degrees Celsius and internal temperature was 39 degrees; it was calculated that this would have reached 49 degrees without air conditioning. The quality of the photos was good, and MAT4 recommended on 23 February to qualify the system as operational.

The GOVERNOR air conditioner did not achieve the original objectives, but AIR2 qualified the system on 8 May 1966. This qualification was unlimited at high altitude and limited to 500 knots at an altitude of up to 1,000 feet and an external temperature of 37 degrees Celsius. The following day AIR3 Chief Rafi Harlev questioned this qualification:

> The limitations of the air conditioning system at low altitude are maximum speed 500 knots and maximum external temperature of 37 degrees Celsius. To the best of my knowledge there were no temperature limitations at high altitude. Since the objective of this installation was to enable high speed at low altitude, I cannot see how this has been achieved. If this is the case then what is the point in the installation of this system?

AIR2 responded on 15 May 1966. As already mentioned, GOVERNOR, without air conditioning, was qualified for operations on 29 April 1964. The limitations without air conditioning were 500 knots at an external temperature of up to 23 degrees, 450 knots at 23 to 27 degrees, and 400 knots at 27 to 30 degrees. An extrapolation of this set of limitations would yield a maximum speed of 330 to 350 knots at an external temperature of 37 degrees, whereas with air conditioning this limitation was not lifted entirely but improved to 500 knots.

Nighttime Napalm

Isaac Nir, flying Vautour 14, illuminated for Shlomo Keren and Micha Gil on the evening of 12 May 1966 during a test of a nighttime napalm attack. The scenario was close air support and the target was a hill. The IDF marked the hill with white smoke bombs and projectors; otherwise it would have been impossible for the illuminating Vautour to spot the target. The illuminator then dropped a flare, but the attackers were unable to spot the hill under the light of a single flare. Thus, the illuminator dropped four flares: one was a dud, and the attackers dropped four napalm bombs; all four napalm bombs hit the hill.

Gunnery Training

The fire power of the Vautour's four 30 mm guns was enormous, but the pilots rarely fired all four guns. Economy dictated that only a single gun would be loaded with rounds during training. Squadron 110 Commander Levy Zur argued on 6 June 1966 that such a policy was counterproductive, and that two guns should be fired during training:

> Gunnery aiming with one, two and four guns was evaluated during this training term. Aiming was right on target when two or four guns were fired. When a single gun was fired the pilot had to offset aiming in order to hit the target. Firing a single gun therefore results in improper training and damage to the guns pack due to asymmetrical firing.

Wing 1 Commander Ezekiel Somech supported this view in a 12 June 1966 letter to AIR5 and MAT5. The latter branch was also in favor of symmetrical firing; this MAT5 view was expressed in a 12 July memo:

> When a single gun is fired the forces are asymmetrical and deform the guns pack and the barrels installation as evident in all aircraft and in all packs. Such impact is not immediate but surface after three or four years and depending on the number of rounds fired. The zeroing of the guns pack is

A Vautour dropped napalm bombs in a range along the Mediterranean Sea coastline north of the Ashdod power station; the delivery pattern of napalm was a high speed, low altitude run contrary to dive attacks with high explosive bombs. (PC)

designed for minimum overlap for maximum coverage. For this reason, a pilot firing a single gun must offset aiming and this offset aim is different per every gun.

Our recommendation is to switch to firing from a pair of guns (upper or lower). It is also suggested that every pilot will fire all four guns at least once every training term in order to get accustomed to the impact on the aircraft and to realistic aiming.

AIR5 responded to the combined pressure of Squadron 110 and MAT5 on 19 July 1966. Though admitting that firing of two guns was preferable from training and maintenance perspectives, the overwhelming consideration was economical. However, if repair of damage would cost more than savings in rounds, then firing with two guns would be more worthwhile. Thus, Air Department Chief ordered AIR2 to prepare a study that would rule which option was more economical. The subject was still being studied on 30 December 1966 when AIR2 officer Ephraim Ashkenazi forwarded to MAT5 a request for information concerning the cost of repairing a Vautour that suffered damage due to firing of a single gun.

A unique formation flown on 7 July 1966 over Tel Nof during the wings parade of Flying School Class 50; Isaac Golan with Elhanan Raz flew lead in Vautour 33; the second Vautour was an N aircraft and the third was an A aircraft. (PC)

Zeiss 12" Camera

On 19 June 1966, AIR2 issued an order to test a Zeiss camera with a 12" lens. The objectives were to qualify the flight envelope with the camera and to evaluate the quality of the photos. The three phases of the test were: exploration of flight envelope at up to 500 knots and 5 G; photography in straight and level flight from five altitude blocks between 25,000 and 45,000 feet, all at 500 knots; and photography in 2.5 to 3.5 G turns at 10,000 feet.

Pilot Boaz Gafni and navigator Elhanan Raz tested the Zeiss 12" in Vautour 33 on 23 June 1966. The test complied with all objectives, and MAT4 recommended on 6 July to qualify the Vautour B Zeiss 12" combination as operational. AIR2 duly accepted this recommendation and qualified the Vautour B Zeiss 12" on 29 August 1966.

Loft Bombing Test

Combine failed to live up to expectations during the first round of its testing onboard Vautour 70. Development of Combine continued and tests would be resumed in due course. Meanwhile, IAI developed a bombing computer and an altimeter for integration with Combine to enable loft bombing. The altimeter and Combine fed altitude and range data, respectively, to the computer that calculated the release point.

On 5 July 1966 MAT8 issued a request to test the bombing computer. First phase included verification of parameters, pri-

Vautours 09 and 26 flying over Tel Nof during the firepower demonstration on 7 July 1966; two Flying School Class 50 graduates – Moshe Teman and Jonathan Zorea – were assigned to the IAF Year 1967 Term 1 Vautour Conversion Course. (PC)

marily altitude. Second phase covered actual bombing, and the suggested pattern was an approach to the target at 250 meters altitude and a speed of 350 knots, with bomb release at a range of three kilometers from the target and actual automatic release by the computer at the calculated points. Eight such bombing runs would be flown in two sorties, with one 250 kg sand bomb to be released on every run.

AIR2 then ordered, on 26 July 1966, a test in line with MAT8 guidelines. Two first phase flights were scheduled for 3 and 5 August; assigned pilots were IAF Chief Test Pilot Daniel Shapira and AIR2 Chief Adam Zivoni, while the assigned navigator was AIR2 electrical engineer Itche Erez. The second phase of actual bombing with eight bombs in two sorties was to be flown only after analysis of data from the first phase of the test.

Shafrir 2 for Vautour

Development of the Israeli air-to-air missile progressed well, and service introduction was expected in 1967. Shafrir 2 was developed for the Mirage, and AIR2 defined on 6 July 1966 a two tier requirement for production of Shafrir 2 missiles. The first tier included 60 missiles for Mirage air-to-air readiness. The second tier covered production of 380 missiles to arm all IAF Mirages (including the Mirage 5 interdictors scheduled for delivery from 1967), half of the 48 Skyhawks (scheduled for delivery from 1967), 13 Super Mysteres, and ten Vautours.

This AIR2 plan was not adopted; Air Department Chief Menahem Bar ruled on 29 July 1966 that only the first tier of 60 missiles for Mirages would be produced. Follow up production was to be decided at a future date.

Operation SMALL SILT

Water War clashes started to escalate in frequency and scale. An IDF vehicle drove onto a mine on 13 July 1966; two soldiers and a citizen were killed. The next morning AIR3 distributed Operation Order SMALL SILT. The objectives were to destroy Syrian engineering equipment – engaged in diversion work of water resources – and to attack surrounding anti-aircraft gun emplacements. Four Squadron 109 Mysteres were to bomb at 16:00, with four Vautours following up with a bombing run (each aircraft was to be loaded with four napalm bombs) and a strafing pass. Five back up 4-ship formations, including one from Squadron 110, were to be on hold, flying racetracks west of the Israeli-Syrian border and waiting for orders to attack Syrian targets. Two Mirage 4-ship formations were to fly combat air patrols; two Mirages and one Vautour were to be on alert for post combat BDA photography.

This was the first aerial retribution operation against Syria since 1951; all IAF strike operations against Syria during the time frame from 1951 to 1966 were air support missions. The IAF switched to Readiness Level C from 11:15, and the IDF issued Mission Order WIND at 13:00 to cover the IAF action, as well as IDF readiness.

The first Squadron 110 formation – call sign BALLET – took-off at 15:51. Squadron 110 Commander Levy Zur (with regular navigator Elhanan Raz in Vautour 61), Wing 1 Commander Ezekiel Somech, reserve pilot Gideon Magen, and emergency posting pilot Daniel Ilan attacked right after the Squadron 109 Mysteres.

The second 110 formation – call sign WALTZ – departed Ramat David at 15:56 as a 3-ship. Deputy A Isaac Golan in Vautour 05, emergency posting pilot Giora Goren, and reserve pilot Jacob Rafaeli attacked right after BALLET. The seven Vautours dropped a total of 24 napalm bombs and were the only IAF attack aircraft to strafe; the Squadron 109 Mysteres bombed and then cleared the area for the Vautours. As a matter of fact, the IAF attack action was called off right after the WALTZ attack

Levy Zur with Elhanan Raz flew Vautour 61 to attack Syria on 14 July 1966; the silver Vautour 61 was photographed on 25 April 1966 beside the camouflaged Vautour 66 that Shlomo Keren with Achikar Eyal would fly over Egypt on 15 August 1966. (PC)

because Syrian interceptors were scrambled. IAF ground control ordered the remaining four attack formations (Ouragans, Mysteres and Super Mysteres) to return to base, and the clash continued as an air-to-air action.

Squadron 110 not only flew the lion's share of the actual attack sorties, but also ended the clash when Boaz Gafni and Alex Meltzer flew a BDA photography mission. The BDA photographs revealed that the air strike destroyed six Syrian tractors as well as vehicles and anti-aircraft guns.

Gunnery Camera Requirement

The IAF trained for a future air superiority campaign; training focused on air base attack patterns, including bombing of runways and strafing of enemy aircraft. A major issue in such a campaign was BDA. An integral element of the IAF air superiority offensive plan was post strike reconnaissance missions, but for finer planning purposes it would have been useful if every attack formation would return to base with its own BDA images. Obviously pilots returning from attack missions would be debriefed, but such debriefs could only present claims. What if all four pilots of an attack formation strafed the same aircraft? They would report separate claims for destruction of four enemy aircraft. What if a strafed aircraft claimed as destroyed was not actually destroyed, but only damaged or completely missed? The solution was to equip every aircraft with a gunnery camera that would automatically photograph during firing of the guns.

An effort to install a gunnery camera in the Vautour stalled in 1964; the prototype installation produced blurred photos and MAT2 recommended – on 6 May 1964 – to relocate the gunnery camera from the nose cone to the leading edge of the right wing or to the vertical stabilizer. Squadron 110 Commander Levy Zur resurfaced this issue on 22 July 1966:

> A lesson from recent operations is that lack of gunnery camera degrade the combat efficiency of the Vautour. The Vautour's sight camera is useless during firing because of vibrations that completely blur the photos. The installation of gunnery camera in the nose is inefficient due to soot and vibrations. I therefore suggest that a gunnery camera will be installed in the leading edge of the wing.

PILLAR DAWN 119

A Squadron 110 Vautour was again utilized to tow a Del Mar target for Shafrir 2 test PILLAR DAWN 119 from 24 July to 12 August 1966. An innovation planned for PILLAR DAWN 119 (but not actually deployed) was the equipment of the Del Mar target with a receiver and of the launched missiles with a transmitter so that exact miss distance would be measured.

PILLAR DAWN 119 was crammed with technical malfunctions, and participating aircraft had to take-off on 19 occasions to launch the planned allocation of four missiles. Squadron 110 Deputy A Isaac Golan flew seven sorties for Del Mar towing in Vautour 67, two with Menahem Eini on 7 August 1966 and five with Elhanan Raz from 8 to 10 August. When everything functioned the test was successful; all four missiles passed within a destruction distance of two to five meters from the target.

A lesson from PILLAR DAWN 119 was an AIR2 requirement, issued on 31 July 1966, to wire more than one Vautour N for towing of the Del Mar target; the AIR2 requirement was that all Vautour N aircraft would be thus wired.

WART

Israel acquired a French CSF electronic warfare system – IAF code WART – that was a stand off jammer optimized against SA-2 systems. WART was packed into two pods, and was intended for the Vautour N with an operator in the back seat. AIR2 electrical engineer Itche Erez signed the WART operational requirement on 4 August 1966. The objective was to wire for WART three Vautour N aircraft, including 70 (the other two would be 65 and 66). Required configurations were full internal fuel, two 1,300 liter drop tanks, and one or two WART pods. Set schedule was to complete design, installation, and test flights by 15 September 1966; as usual, the actual schedule took much longer.

On 4 December 1966 MAT4 issued a request to test a "universal pod" on Vautour 70; the objective was to qualify the Vautour N WART flight envelope with the pod attached to an outer wing station. Correspondingly, AIR2 ordered on 5 January 1967 that four test sorties would be flown from 15 January 1967.

Deputy A Shlomo Keren and regular navigator Elhanan Raz flew the three WART "universal pod" test flights from 9

The first iteration in the IAF quest for a Vautour gunnery camera was this installation in the nose cone of Vautour 18 that dated back to 1962 and was abandoned in 1964; the gunnery camera issue was revived in 1966. (PC)

Isaac Golan flew seven PILLAR DAWN 119 target tow sorties in Vautour 67; this photo of Vautour 67 was taken in June 1964. (IDFA)

A demonstration of an incorrect WART pod loading onto the outer wing station of Vautour 70. (PC)

to 10 March 1967. Vautour 70 with WART was cleared to fly at a flight envelope limit of up to 500 knots, maximum Mach 0.98, maximum 3.5 G, and maximum roll rates of 90 and 70 degrees per second at 350 and 450 knots, respectively. MAT4 therefore recommended on 26 March to qualify the Vautour N WART combination.

Joint Reconnaissance Egypt

Deputy A Isaac Golan, with navigator Menahem Eini, flew three GOVERNOR test flights in Vautour 66 from 3 to 4 August 1966 prior to qualification of the Vautour N GOVERNOR aircraft. Ten days later, on 14 August, two Vautours deployed south to Tel Nof for a very special joint reconnaissance operation over Egypt. The objective of this operation was to photograph SA-2 batteries defending the Suez Canal. Each of the GOVERNOR equipped Vautours was to lead a Squadron 119 reconnaissance Mirage; each pair was tasked to photograph the missile batteries from low altitude.

The mixed pairs – Deputy A Isaac Golan with Menahem Eini in Vautour 31 and Deputy B Shlomo Keren with Achikar Eyal in Vautour 66 were the leaders – departed Tel Nof on 15 August 1966 and photographed nine of the 12 SA-2 batteries deployed along the Suez Canal: six batteries west of the canal and three batteries east of the canal. As a bonus, images of Abu Sueir and Fayid air bases were also taken. The mission was so meticulously planned and executed that the four reconnaissance aircraft flew such a daring mission without encountering a single threat; no anti-aircraft fire, no reaction from the SA-2 batteries that they photographed, and no enemy interceptors.

Simi's 3-Ship Scrambled to Syria

Once again Squadron 110 was in the thick of a Water War clash on 15 August 1966. Several IAF formations were

Superb view of Vautour 33 - the convex leading edge wing is vividly apparent - during a display of reconnaissance assets at Hatzor on 20 September 1966; Isaac Golan flew Vautour 33 from Ramat David to Hatzor that day. (PC)

The Squadron 110 badge was only painted on the vertical stabilizers of the Vautours from the spring of 1966. This photo was taken in January 1967. (PC)

scrambled, but only the Vautours attacked. The clash started when an Israeli Navy boat ran aground on the northeast coast of the Sea of Galilee. More boats then came to the aid of the stranded boat; this action focused the attention of nearby Syrian fortifications, which opened fire. Air power involvement started at 06:10, when Squadron 109 Mysteres flew visual reconnaissance; Syrian MiG-17s attacked the boats at 09:00 and combat air patrol Mirages engaged the escorting MiG-21s.

It was then that the Squadron 110 readiness formation was scrambled. Emergency posting pilot Simi Saar in Vautour 61 and Absalom Friedman and Boaz Gafni took-off at 09:25. Each Vautour was armed with two 500 kg bombs under the wings and six 250 kg bombs in the bomb-bay. The three pilots bombed Syrian fortifications; shortly after this strike fire ceased courtesy of UN mediation.

The Israeli Navy boat was still stranded when fire ceased on 15 August 1966; recovery attempts continued until 26 August, when the boat was finally recovered. The IAF was on readiness to support this recovery effort if the Syrians opened fire. Emergency posting pilots reported to their squadrons and a maximum number of combat aircraft were on readiness; Squadron 110 fielded 18 qualified pilots and 13 serviceable aircraft on the evening of 23 August, but no more action was needed, as the Syrians did not open fire.

Gun Round Counters

Cockpit standardization was never achieved, but a prototype modification of the top priority issue – gun round counters – was installed in Vautour 26. MAT5 reported this installation on 19 October 1966 and asked AIR2 to qualify this modification, while Squadron 110 Gunnery Officer Michael Dvir reported on 28 October:

> The prototype guns rounds counters installation in Vautour 26 was evaluated and fully complies with the requirement.

AIR2 duly qualified the Vautour gun round counters on 4 November 1966, and ordered MAT5 to apply this modification to all Vautour A aircraft. This directive was not yet implemented by 27 February 1967, when Squadron 110 Commander Levy Zur requested if the process of gun round counters modification would indeed progress from successful prototype fit in Vautour 26 to standard installation in all Vautours armed with guns.

Joint Reconniassance Syria

Another joint Squadron 110 and Squadron 119 reconnaissance operation was flown over Syria on 30 October 1966; the objec-

The cockpit of Vautour 26 with the gun rounds counters visible to the right of the stick's hand grip. (PC)

tive was to photograph all active SAF air bases in order to enable the update of IAF air superiority offensive plans.

This time the two Squadron 119 Mirages and two Squadron 110 Vautours flew in homogenous pairs; Levy Zur and Elhanan Raz led the Squadron 110 pair in Vautour 33.

Vautour 70 to Squadron 110

On 30 October 1966, AIR2 Chief Adam Zivoni chaired a discussion concerning the status of Vautour 70. From arrival in Israel during March 1964, Vautour 70 was operated from IAI as a non-operational aircraft. Squadron 110 objected to this arrangement and demanded that Vautour 70 be assigned to the squadron as an operational aircraft with secondary test assignment. AIR2 accepted this view. Vautour 70 would be permanently assigned to Squadron 110. Whenever needed, Vautour 70 would be temporarily deployed for tests at IAI. Finally, AIR2 directed that Vautour 70 would be transferred from IAI to Squadron 110 on 18 November 1966.

Quadruple Bomb Rack

The qualification of the Vautour Light Bomb Rack – capable of lifting two bombs – was merely a stopgap solution pending AIR2 qualification of the Vautour Quadruple Bomb Rack (capable of lifting four bombs) on 2 November 1966.

The first round of QBR tests lasted from 8 December 1965 to 9 January 1966. Pilots Levy Zur and Yigal Yanay and navigator Alex Meltzer flew 13 flights in two configurations: four QBR units with 16 bombs or two QBR units with eight bombs and two 1,300 liter drop tanks. Weapons tested on QBR were flares, smoke bombs, and 50, 70, 90 and 100 kg high explosive bombs. A 100 kg bomb from the inner QBR hit the right engine of Vautour 62 during the ninth test flight. This mishap delayed qualification of QBR; MAT5 recommended on 11 February 1966 to qualify QBR for all tested weapons but 100 kg bombs and to continue testing with 100 kg bombs.

Vautour 70 – flown by Simi Saar right behind the leader of this 14-ship formation – was photographed during a rehearsal prior to the 25 April 1966 Independence Day fly-past, when it was not yet assigned to Squadron 110; an assignment decision was made on 30 October 1966.

Vautour 62 with two QBR units under its left wing during the first series of test flights, December 1965 to January 1966. A denotes test camera installation; B indicates where the 100 kg bomb hit the engine (albeit the right engine and not the photographed left engine). (IDFA)

In the wake of the 2 November 1966 qualification of QBR, small bombs were added to the Vautour's variety of weapons, as this November-December 1966 ordnance display at Ramat David clearly illustrates. (PC)

At the conclusion of the follow up series of test flights, QBR was qualified for inner wing stations (with flares and 50 or 70 kg bombs) and outer wing stations (with flares and 50, 70 or 100 kg bombs). Qualified configurations were QBR under all four wing stations or QBR under outer wing stations and 1,300 liter fuel tanks under inner wing stations; both options were in addition to qualified bomb-bay loads.

The qualification of QBR resurfaced the issue of cockpit standardization. While all Vautour A aircraft were indeed compatible with QBR, MAT5 reported on 6 November 1966 that Vautour 66 was the only N aircraft capable of operating QBR. This was because the original Vautour N bombing system was not compatible with QBR, and only Vautour 66 had been modified with a standard Vautour bombing system; all other Vautour N aircraft were to be thus modified as well.

Another issue that surfaced due to QBR introduction was that two configurations were out of normal center of gravity limits: eight 100 kg bombs, two 1,300 liter tanks and bomb-bay fuel tanks in the Vautour N; eight 100 kg bombs, two 1,300 liter tanks and six 250 kg bombs in the Vautour A.

PILLAR DAWN 119A

The PILLAR DAWN 119 lesson concerning the Vautour, or modification of all N aircraft as Del Mar tow aircraft, was only partially accomplished before PILLAR DAWN 119A from 27 November to 29 December 1966. MAT5 reported on 22 September that Del Mar tow modification was applied to Vautours 66 and 67, while Vautour 62 was the next aircraft to be thus modified. It was recommended not to utilize Vautour 66 for Del Mar towing duties because this aircraft was fitted with GOVERNOR, but availability issues prevailed and Deputy A Isaac Golan, with navigator Elhanan Raz, flew two Del Mar tow sorties in Vautour 66 on 25 and 26 December, as well as one such sortie in Vautour 61 on 29 December, in support of PILLAR DAWN 119A. Besides the Del Mar difficulties, PILLAR DAWN 119A was a success for the Shafrir 2 and another failure for Shafrir 1 (two missiles were launched and both missed).

Conversion Course Accident

Squadron 110 again experienced a gap in assignment of new young pilots. This gap lasted for two terms, from assignment of Flying School Class 46 graduates Absalom Friedman and Ran Zur at the start of IAF Year 1965 Term 3 to the assignment of Class 49 graduates Ilan Lior and Nahum Roman at the start of IAF Year 1966 Term 3.

It was during the Vautour Conversion Course preparatory flights in a Meteor trainer that tragedy struck on 28 November 1966. Instructor Boaz Gafni and student Nahum Roman were practicing single engine landings when Meteor 14 crashed; Gafni and Roman were killed.

Shafrir 1 for Vautour N

The Vautour B reconnaissance aircraft were qualified to launch Shafrir 1, but Vautour 66 was also flying reconnaissance missions, so the next logical step was to fit Shafrir 1 to the Vautour N. This modification was integrated with the installation of a standard Vautour A bombing system – essential for compatibility with the Quadruple Bomb Rack – in a Vautour N, with Vautour 61 as the test-bed. AIR2 qualified the integrated Shafrir 1 and bombing system modification on 13 December 1966, and two days later Vautour N Shafrir 1 was also qualified.

Gunnery Camera Test

The suggestion to install a gunnery camera in the leading edge of the wing was not accepted; the preferred fit was in the leading edge of the vertical stabilizer, in place of the VHF antenna (the long term plan was to replace the remaining VHF transceiver with a second UHF radio). AIR2 ordered a test on 29 December 1966, and Vautour 07 had its VHF antenna removed and was fitted with prototype installation of the gunnery camera.

Micha Gil flew the gunnery camera test flight in Vautour 07 on 1 March 1967; Gil flew six strafing passes. The first two passes were dummy, while during the remaining four passes all four guns were fired (each gun consumed a total of 50 rounds). Then the aircraft simulated a bombing run, turned at up to 5 G, and rolled at the maximum rate (120 degrees per second). The gunnery camera functioned flawlessly; the photos were clear and sharp, and the camera was still operational after landing, whereas the sight camera malfunctioned due to vibrations during strafing.

In line with the results of this successful test, AIR2 officer Ephraim Ashkenazi announced qualification of the Vautour gunnery camera on 6 April 1967. However, the gunnery camera was to be installed in all Vautour A and N aircraft only upon replace-

Vautour dummies at Tel Nof in November 1966. (PC)

Vautour 61 was utilized to qualify the Vautour N Shafrir 1 combination; this photo was taken in January 1967. (PC)

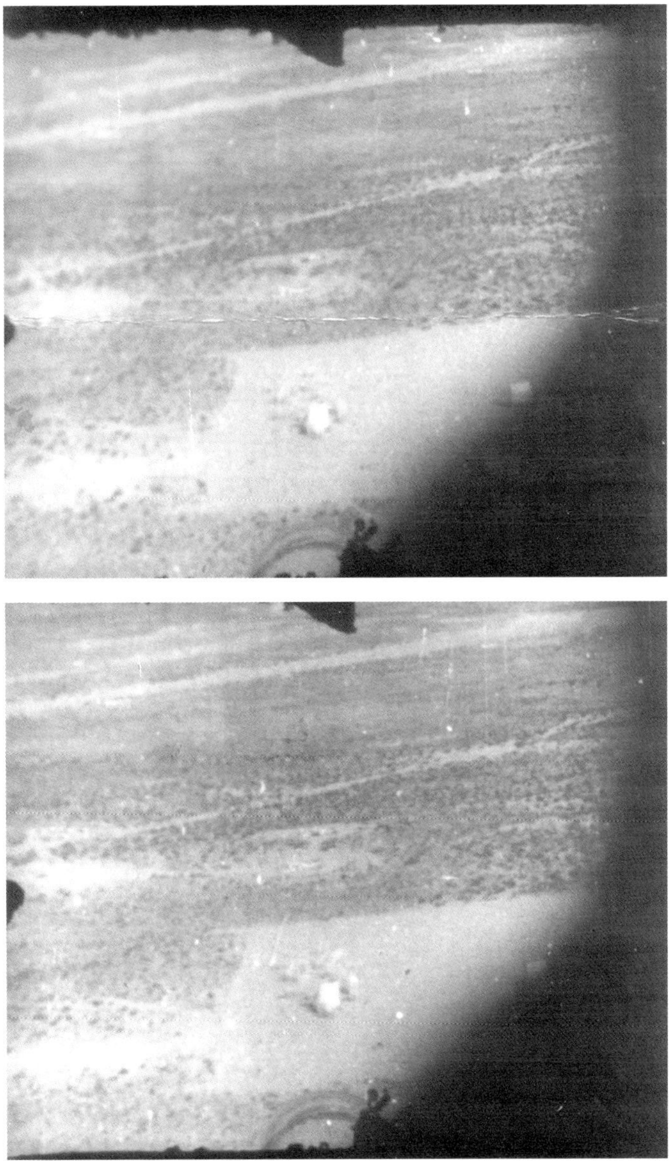

Two frames from the gunnery camera of Vautour 07 during strafing on 1 March 1967; the top of the cockpit canopy is visible at the bottom of the frames, just below the strafing target, with surrounding dust and smoke due to the hit of rounds. (IDFA)

ment of the VHF transceiver with a UHF radio; a replacement that would render redundant the VHF antenna in the leading edge of the vertical stabilizer.

Vautour 70 Standard

The Vautour standardization issue was obviously exacerbated after delivery of Vautour 70 to Squadron 110. On 13 March 1967, AIR2 officer Ephraim Ashkenazi issued a memo that summed up a 27 February discussion concerning Vautour 70; a discussion that was attended by Squadron 110 Commander Levy Zur, Technical Flight Commander Nathan Segall, and representatives from AIR2 and from MAT2 to MAT8.

The conclusion of this discussion was that, although Vautour 70 was assigned to Squadron 110 for service as an operational aircraft, it was also non-standard from the outset and fitted as a test platform. It was agreed that Vautour 70 would never be brought to a similar standard as other Squadron 110 aircraft. The main non-standard features were lack of guns, a different nose, and inability to be fitted with a fuel tank in the front suite of the bomb-bay; these non-standard features were to remain unchanged. Recommended changes were to standardize the pilot's cockpit, to replace the Mirage CSF-95 sight with the Mark IVE sight, to paint the aircraft in a standard camouflage scheme, and to evaluate qualified configurations because, due to internal test equipment, the center of gravity limitations were exceeded long before the aircraft was loaded up to its maximum weight.

Vautour 70 was therefore to remain in non-standard configuration, so it was suggested to adapt it to fly special missions, such as reconnaissance, Del Mar target towing, and as a Combine equipped maritime attack aircraft.

A follow up to this discussion was an AIR2 operational requirement, issued on 2 May 1967, to modify Vautour 70 for reconnaissance with the GOVERNOR camera.

Friday the 7th

Friday, 7 April 1967, mirrored the action of Friday, 13 November 1964. A small scale clash escalated to the point that air strike readiness was authorized at 12:14 and attack aircraft were then

Non-standard equipment in the rear cockpit of Vautour 70; the black-yellow stripped right ejection handle is at the bottom, to the right of the seat. (PC)

scrambled. Squadron 110 launched four Vautours at 12:45: Levy Zur with Oded Erez in Vautour 62, and Adam Zivoni, Gideon Magen, and Ran Goren in Vautour 17. The attack aircraft orbited on hold until ordered to strike, and the Vautours were the first to attack from 13:32. Their target was the Imarat Az El Din fortification.

Ouragans, Super Mysteres, Mysteres and Mirages followed the Vautours. A second Squadron 110 formation – Deputy B Shlomo Keren with Shabtai Ben-Shoa, Micha Gil in Vautour 12, Simi Saar, and Michael Dvir – also took-off and was waiting for its turn to attack when Syrian MiG-21 interceptors appeared. All IAF attack formations were called back to base and combat air patrol Mirages engaged the MiG-21s.

The Mirages claimed two MiG-21s shot down and IAF air strikes were resumed at 15:25. Two Squadron 110 formations took part in this phase of the action. The first was a 3-ship (Isaac Golan in Vautour 05, Isaac Nir in Vautour 67, and Jacob Tal) and a 4-ship formation (Gideon Magen with Alex Meltzer, Herzle Bodinger, Ran Goren in Vautour 17, and Giora Goren); the former attacked mortars and anti-aircraft guns, while the latter attacked anti-aircraft gun emplacements.

Again Syrian interceptors appeared from 15:52, but this time IAF strikes were not called off, and the Mirages engaged the MiG-21s while attack aircraft continued their attacks; the Mirage pilots claimed four more MiG-21s shot down. By the time action ended, the IAF launched 171 combat aircraft sorties against 34 SAF combat aircraft sorties. IAF actual attack sorties count was 84 and total weight of bombs dropped was 65 tons. Squadron 110 launched 15 sorties, but only 11 sorties actually attacked (13% of the IAF total number of actual attack sorties) with 62 bombs weighing 24 tons (37% of the IAF total tonnage), again emphasizing the value of the Vautour heavy attack aircraft.

Arab Solidarity

The Israeli claim that six Syrian MiG-21s were shot down on 7 April 1967 was a blow to Soviet prestige, while the scale and weight of the IAF attack was a test to Arab solidarity. EAF Commander was dispatched to Syria on 10 April, but

A Vautour that flew one mission on 15 August 1966 and two missions on 7 April 1967; the squadron badge was painted on the aircraft from 1966, and the VHF antenna is to the left of the 7 April 1967 mission markings. (PC)

for a while all seemed quiet. Then something happened. A Syrian reconnaissance mission was flown over Israel along the Israeli-Syrian border on 14 May, and the next day Egypt reacted.

The Egyptian army marched into Sinai on 15 May 1967; it was Israel's 19th Independence Day. Israel did not foresee the Egyptian march into Sinai. Israeli Intelligence evaluations did not expect the Egyptian armed forces to be ready for war before some time in between 1968 and 1972. Contrary to the February 1960 RETAMA crisis, the Egyptian deployment to Sinai in May 1967 was public. The next day, Egypt demanded the withdrawal of UNEF from their observation posts; UNEF began to withdraw from Sinai on 19 May. The Government of Israel responded to a UN Secretary-General U Thant appeal, exercising restraint, but also warning that resumption of Palestinian terror activities along the borders or an Egyptian closure of the Strait of Tiran to Israeli shipping would be considered as *casus belli*.

The IAF prepared for war. Emergency posting service persons reported to their units and reserves were mobilized.

Imarat Az El Din fortification under Vautours' attack on 7 April 1967. (PC)

In line with IAF "wartime" procedures, unqualified pilots were "slipped back". Jonathan Zorea - photographed after returning from his Vautour solo flight on 10 April 1967 and prior to the ceremonial baptizing – was "slipped back" to Squadron 116 Mystere and was killed on 5 June 1967 when his Mystere was lost in combat over Syria. (PC)

The emergency situation "slip back" policy was applied. A very short while prior to this crisis, Squadron 110 Deputy A Isaac Golan succeeded former Squadron 110 Commander Yossi Sarig as AIR3 Section A (attack and reconnaissance mission planning) Chief; Squadron 110 Deputy B Shlomo Keren succeeded Golan as Deputy A, and Wing 1 Operations Officer Jacob Gal entered office as Deputy B. Then the Egyptian forces deployed to Sinai, and Sarig was called back to Israel from the USA – where he was due to start a Skyhawk Conversion Course prior to his next assignment as a Skyhawk squadron commander – and returned to the office of AIR3 Section A Chief, Golan returned to Squadron 110 as Acting Deputy A, Keren to Acting Deputy B, and Gal to Wing 1 Operations Officer. A similar "slip back" policy was applied to the three regular Squadron 110 pilots not yet qualified as Vautour operational pilots; Ilan Lior, Moshe Teman, and Jonathan Zorea were "slipped back" to the previous type of aircraft on which they were combat rated, and this type was the Mystere.

In a show of force, an Egyptian paratrooper battalion was air dropped over Sharem El Sheikh on 20 May 1967. Then, just before planned arrival in Cairo of UN Secretary-General U Thant on 22 May, Egypt President Gamal Abed El Nasser announced the closure of the Strait of Tiran. Adding insult to injury, the Egyptian President – laughing and mocking with Egyptian fighter pilots during a visit to an EAF air base – made an immortal statement: "if Israel wants war then 'Ahalan Va-Sahalan' (welcome)". The die was cast.

Squadron 110 staged a reconnaissance mission over the Golan heights on 30 April 1967; Isaac Golan with Elhanan Raz flew lead in Vautour 33, and Micha Gil with Menahem Eini flew wing in 31. This photo was taken immediately after the mission, from left to right: Raz, Gil (who flew his first reconnaissance mission over enemy territory) and Eini. (PC)

5

THE SIX DAY WAR

JUNE 1967

THE MIDDLE EAST WAS CLEARLY HEADING FOR WAR. Israel mobilized reserves to face Egypt, and the IAF was busy returning aircraft undergoing maintenance to serviceable status, training reserve crews, and refreshing operational orders; the cornerstone of the latter effort was the air superiority offensive, Plan FOCUS.

Plan FOCUS

AIR3 Chief Rafi Harlev signed the Operation FOCUS master plan on 16 March 1967. The plan was issued in 54 copies to senior staff officers (copies 33 to 51), air base commanders (copies 1 to 6), squadron commanders (copies 7 to 25), and control unit commanders (copies 26 to 32), plus three reserve copies

Vautour 21 – with suffix digit "0" – on display at Tel Nof a short while before the war. The maximum offensive load is in the center: four 500 kg bombs, six 250 kg bombs and the guns pack. The alternative under wing offensive loads on the sides are: 24 T-10 rockets or two rocket pods or 16 small bombs (apparently 20 are on display) or (as loaded under the wing) two napalm bombs and two 1,300 liter drop tanks. (PC)

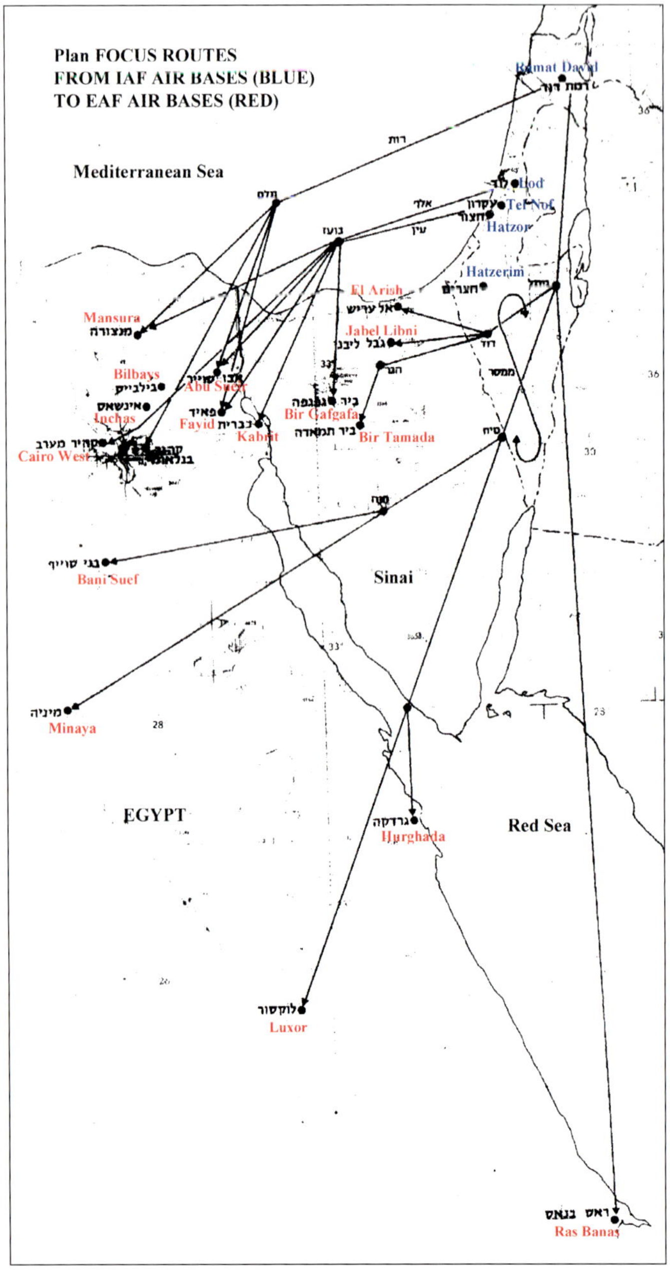

The Plan FOCUS routes from IAF air bases (blue) to EAF air bases (red). (IDFA)

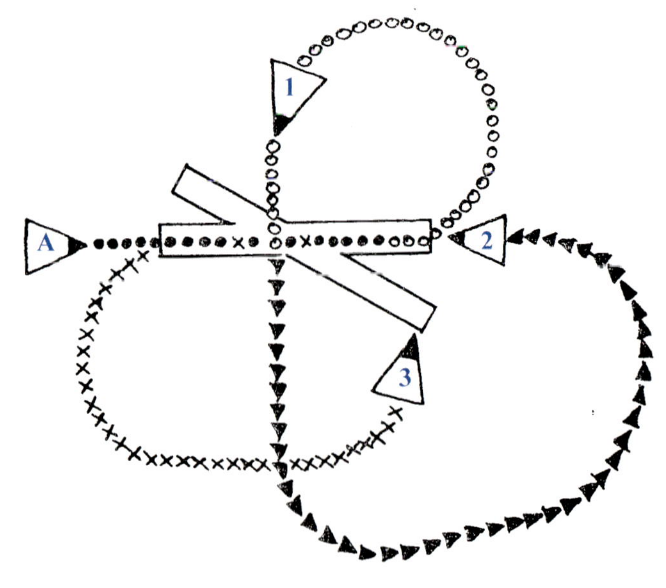

The simplest of the IAF 1960s air base attack patterns was the "270" pattern, which included a bombing run (A), a 270 degree turn, first strafing pass (1), a 270 degree turn, second strafing pass (2), and a 270 degree turn and third strafing pass (3). (PC)

(52 to 54). FOCUS was to be stored in safes until activation. The objective of FOCUS was to achieve air superiority.

FOCUS covered four scenarios: Plan A against Egypt; Plan B against Syria; Plan C against Egypt and Syria, or Egypt, Syria, and Jordan; and Plan C plus Iraq and Lebanon. Mission definition was "neutralization of enemy air force through damage to runways and destruction of maximum number of aircraft on the ground". Herzle Bodinger recalled:

> I graduated Flying School Class 40 in March 1963 and ever since then we practiced air base attack missions. The IAF Year was divided into three training terms and a significant slice of the training cycle covered air base attack. We learnt all air bases in the region; pilots could approach the board at the squadron's briefing room and sketch the runways layout and aircraft parking areas of any given air base. Obviously we better learnt the air bases that were assigned as targets of our squadron and in this context I must mention the quality of our precise intelligence. When I finally arrived over the targets that I learnt during the preceding four years, I felt as though I was on a revisit rather than being there for the first time.

Navigation aids of IAF combat aircraft were limited to a compass, map, and watch, so professionalism and skills of formation leaders were crucial to ensure the success of FOCUS. Surprise was critical; routes to the targets were to be flown at extremely low altitude to escape enemy radar detection; no altitudes were specified, as the pilots were to fly the lowest they dared. Even the long range Vautours could not fly a low level mission from Ramat David to the farthest air bases, such as Bani Suef, Luxor, and Ras Banas. Attacking such far away air bases was crucial, as these were expected to host elements of the Egyptian bomber force, supposedly safe beyond the reach of Israel's air power. Therefore formations tasked to attack the far away air bases were to take-off after H-Hour.

Returning formations were to disengage at low altitude. A climb to economical cruise altitude was recommended once safe separation from the enemy's air defense threat was achieved. Formations flying into SA-2 defended zones were to climb only after flying out of the engagement envelope of the missiles. To aid planning, intelligence prepared a map of known SA-2 battery locations and an engagement envelope circle drawn around every battery. Substantial overlapping between SA-2 battery engagement circles resulted in the coverage of the entire Suez Canal Zone and Nile Delta, though obviously this was based on optimum high altitude maximum range performance of the SA-2 batteries. Actual coverage was much more limited, especially against low flying aircraft.

A time frame of 15 minutes was assigned per attack formation. During this time frame the Vautours were to bomb and strafe. The Vautour was the only IAF attack aircraft authorized to fly two bombing runs, because the Vautours were armed with eight bombs in air base attack mission configuration, whereas all other combat aircraft lifted only two bombs, except for the Ouragan, which was armed with two bombs and rockets but attacked with both in a single pass. The bombing runs were flown in line astern, with the second pair bombing immediately after the first pair. Since approach was at low altitude, the aircraft were to pull up to a minimum altitude that would guarantee a safe 35 degree dive bombing. For the strafing passes, the 4-ship formation was to split into two pairs that would attack from a different direction each time. When a pair strafed the air base was divided into two sectors, with the leader attacking targets in the left half of the air base and the wingman strafing targets in the right half of the air base. Strafing passes were also to be flown at low altitude; the AIR3 recommendation was lower than 1,500 feet above ground level over targets defended by radar guided anti-aircraft artillery. Formation leaders were to aspire to fly four strafing passes depending upon their real-time evaluation of local anti-aircraft fire and their fuel state. The only pre-planned constraint was that a formation would finish its attack within 15 minutes before the next formation would appear over the target.

The Vautours were to bomb with General Purpose bombs that were equipped with two types of detonators: impact fuse inside the bomb's nose cone and "7-15" short delay fuse inside the bomb's tail section. Long delay fused bombs – set to explode an hour or more after bombing in order to disrupt the enemy's battle damage repair effort – were also included. In the Vautour's case, one in every four bombs was equipped with a long delay fuse. Detonators were also tailored for specific targets. A few enemy air bases had concrete runways, so formations assigned to attack these air bases were equipped with 0.1 second delay fuses inside the bomb's nose cone. Runways prepared for heavy bomber operations were naturally hardened, so Vautour bombers planning to attack these air bases had their bombs equipped with a hardened penetration nose cone and only "7-15" delay fuse in the tail section.

Only the first wave of FOCUS was pre-planned in detail; tentatively, the second wave plan was to mirror the first wave. Obviously AIR3 was to utilize BDA data to plan the second wave, but if no new AIR3 orders would be issued then a formation that returned from a first wave air base attack mission would fly a replay mission to the same target along the same route during the second wave.

FOCUS Plan A

Egypt was Israel's principle enemy during the 1960s, and the EAF was the most powerful air arm facing Israel. The EAF was numerically larger than the IAF; AIR4 evaluated in March 1967 that the EAF fielded 16 squadrons of frontline combat aircraft versus the IAF's nine squadrons of combat aircraft. Since the standard of an IAF squadron included more aircraft than the standard on an EAF squadron, and since IAF serviceability was higher than EAF serviceability, the actual gap between EAF and IAF fighter elements was even smaller.

An Egyptian advantage was the shear size of the Nile nation. The EAF had many more air bases than the IAF, and some were beyond the low altitude mission profile scope of IAF combat aircraft. Thus, the Vautours were tasked to attack the most distant EAF bases and/or the EAF bomber bases. Additionally, Squadron 110 was also tasked to fly a BDA photography mission over the Sinai air bases at the end of the first wave.

The regular cadre at the start of IAF Year 1967 Term I, standing left to right: Shabtai Ben-Shoa, Micha Gil, Squadron 110 Commander Levy Zur, Deputy B Shlomo Keren, Ramat David Wing Commander Ezekiel Somech, Deputy A Isaac Golan, Giora Goren, and Abraham Vilan; kneeling left to right: Jonathan Zorea, Ilan Lior, and Moshe Teman. (PC)

The first known FOCUS task assignment program, prepared on 10 April 1967, included five Squadron 110 missions. Missions 1/110 and 2/110 were to attack El Minaya (supposedly an Il-28 bomber base) at H+30 and H+45. Missions 3/110 and 4/110 were to attack Luxor (supposedly an Il-28 bomber base) at H+30 and H+45. Missions 2/110 and 4/110 were to include a reconnaissance Vautour that would bomb but would not strafe; instead, it would photograph with GOVERNOR the attacked air base for BDA. An alternative was that two of these formations would be tasked to attack Ras Banas (supposedly a Tu-16 bomber base) at H+45 and H+90. Mission 5/110 tasked a single GOVERNOR Vautour to photograph the Sinai air bases from H+55 to H+90.

FOCUS Plan B

Syria was the most active and aggressive enemy of Israel during the first half of the 1960s. The SAF was smaller than the IAF and generally treated as no match. The March 1967 AIR4 evaluation referred to only four Syrian frontline squadrons against the nine IAF fighter squadrons.

In FOCUS A, the AIR3 planners had to spread the attack against 13 Egyptian air bases, but in FOCUS B, the attack was concentrated against only five Syrian air bases. Because of this excess power issue, only Plan B included attacks against air defense targets and a massive air-to-air element, including offensive combat air patrols over Syrian air bases. More formations were available per air base and a tighter schedule was set accordingly. If four formations were planned to attack Cairo West during the first wave of Plan A, then no less than nine and up to 11 formations were planned to raid Damascus Elmaza during the first wave of FOCUS B.

Squadron 110 was assigned with four FOCUS B attack missions: Mission 1/110 to Saikal at H+3; Mission 2/110 to Dmer at H+6; Mission 3/110 to Elmaza at H+6; and Mission 4/110 to Dmer at H+16. Because of the shorter distances, the standard Plan A offensive load of eight 70 kg bombs was upgraded to eight 70 kg bombs plus six 250 or 400 kg bombs in the bomb-bay for Vautour A aircraft, or 16 70 kg bombs for Vautour N aircraft, which were unable to bomb from the bomb-bay but were

The prewar "look" of the Vautours included an individual brown/blue camouflage pattern on the upper surfaces and gray undersurfaces; the nose cone of the A aircraft was red and the nose cone of the N aircraft (except the non-standard 70, still in silver) was black; the squadron badge had been painted on the vertical stabilizers since 1966 and yellow mission markings were also painted on the vertical stabilizer (in this photo only Vautour 18 is adorned with mission markings for missions flown in November 1964 and August 1966, while 09 and 12 did not fly offensive missions by the time this photo was taken in January 1967); the tradition to name individual aircraft started shortly before war. (PC)

capable of flying with two internal 1,460 liter fuel tanks in the bomb-bay as a fair substitute to the 1,300 liter external fuel tanks of the Vautour A aircraft. Additionally, there were two Plan B reconnaissance tasks. Mission 5/110 was a Vautour B equipped with K-38 and GOVERNOR cameras tasked to photograph Saikal, Dmer, Elmaza, and Blei from high altitude (the Syrians had no SA-2 batteries) starting at H+105. Mission 6/110 was another, similarly equipped Vautour B on standby from H-Hour to fly a reconnaissance mission over Syria.

FOCUS Plan C

Israel's nightmare was war against an Arab coalition. Tiny poor Israel, facing alone the mighty rich Arab world, was an almost Armageddon scenario in the eyes of the Israeli leadership of the 1950s and 1960s. When Israel's first Prime Minister and Minister of Defense David Ben-Gurion coined the Israeli defense strategy, he referred to a David versus Goliath scenario. The three simple elements of Ben-Gurion's brilliant defense concept were brief conflicts to be fought on enemy's soil with swift operational decision.

The logic was that Israel's modest economy and small population could not sustain a lengthy war. Israeli territory had no "strategic depth" to withstand fighting on its own soil, while overwhelming and swift operational decision were essential to deter Israel's enemies from pursuing another war in the future.

FOCUS Plan C was the AIR3 answer to a coalition that was to include the four nations sharing a border with Israel (Egypt, Jordan, Lebanon, and Syria), plus Iraq. From an air power perspective, Egypt was the main enemy. Jordanian and Lebanese air power were negligible. Iraqi air power was relatively substantial, but too far away to matter. Syrian air power

The standard IAF air base attack formation was a 4-ship, but 110 was obliged to fly air base attack missions in 3-ship formations due to availability and serviceability of the Vautours (actually four of the squadron's 13 Six Day War Day 1 air base attack missions took-off as 3-ship formations and another two took-off as 4-ship formations but attacked as 3-ship formations); this January 1967 image of Vautours 09, 12, 14 and 17 was taken over the Mediterranean coastline near Caesarea. (PC)

was a quarter of EAF strength and more tactically oriented than the EAF, which fielded a substantial bomber force. AIR3 decided to ignore Lebanese and Iraqi air bases; the Plan C targets were 13 air bases in Egypt, three in Syria, and two in Jordan.

Squadron 110 missions in Plan C were exactly the same as in Plan A, except for optional orders per every formation: attack Ras Banas at H+20 for 1/110, Abu Sueir at H+45 for 2/110, Ras Banas at H+30 for 3/110, and Ras Banas at H+45 for 4/110. The Plan C Mission 5/110 was also the same as in Plan A.

Preparing for War

Less than two months after FOCUS was signed and distributed the plan was pulled out of safes in the wake of the Egyptian march into Sinai from 15 May 1967. The IAF prepared for a preemptive strike and FOCUS was evolved into a detailed Mission Order. Herzle Bodinger recalled:

> I served as a Flying School instructor at Tel Nof; at the time about half of the Flying School was transferred to Hatzerim but the other half was still at Tel Nof. Flying School instructors had their emergency postings and we were called to our squadrons. All aircraft were armed and ready for war as though Operation FOCUS would be launched the next day. Static aircraft were prone to malfunctions so we occasionally refreshed the aircraft in flight. Moreover, when so heavy aircraft were parked for prolonged time frames, the tires turned less and less elastic, 'squared tires' we used to call this phenomenon that was especially noticeable during taxi as the ride was bumpy rather than smooth as usual.
>
> Therefore every aircraft was flown once every several days but instead of simply flying the aircraft to make sure that it was serviceable, we trained for realistic air base attack mission scenarios. Additionally, the aircraft configurations were different than usual. To enable the Vautour to fly the planned long range Operation FOCUS missions, the bombing gear that normally could haul six 400 kg bombs was removed from the bomb-bay and two fuel tanks were inserted. For maximum fuel load we had external fuel tanks on the inner wing stations so to enable us to lift more than two bombs we were issued with Quadruple Bomb Racks that were loaded onto the outer wing stations. The eight bombs we thus lifted were also special. Our primary targets were Tu-16 air bases that had hardened runways so our bombs had penetration steel nose sections – similar to bombs optimized to attack warships – and only tail fuses.
>
> We never before flew in such a configuration so during those training flights we actually tested fuel consumption and collected data for our planned long range missions. We took-off from Ramat David; flew south to Eilat, turned north to Metula and again south to Beer Sheba where we dropped to low altitude and practiced the air base attack pattern – two bombing runs and three strafing passes – over the Halutza range. We thus simulated long range missions within the limited boundaries of Israel.

Israel figured that the situation was so desperate that war was practically inevitable, so an emergency aid request was forwarded to France on 18 May 1967. The Israeli wish list included 25 combat aircraft – Mirages, Super Mysteres, and Vautours – of whatever mix possible; but no more Vautours would arrive from France prior to the outbreak of war.

FOCUS Update

Egypt's military march into Sinai, the withdrawal of UNEF, and the 18 May 1967 flight of EAF MiG-21s over Dimona certainly escalated the crisis, but the Rubicon of *casus belli* was only crossed on 22 May, when Egypt closed the Strait of Tiran to Israeli shipping. The following day, AIR3 updated the task assignment program of FOCUS Plan A.

The main target of Squadron 110 was changed to Bani Suef, a Tu-16 bomber base in Upper Egypt, west of the Nile. Missions 1/110 to 3/110 were to attack Bani Suef at H+40, H+55, and H+70. Mission 4/110 was tasked to attack Abu Sueir – in the Suez Canal Zone, west of Ismailia, an Il-28 bomber base – at H+40. The reconnaissance Mission 5/110 remained unchanged, but another special mission was added. Mission 6/110 tasked a single Vautour to fly a WART electronic warfare profile from H-Hour.

Vautour 17 was photographed at Hatzerim – the IAF's new fourth air base in the Negev Desert – on 17 May 1967. (PC)

Herzle Bodinger flew Vautour 26 on 22 May 1967 during a 2:15 flight hours sortie that simulated an air base attack mission against Bani Suef; this simulation proved the feasibility of such a strike, and the following day AIR3 changed the FOCUS task assignment program. (PC)

War is Imminent

Israeli Intelligence concluded on 25 May 1967 that Egypt would attack Israel the following day. If this was the case, then the IDF had very little time left to prepare for an IAF preemptive strike. Thus, four Vautours deployed from Ramat David to Tel Nof during the late afternoon hours of 25 May. Gideon Magen was the leader of this formation, and among the pilots was Herzle Bodinger in Vautour 09.

The deployment to Tel Nof was aimed at securing the success of Mission 1/110 to attack Bani Suef, because the Vautours were unable to fly a low altitude mission profile from Ramat David to Bani Suef. The solution was that the first formation to attack would deploy south to Tel Nof. This formation would fly at low altitude all the way to the target and launch the attack against Bani Suef. The second formation would follow from Ramat David in a complicated profile of low altitude until H-Hour, climb to a more economical high altitude cruise, and descend back to low altitude just prior to the attack.

Egypt did not attack Israel on 26 May 1967 and the four Vautours returned from Tel Nof to Ramat David on 29 May.

Reconnaissance for War

Egyptian MiG-21s penetrated Israel on seven occasions from 17 May to 4 June 1967. Israel banned fighter pilots from crossing the borders during pursuits after intruding MiG-21s, in order to avoid an escalation, but collection of intelligence information in reconnaissance flights was essential.

Shlomo Keren with Achikar Eyal, in Vautour 31, flew a nighttime reconnaissance mission over Sharem El Sheikh on 24 May 1967. Isaac Golan with Alex Meltzer flew a reconnaissance mission over south Sinai in Vautour 31 on 2 June. The same day Levy Zur, with Elhanan Raz, flew a reconnaissance mission over "enemy territory", also in Vautour 31. The final prewar Vautour reconnaissance mission was flown at noon on 4 June, when Simi Saar, with Achikar Eyal in Vautour 33, and Ovadia Nachman in Vautour 31 scanned Sinai to pinpoint the exact location of the Shazly Group; an elite brigade size task force in Sinai. When Simi Saar returned from this mission he was summoned to the office of Squadron 110 Commander Levy Zur. Simi Saar recalled:

> Levy Zur told me that war may start tomorrow's morning. If this was the case then my target would be Bani Suef with take-off from Tel Nof; our mission was to destroy the Egyptian Tu-16s based at Bani Suef. I wrote a note to my wife and asked from the secretary at the squadron's operations cell to arrange its delivery to our home at the base neighborhood. I wrote that I would not return home that evening because I would be on a deployment, nothing further. The note was indeed delivered and I gathered the pilots of my formation. My navigator was Alex Meltzer; the pilots were Giora Goren, Jacob Rafaeli and Herzle Bodinger. We ferried the aircraft at dusk, in radio silence, in singles; our aircraft were full of fuel and fully armed for our mission. Landing such a heavy aircraft was a challenge but we all landed safely and taxied to a Flying School line that was nearest to runway 18.
>
> Later that evening, circa 22:00, the Tel Nof Air Base Commander Samuel Shefer summoned all unit commanders to his office and I reported to this meeting as a commander of a force in the base. It was actually a briefing for war. Shefer explained the background and ended his presentation with a statement that war would start tomorrow's morning; that this was a final decision. We were banned from telling anything to the pilots but we had to brief the technical officers so that our aircraft would be fully fueled early in the morning when it is cool and the fuel density is maximal.
>
> I returned to Alex Meltzer's home. Meltzer served in the Flying School [as Navigators Training Squadron Command-

Reservists push a 1,300 liter drop tank past Vautour 65 during the build up to war in late May or early June 1967. Because Vautours were pulled out of periodic maintenance, there was not enough space for all aircraft in HAS or pens, so some aircraft had to be parked in dispersal points at the line. In order to ease stresses, the bombs were placed underneath the wing stations. (PC)

Squadron 110 maintainers at HAS Number 8 in front of an armed Vautour and beside a mobilized car with blacked out head lamps; Technical Flight Commander Nathan Segall stands third from left, while Shlomo Kahana crouches first from left. (PC)

er]; when the emergency started, he reported to Squadron 110 at Ramat David and his wife went to stay with her parents so their home was empty and there we stayed overnight, all five of us. I gathered them for a briefing but I did not tell them about the war. We arranged an early rising for 03:00 and we went to sleep. I knew that the die was cast and I just could not sleep; I did not sleep at all that night.

We woke up at 03:00 and gathered for a [formation] briefing at 03:30 and then I told them that it was war, that nothing would stop us. I presented H-Hour [07:45] and said that we would take-off in complete radio silence. Our target was the most distant target so we actually took-off first; our Vautours were the first [IAF] aircraft to take-off [on an Operation FOCUS mission].

Vautour 61 – photographed departing Ramat David in January 1967 - was the only aircraft undergoing maintenance that could not be readied in time for the Six Day War (Vautour 61 is not known to have flown after 2 April 1967 and before 25 March 1968), and for this reason it was never painted in the IAF pre-Six Day War scheme, but transformed from silver directly to the post Six Day War scheme. (PC)

Squadron 110 on the Morning of 5 June 1967

The IAF order of battle on the morning of 5 June 1967 – Six Day War Day 1 – included 21 Vautours: 13 A, two B, and six N. The actual number of A aircraft was only 11. Additionally, Vautour 61 was undergoing D-Level maintenance and it was impossible to finish this work in time. The actual Squadron 110 order of battle therefore included 18 Vautours: 11 A, two B, and five N. All of these were made serviceable for FOCUS, and 17 actually took-off on FOCUS first wave missions.

The squadron's pilots and navigators were composed of three populations: regulars, emergency postings and reserves.

The regulars included Squadron 110 Commander Levy Zur and Acting Deputy B Shlomo Keren, plus three pilots (Micha Gil, Operations Officer Isaac Nir, and Michael Dvir) and two navigators (Elhanan Raz and Shabtai Ben-Shoa).

Emergency postings included Acting Deputy A Isaac Golan, 11 pilots and five navigators. The 11 emergency posting pilots included Wing 1 Operations Officer Jacob Gal, seven Flying School instructors (Herzle Bodinger, Daniel Ilan, Giora Goren, Jacob Tal, Abraham Vilan, Absalom Friedman, and Ran Zur), two staff officers (AIR2 Chief Adam Zivoni and AIR5 Aerial Training Section Chief Simi Saar), and Ovadia Nachman, who was on a preretirement leave. The five emergency posting navigators included Flying

School Navigators Training Squadron Commander Alex Meltzer and three staff officers (Achikar Eyal, Uri Talmor, and Menahem Eini) and Oded Erez, who returned to Israel from his assignment in France on 6 June 1967 and arrived at the squadron on 7 June.

The reserves numbered eight; all were pilots, including two El Al pilots (Uri Margalit and Simon Zimon), one agricultural pilot (Beni Zohar), an IAI test pilot (Zeev Tavor), an industrialist (Yoash Tsiddon), and three Kibbutz members (Jacob Rafaeli, Gideon Magen, and Ran Goren).

The total number of pilots and navigators available to Squadron 110 on the morning of 5 June 1967 was therefore 25 and six, respectively. The "fighting executive" post – the IAF term for squadron officer in command – was rotated between the three squadron executives (Squadron 110 Commander Levy Zur, Acting Deputy A Isaac Golan, and Acting Deputy B Shlomo Keren) and senior leaders, including Simi Saar, a former Squadron 110 Deputy A. The "fighting executive" on the morning of 5 June, and throughout the duration of FOCUS first wave, was Isaac Golan.

Previously, on the evening of 4 June 1967, AIR3 prepared the final iteration of the FOCUS task assignment program. Squadron 110's missions remained unchanged. D-Day and H-Hour were set to 5 June 1967 and 07:45. By the early morning hours of 5 June all aircraft were ready to go. Levy Zur recalled:

> It was still before sunrise on the morning of 5 June, when all pilots gathered inside the briefing room. I stepped to the black board and wrote, in large bold letters, 'H-Hour 07:45'; I waited a few seconds for the meaning to sink in and then I started the briefing.
>
> The pilots were all trained in air base attack missions; it was our primary mission and we practiced this mission for years. Leaders knew their targets and pilots were well aware of their roles. I emphasized two issues: very low altitude flight to the targets and absolute radio silence, not a word prior to H-Hour even if a pilot would run into some kind of emergency situation.

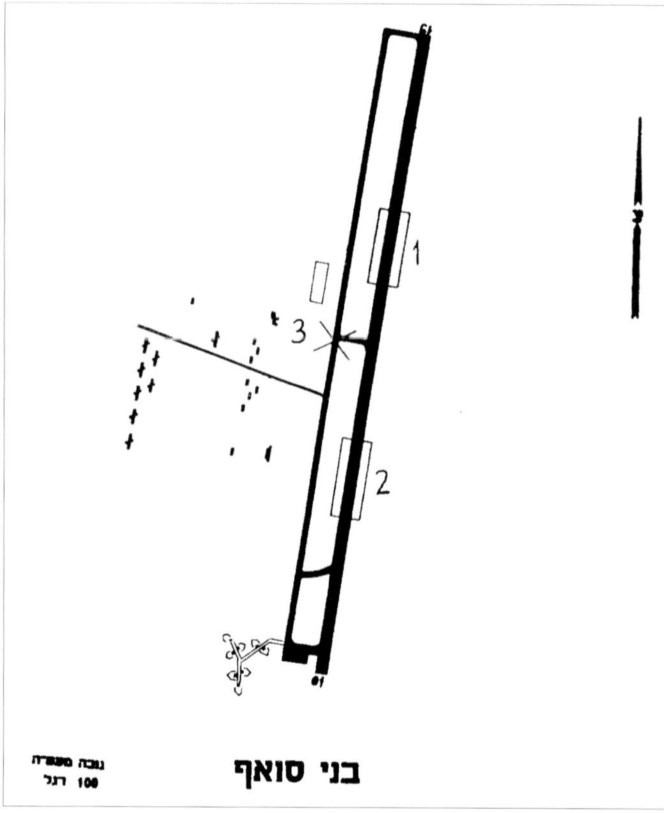

The Operation FOCUS scheme of Bani Suef; primary targets 1 and 2 were along runway 01/19, secondary target 3 was along the parallel taxiway; the Tu-16s were in pens west of the south end of the taxiway. (IDFA)

The silhouettes of three Tu-16s painted on the vertical stabilizer; each of the three TIGRIS Vautours – 62, 26 and 09 – destroyed three Tu-16s at Bani Suef during Mission 1/110. (PC)

Take-Off 07:17 – TIGRIS to Bani Suef

The farthest EAF air base to be attacked during FOCUS first wave was Bani Suef; a Tu-16 bomber base. Bani Suef's layout was rather simple, with a single runway (01/19) and a parallel taxiway. The primary aiming points 1 and 2 were along the runway, and the secondary aiming point 3 was halfway along the taxiway. The geographical elevation of Bani Suef was 100 feet above sea level. AIR4 evaluated that 15 Tu-16s were present at Bani Suef on the morning of 5 June 1967.

Mission Order 1/110 specified a configuration of eight 70 kg bombs, two 1,460 liter fuel tanks in the bomb-bay, and two 1,300 liter external fuel tanks. The task was to bomb aiming points 1 and 2 at 08:10 – 25 minutes after H-Hour – and the pilots were briefed to activate OFRA communication jammers during the attack. The formation assigned to fly Mission Order 1/110 was Simi Saar with Alex Meltzer in Vautour 62, Giora Goren in 12, Jacob Rafaeli in 26, and Herzle Bodinger in 09; their radio call sign was TIGRIS. Herzle Bodinger recalled:

> We could not fly a low altitude profile mission from Ramat David to Bani Suef so we had to deploy south to Tel Nof.

I was selected for this mission because I was a Flying School instructor, I lived at Tel Nof and my presence there was not expected to attract any exceptional attention. We were summoned to the wing commander and ordered to ferry our Vautours to Tel Nof in singles, flying at low altitude, without lights, in radio silence. When we arrived at Tel Nof we taxied to a remote apron where we parked. We then returned to our homes, the cover story was that we were on leave to visit our families. I obviously realized that there would be war tomorrow but I said nothing to my wife. Only in the morning, just before I left to the briefing room at about 05:00, I suggested to her to take our baby son – now Ramat David Air Base Commander – to visit her parents at Ramat Gan but I did not tell her why.

We were told during briefing that the future of the State of Israel was dependent upon the success of FOCUS. Preservation of surprise was emphasized. We had two radio sets – formation channel and control channel – and one of these was switched OFF while the other set was switched ON in case of 'abort mission' message but the transmission selector was switched to the OFF radio set, ensuring that accidental clicks would not violate the electromagnetic silence.

The canopy of Giora Goren's Vautour failed to lock and we could not wait for the maintainers to fix the problem; we had to take-off right on time so we departed as a 3-ship. We flew south, very low, and over Hatzerim we saw the deception Fouga trainers. When we crossed the border, I realized that it was real and shortly after crossing the border we flew over an Egyptian military convoy. The Egyptian soldiers waved and this filled me with joy, surprise has been achieved. We flew towards Bir Tamada where we turned, crossed the Gulf of Suez and flew into a huge canyon that had to lead us right to the target.

As we approached the Nile, morning mists began to cover the sky from 700 feet above ground level so we flew under a canopy of a thin layer of clouds. This was a worrying phenomenon as we had to climb, visually acquire the target and dive to bomb. We flew over rural cultivated area and crossed the Nile, a huge wide river with sailing boats. The Egyptians did not construct air bases on land that could be cultivated, as this was a rare resource in a mostly desert nation, so Bani Suef was located in the dunes further inland beyond the edge of the rural area. As a result, the morning mist was mostly over the humid cultivated land but not over the dry dunes. We climbed and there it was, spread out in front of us. A new large air base with long wide runways and the aircraft were parked in open pens.

We flew the first bombing run and got all the bombs right on target; this success was repeated during the second bombing run, I was trailing behind so I saw the hits of the first two aircraft. The bombs were set to explode 7-15 seconds after hitting the surface so that all aircraft would clear the area before the explosions. I saw the holes of the hits ahead of me and the explosions when I looked back over my shoulder.

We then flew three strafing passes. The Vautour's 30 mm guns pack covered an area as large as 20 meters wide and 20 meters long. During training we only fired the lower right gun that had the central aiming point, with the aiming points of the other three guns scattered around, but during the war we fired all four guns for the first time. We just could not miss such a large target as a Tu-16 and every one of us destroyed one bomber per strafing pass. We were three aircraft and we flew three passes so we destroyed nine Tu-16s.

We disengaged but flew east at low level, an SA-2 battery was located south of the air base. The SA-2 was not effective against aircraft flying lower than 3,000 feet and most of the time we were really low. Only during pop-up [to bomb] we climbed to 6,000 feet so we endeavored that the time frame that we flew higher than 3,000 feet would be limited to less than 30 seconds. They launched a surface-to-air missile but it did not lock on to us. By the time we climbed we began to hear reports from other formations and it was crystal clear that FOCUS succeeded though losses were indeed sustained.

During disengagement I also suffered an electrical trim control malfunction and I had to fly slower, I never inquired into it so I do not know if I got hit but I reported to my leader and still managed to fly all the way to Ramat David.

The three TIGRIS Vautours landed at Ramat David from 09:10 to 09:14. Vautour 12 was already at Ramat David; the canopy malfunction was repaired and Giora Goren ferried the aircraft back home, departing Tel Nof at 08:20 and arriving at Ramat David at 08:51.

A WART pod is loaded – correctly – onto the right outer wing station of Vautour 70. (PC)

Take-Off 07:21 – ARNON Flies WART

The first Vautours to take-off from Ramat David flew Mission Order 6/110. Call sign ARNON, Adam Zivoni with Menahem Eini in Vautour 70 and Uri Margalit with Yoash Tsiddon in 65 were tasked to fly at low altitude to a position over the Mediterranean Sea, north of Port Said. Exactly at H-Hour they would climb to 20,000 feet and activate their WART electronic warfare system. The two crews completed their eventless mission as planned and returned to Ramat David at 08:46 (Vautour 65) and at 09:29 (Vautour 70).

Take-Off 07:25 – JORDAN to Bani Suef

Mission Order 2/110 was tasked to bomb aiming points 1 and 2 at 08:25. The formation flying Mission Order 2/110 was JORDAN: Levy Zur with Uri Talmor in 67, Abraham Vilan in 17, Micha Gil in 08, and Daniel Ilan in 18. JORDAN was planned to attack 15 minutes after TIGRIS, but departed Ramat David only eight minutes after TIGRIS took-off from Tel Nof; this was because the two formations flew different mission profiles. TIGRIS flew at low altitude all the way to Bani Suef; JORDAN started at low altitude, climbed to cruise, and descended back to low altitude before the attack.

Exactly at H-Hour (07:45) the JORDAN Vautours initiated a steady climb, and for this reason the second Vautour formation to raid Bani Suef was able to fly the mission from Ramat David, while the first formation had to take-off from Tel Nof. As they crossed the Gulf of Suez SA-2 launches were noted to their right, possibly from a Suez Canal Zone battery. The SA-2 missiles exploded harmlessly in mid-air, higher and far away from the Vautours that by then had started to hear TIGRIS chatter over the radio. JORDAN dropped to low altitude for penetration below the SA-2 shield that defended Bani Suef. As the Vautours flew over the Nile, pillars of black smoke emanating from Bani Suef were observed and the formation pulled up to bomb. The runway was full of holes, so Levy Zur ordered to bomb the parallel taxiway during the second bombing run.

Burning Tu-16s generated huge pillars of black smoke, and this smoke screen complicated JORDAN's scan for valid targets. Levy Zur aimed at a seemingly intact Tu-16 during his first strafing pass; the resulting mushroom like explosion was so huge that the Squadron 110 Commander had to pull up abruptly in order to avoid flying into the inferno of fire and smoke. Meanwhile, the local SA-2 battery was active; one SA-2 launch was noted right after the first bombing run and a second SA-2 homed onto the Vautours after the first strafing pass. Thus, JORDAN 1 targeted the SA-2 battery's command and control center during his second strafing pass.

In his third strafing pass, the JORDAN leader focused his attention on two Tu-16s that were seemingly undamaged, strafing them one after the other in a single pass. Overall, JORDAN claimed the destruction of seven Tu-16s; the Squadron 101 Mirage formation that followed JORDAN from 08:40 reported that all resident Tu-16s had indeed been destroyed by the two Vautour formations. Mission completed, the JORDAN Vautours returned to Ramat David at 09:23 and Levy Zur replaced Isaac Golan as "fighting executive" for the duration of the second wave.

Take-Off 07:49 – BANIAS to Abu Sueir

Located halfway along the 160 km long Suez Canal – some 15 km west of the city of Ismailia, along the highway from Ismailia to Zakazik, Banha, and Cairo – Abu Sueir bore the brunt of the IAF preemptive strike; AIR4 evaluated that the large air base hosted 27 Il-28 bombers and 19 MiG-21 fighters on the morning of 5 June 1967.

Seven formations were tasked to raid Abu Sueir from 07:45. Squadron 110's was the fourth formation to attack; Mission 4/110 was tasked to bomb aiming points 1 and 3 along runway 04/22 from 08:25. The secondary target of the Vautours was a nearby SA-2 battery. Originally, the mission included five Vautours; the fifth was a B reconnaissance aircraft that was planned to bomb Abu Sueir and then collect BDA images while the other four Vautours would fly their strafing passes. However, there were two Vautour B reconnaissance aircraft and one was assigned to Mission 5/110, so the assignment of the other Vautour B was changed from Abu Sueir to Bani Suef. As a replacement, the GOVERNOR Vautour 66 was assigned to fly Mission 4/110. The BANIAS manning was Shlomo Keren with Shabtai Ben-Shoa in 66, Ran Zur in 05; Beni Zohar in 07, and Michael Dvir in Vautour 14. Beni Zohar recalled:

> (Wing 1 Commander Ezekiel) Somech briefed all air crews a day before the war. He did not say 'there will be war

Levy Zur with Uri Talmor flew Vautour 67 on Mission 2/110 to Bani Suef; Vautour 67 was photographed in April 1966 with a recently applied squadron badge on the vertical stabilizer, albeit pointing backwards. (PC)

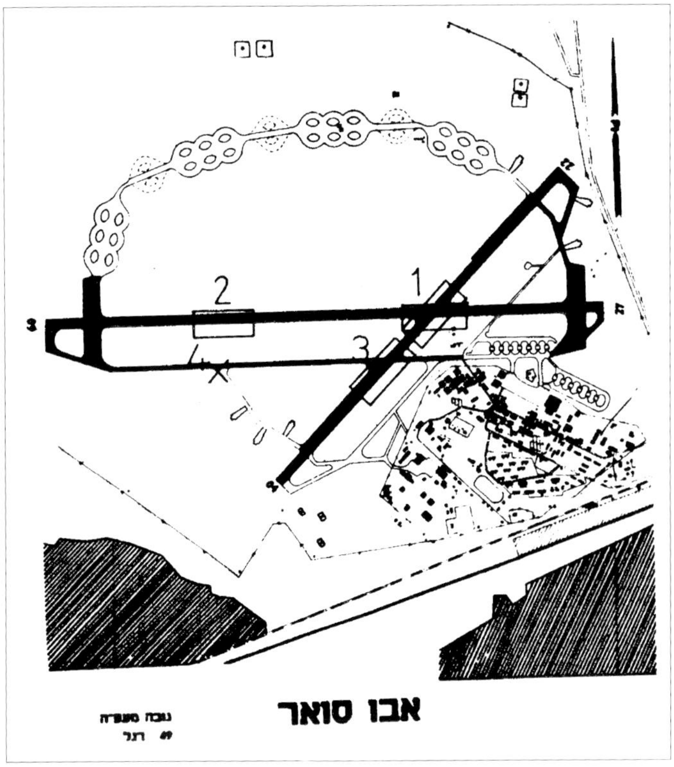

Abu Sueir's aiming point 1 was the intersection of runways 09/27 and 04/22, aiming point 2 was along runway 09/27, aiming point 3 was the intersection of 04/22 and the taxiway, and aiming point 4 was along the taxiway. A prominent feature peculiar to Abu Sueir was the pattern of the aircraft dispersal areas to the north of 09/27. There were four dispersal areas, each with three double dispersal stations that resembled the shape of the digit "8". The "eights" – as IAF pilots referred to these dispersal areas – were connected to the runways via a crescent shaped taxiway. (IDFA)

tomorrow', but he presented guidelines. We had to take-off in radio silence, a pretty complicated procedure that we could not [fully] practice as this would have revealed our secret so we only practiced it in taxiing. He told us that during the war we should not pay attention to our personal safety or to the safety of the aircraft. For example, if a pilot experienced a malfunction during the radio silence take-off, he should do everything to keep the runway operational even at the risk of being killed. We assumed that we would have 20 to 30% losses during the first wave. Again, his directive was to ignore risks and press ahead with the mission regardless of consequences.

My first mission was to Abu Sueir. We flew to the target at low altitude and as we climbed [to bomb] I saw additional aircraft trailing us. There was a possibility that Mirages would fly a combat air patrol to protect us. At first glance I thought that these were Mirages but then I saw, 1,000 to 1,500 meters in front of me, an air-to-air missile. I could actually see its spiral trajectory chasing the leading pair that was exactly diving to bomb but it was crystal clear that this missile could not harm them. Only then I realized that these aircraft were MiG-21s, not Mirages, so to comply with [Somech's] briefing I had to do everything to ensure that the attack would be pressed on, at least three of us would attack.

I transmitted over the radio that I would engage the MiG-21s. They came in a not very threatening manner, more or less at our speed, a little faster. They slowly closed on us with the two [MiG-21] pairs trailing each other rather than splitting, lighting their afterburners and chasing us, it actually resembled an escort!

I ordered my wingman to bomb and I started to turn. If those MiG-21s had followed the bombers I would have followed them but all four of them followed me, turning. Not a single one of them really threatened me; I did not even jettison my bombs. They were not turning very well and I observed all the mess down there at the air base: smoke from previous air raids, dust from the ongoing attack. I had to get rid of the bombs but why to drop these bombs without a purpose?

I lowered my nose, entered the smoke but as I exited the smoke I noticed that I was in between the runways, there was

The only Six Day War mission that Vautours 05 and 07 flew together was 4/110 to Abu Sueir on 5 June 1967; this photograph of 07 and 05 in a low altitude pass along a Ramat David runway was taken in the early 1960s. (PC)

A view of the westernmost "eights" complex at Abu Sueir during the FOCUS first wave attack; five intact Il-28 bombers are visible, while a sixth Il-28 is already burning in the background. (PC)

nothing I could do [to correct the aiming of bombing]. I released the bombs, climbed back and then I saw only three [MiG-21s]. I reported that I was mixing it with only three and that the fourth was with the attackers, he bothered them a little, they threatened him and he disengaged. I turned with the MiG-21s so the other Vautours could bomb and strafe according to plan, albeit in a 3-ship formation. [When they finished the attack] they reported 'we are going home' and only then I noticed that I had a fuel problem. One of the 1,200 kg belly tanks did not feed fuel, I was stuck with that weight; I could not drop it, I could not use it.

I disengaged when I was in a 90 degrees advantage over one MiG and the pair was 180 degrees from me. If I had time and fuel I might have shot him down, but this was not the case and during the turn, when I was heading towards Ismailia, I simply dropped my nose and dived from about 6,000, 8,000 or 10,000 feet. I did not even look back at them, as an agricultural pilot I simply shaved the earth.

I regrouped with the formation and as we crossed the coastline we strafed an SA-2 battery. I thought that we had to damage the battery because soon I had to climb. Up to an altitude of 2,000 feet we were pretty safe and below 5,000 feet the SA-2 was not very effective but I had to climb to save fuel. The others continued a bit longer at low altitude just in case but I climbed immediately. I shut down one engine, left the other at full power and it kept me going at 30,000 feet. I figured that I would have to eject over the sea but the closer [to Israel] the better. Then I reached a point where I evaluated that if I would switch off both engines and glide, there would be a chance that I might land. I was permitted direct approach, restarted one engine but left it on idle just in case and landed.

The people at the HAS told me that when I exited the cockpit they could not understand me, they only heard 'missile... fighters...air combat...fuel...one engine...' and although I do not remember that I was so excited, we actually returned from a mission from which we thought that we would not return alive.

BANIAS claimed destruction of five Il-28s, one MiG-17, and one MiG-21. Beni Zohar landed at 08:59, and the three other BANIAS Vautours followed at 09:01. For this action, Beni Zohar was awarded with the IDF Gallantry Medal in April 1973.

Take-Off 07:57 – VISTULA to Bani Suef

The fourth and final FOCUS first wave formation to attack Bani Suef was VISTULA. These four formations included the three

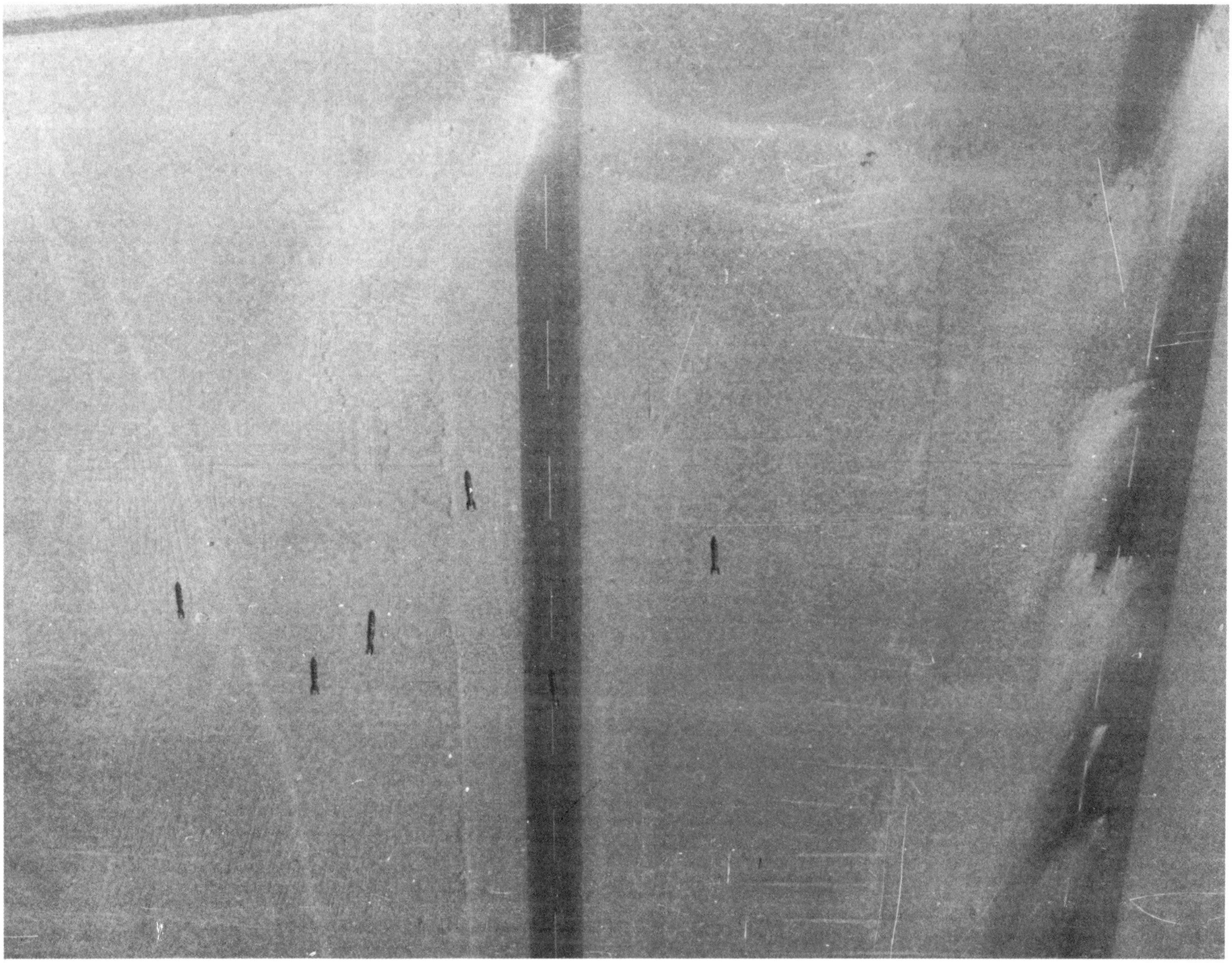

Vautour 33 bombing Bani Suef, with runway 01/19 on the right and the taxiway in the center. (PC)

Squadron 110 Vautour formations plus a Squadron 101 Mirage formation. VISTULA was tasked to raid Bani Suef from 09:00, but departed Ramat David only three minutes before the Mirage formation – tasked to attack at 08:40 – took-off from Hatzor. This was because the Vautours were slower and had to cover more distance; the VISTULA flying time from Ramat David to Bani Suef was 63 minutes, while it took the Mirages only 40 minutes to fly from Hatzor to Bani Suef.

VISTULA flew as a 3-ship formation: Gideon Magen with Elhanan Raz in 33, Ran Goren in 03, and Jacob Tal in 21. They were tasked to bomb aiming points 1 and 2, operate OFRA jammers during the attack, and collect BDA imagery. Because of the latter, VISTULA 1 flew in a Vautour B reconnaissance aircraft.

Having taken off after H-Hour, the VISTULA Vautours cruised over Sinai at 36,000 feet. When they flew over the Gulf of Suez they started to hear the radio chatter of fellow squadron members reporting SA-2 launches over Bani Suef and MiG-21s over Abu Sueir; Gideon Magen was relieved, as due to the late departure of VISTULA he was afraid that he may miss the thick of the action, but apparently there was still plenty of action for everyone. Ran Goren recounted:

I was called to active service on 24 May 1967 after the Straits of Tiran were closed. We thought that we would attack immediately but we were proved wrong. Every evening I went to sleep feeling that tomorrow would be D-Day but woke-up the next morning into another day of preparations. Yet, on Sunday evening my hunch was stronger than ever, there were perceptible signs of action. Nothing was said but I hardly slept that night until the wake up call at 05:00.

We gathered in the briefing room and [Levy] Zur, the squadron commander, opened, 'Is anyone tired? Is anybody sick?' No one answered even though many of us hardly slept the previous night. He said 'H-Hour is 07:45' and wrote it on the board in large letters and in a frame. I was very excited but calm. Certainty is better than uncertainty. Zur talked about the risks, about tenacity of purpose and about losses. We would loss friends.

The Wing Commander then entered. 'Today we are going to realize the greatest IAF mission. It is not a partial, limited operation. This is the maximum we can do, on our own, alone. We face the ultimate test to the IAF and to the nation. I am convinced that we will prevail. The most important thing is to keep the surprise. Do not say a word on the radio even when you see

> your friend going to collide into a hill. If you have problem in take off just go aside into the ground and clear the runway for the others to take off". His words made clear the dedication that was expected from us and the risks we were expected to undertake.
>
> Each formation then briefed its own mission. We flew as a 3-ship, a leader in front with a trailing pair. The briefings were short; everything had been studied and discussed the previous days.
>
> Immediately after take-off I noted that RPM of the right engine fluctuated but it was war so I pressed ahead. As we climbed, the RPM dropped until it stabilized at 7,900; I lacked power but I could fly. By then we heard the reports of returning formations and I envied them, they finished their mission and we were still to accomplish ours. It was a strange feeling to cross the border into Sinai, a line that we once were so careful not violate. We flew over the desert and scanned the sky, maybe MiGs were lurking out there.
>
> After crossing the Gulf of Suez we flew low over featureless desert with poor visibility due to haze, both issues complicated our navigation when suddenly we saw the Nile; a huge river with sailing boats and palm trees. The town of Bani Suef appears to our right, instead of left as planned, so we turned steeply to the right and pulled up. The air base was wrapped in smoke but anti-aircraft fire was active and all around us. I made myself concentrate on aiming at the parallel taxiway, released four 70 kg bombs that hit their mark and flew away so low that I was unable to strafe an anti-aircraft artillery position down along the runway.
>
> Flying very low over the air base I figured that the scene was like in a movie: pillars of smoke, burning aircraft, explosions in the air of anti-aircraft artillery shells and explosions on the ground of our bombs. Suddenly a delay fused bomb, dropped by a preceding formation, exploded nearby and produced a huge mushroom.
>
> We turned right 360 degrees for another bombing run from south after which we turned left 270 degrees to strafe from west. There were no more aircraft to strafe, all aircraft were on fire. I aimed at an aircraft that looked more complete than others and noted an anti-aircraft gun to its right. I strafed that anti-aircraft artillery emplacement during my second pass and the control tower during my third.

The third and fourth formations to attack Bani Suef did not report SA-2 launches, so Levy Zur's strafing of the surface-to-air missile battery's command and control center may have been effective. VISTULA 1 flew a reconnaissance pass over Bani Suef to capture the important BDA images when VISTULA 2 and 3 flew their first strafing pass. Then the Vautour B cleared the target area and orbited south of Bani Suef while the two attackers flew their second and third strafing passes.

When the VISTULA gunmen reported that they had completed their mission, VISTULA 1 returned for one final photographic pass. By then VISTULA 2 and 3 were well east of Bani Suef, so Gideon Magen throttled to full power and chased his two wingmen, regrouping with them just as they flew out of the estimated SA-2 danger zone and initiated their climb to high altitude cruise back home. On their way ground control vectored VISTULA to intercept enemy aircraft, presumably MiG-21s and Tu-16s, over the Negev Desert, but they actually chased ghosts, as no Egyptian fighters or bombers were flying over Israeli territory at the time, and the three VISTULA Vautours landed at 09:47.

A vulture over carcasses; Vautour 31 photographed these wrecked Il-14s airlifters and Mi-6 helicopter – and its own shadow – at Bir Tamada. (PC)

Take-Off 08:29 – Photographing the Sinai Air Bases

The 14th Vautour to depart Ramat David on the morning of 5 June 1967 was a B reconnaissance aircraft, Vautour 31 flown by Ovadia Nachman with Achikar Eyal, radio call sign RIVER 275. The lone Vautour flew Mission 5/110, tasked to photograph El Arish from H+55, then Jabel Libni, Bir Tamada, and Bir Gafgafa from H+90. Fitted with the GOVERNOR camera, Nachman and Eyal flew this crucial BDA collection mission at low altitude and returned to Ramat David at 09:34.

FOCUS First Wave Summary

The IAF official claim for FOCUS first wave was the destruction on the ground of 189 aircraft; an average of 21 claims per squadron, or 4.2 claims per air base attack formation. The Squadron 110 claim was 23 aircraft: 16 Tu-16 bombers at Bani Suef and seven (five Il-28 bombers plus a MiG-17 and a MiG-21) at Abu Sueir; Squadron 110 averaged 5.75 claims per air base attack formation.

In addition to the three formations that raided Bani Suef and the formation that attacked Abu Sueir, Squadron 110 also launched a pair on an electronic warfare mission and a single Vautour B reconnaissance aircraft. Thus, a total of 17 Vautours (three from Tel Nof and 14 from Ramat David) took part in FOCUS first wave missions. All 14 Vautours flying the four air base attack missions faced enemy anti-aircraft fire; some were targeted by SA-2 missiles and others engaged MiG-21s, but all returned safely to base. FOCUS succeeded, the IAF achieved air

Vautour 31 over Bir Tamada, photographing, left to right: two destroyed helicopters, possibly Mi-4s, and the wreckage of a MiG-17. (PC)

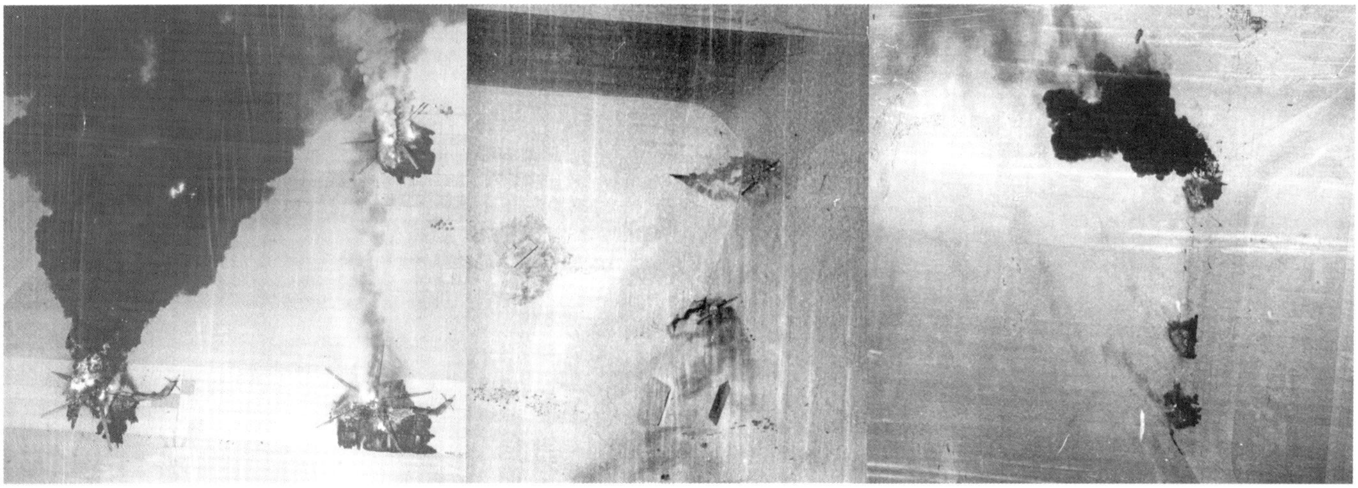

Three images taken over Bir Gafgafa (EAF EL Maliz), left to right: four Mi-6 helicopters destroyed on the northwest apron, two MiG-21s destroyed in front of a pen along the taxiway, three burnt out fighters – most likely MiG-21s – and a burning MiG-15 or MiG-17 at the southeast apron. (PC)

superiority, and Squadron 110 again demonstrated the value of a long range, heavy attack aircraft.

FOCUS Second Wave Launch

All Vautours were at Ramat David for 15 minutes from 09:47 when VISTULA returned from Bani Suef until 10:02, when the first Squadron 110 formation, tasked to fly a FOCUS second wave mission, departed Ramat David. The second wave also targeted Egyptian air bases; the objective was to exploit the success of the first wave and to consolidate IAF air superiority over Egyptian air power.

Originally, FOCUS Plan A called for every formation returning from a first wave mission to turn around its aircraft (service, refuel and rearm) and take-off for a second wave mission against the same target it had attacked during the first wave. In real-time, AIR4 analyzed the debrief reports – and later the BDA images – and this data was streamed to AIR3, which produced a new set of mission orders for FOCUS second wave missions.

Take-Off 10:02 – KISHON to El Minaya then Inchas

El Minaya air base was not attacked during the first wave. Two Mirage formations and a Vautour formation were tasked to attack El Minaya during the second wave; the former formations took-off at 09:48 and 09:51. KISHON followed: Isaac Golan with Alex Meltzer in 62, Absalom Friedman in 05, Isaac Nir in the repaired Vautour 12, and Simon Zimon in 26. For all four pilots this was their first Six Day War mission; Meltzer returned from Bani Suef at 09:11.

Postwar, popular press reports stated that IAF turnaround time was only seven minutes. This was not actually the case, as turnaround was fast, but not that fast. The average turnaround for the KISHON aircraft was 53 minutes: Vautour 62 returned from Bani Suef at 09:11, 05 returned from Abu Sueir at 09:01, and 26 from Bani Suef at 09:14.

Other FOCUS features often emphasized were flexibility and improvisation. When the KISHON Vautours were cruising at 40,000 feet some 50 miles from El Minaya, IAF ground control ordered a change of target. The first Mirage formation to attack El Minaya destroyed all targets – there were only Il-14 airlifters at this base – and the second Mirage formation was unable to track down any "live" targets; there was no point for a third formation to visit El Minaya. so KISHON was therefore ordered to attack Inchas, a MiG-21 fighter base in the Nile Delta that had already been attacked by five formations. Melt-

Inchas prior to arrival of KISHON, as photographed by a post first wave Squadron 119 Mirage BDA mission. (PC)

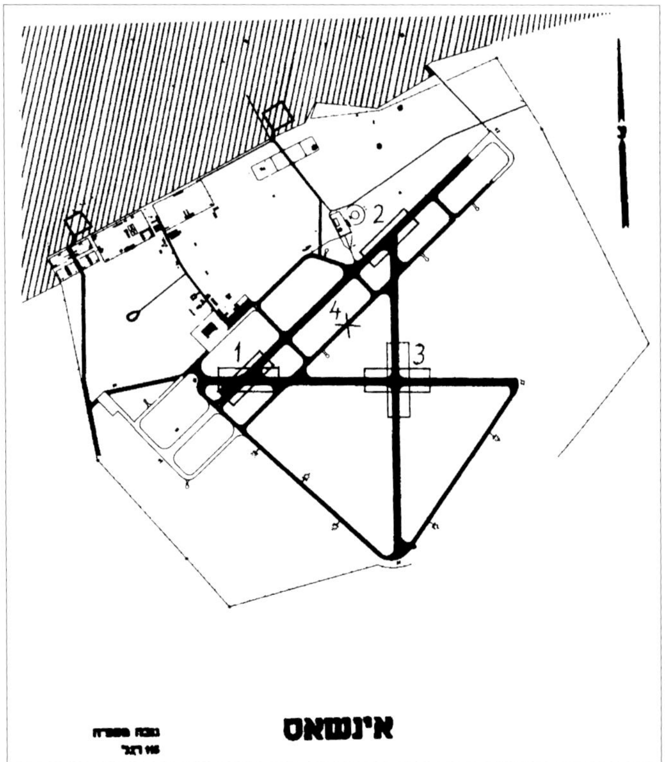

AIR4 estimated that 32 MiG-21s were at Inchas on the morning of 5 June 1967, but many of these had already been destroyed by the time of KISHON's attack. (IDFA)

zer planned a low altitude approach route to the SA-2 defended Inchas and Golan briefed the pilots en route; by then only three Vautours were heading to Inchas. KISHON 4 reported a fuel problem so Golan ordered Zimon to return to base all alone, without an escort, such was the level of air superiority achieved and the tenacity of purpose. On his way back to Ramat David KISHON 4 bombed and strafed targets of opportunity in Sinai prior to landing at 11:37.

The three KISHON Vautours popped up to bomb right into a hellish anti-aircraft fire that damaged Vautour 12; Nir pressed on regardless. Radio chatter heard during the second bombing run revealed that a Squadron 105 Super Mystere formation was "head on", flying their own bombing run opposite to the Vautours. The two leaders then coordinated the attack; the agreed solution was for the Super Mysteres to orbit until the Vautours finished their second bombing run. Isaac Golan recalled:

> Anti-aircraft fire over Inchas was really intense, a lot of fire. The runways were visibly holed but it was very difficult to spot targets because of the smoke.

KISHON's bombs were gone, but the Vautours were still armed with four fully loaded 30 mm cannon, so Golan decided to fly along the Cairo to Ismailia highway in search of the El Manaif SA-2 battery; Golan and Meltzer were well acquainted with this specific battery from their prewar reconnaissance missions. They located the SA-2 battery and strafed the array of antennas. Encouraged by their success, they then looked for another SA-2 battery that was reportedly deployed in Sinai, east of the north sector of the Suez Canal, but this time they did not spot the typical pattern of the SA-2 battery and returned to Ramat David at 11:45 after an eventful long range mission that involved flexibility and improvisation.

Take-Off 10:23 – NILE to Cairo West

The only formation to revisit Cairo West during the second wave was Squadron 110's NILE flying Mission 8/110: Beni Zohar with Uri Talmor in 66, Zeev Tavor in 17, Jacob Gal in 18, and Giora Goren in 08. All three wingmen were flying their first Six Day War mission. NILE leader attributed accurate navigation to the presence of a navigator in the lead Vautour; they crossed the Sinai coastline only two kilometers beside the

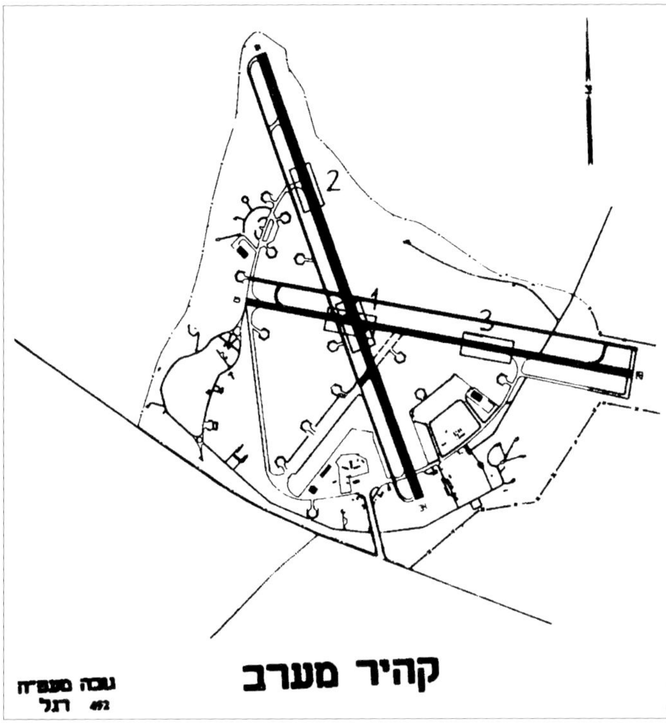

Cairo West was the center of the EAF heavy attack force. AIR4 evaluated that 15 Tu-16 bombers, 15 MiG-21 fighters, and 22 MiG-15 and/or MiG-17 aircraft were at Cairo West on the morning of 5 June 1967. Most of these were destroyed by the time NILE arrived. (IDFA)

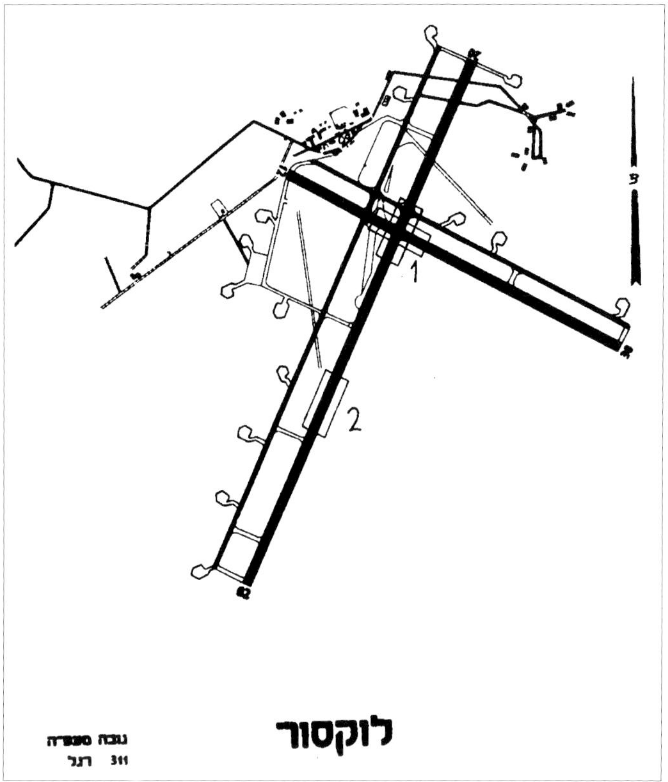

The Luxor layout included 14 pens fit for Tu-16 bombers. (IDFA)

planned crossing point and climbed exactly over the planned pop up point.

The Vautours flew their usual ritual of two bombing runs and three strafing passes in the face of light, and evidently ineffective, anti-aircraft fire. Engineering equipment utilized to repair battle damage to the runways was noted, and some of these vehicles may have been hit during the Vautours' bombardment.

Most targets at Cairo West were damaged or destroyed; some visibly intact targets were evidently damaged upon closer examination. NILE 1 selected a seemingly operational Tu-16 during the first strafing pass, but when they approached the bomber it became crystal clear that it was already full of holes. Overall, NILE claimed destruction of three MiG-21s; all other aircraft that they strafed were already destroyed or damaged. The NILE Vautours landed at Ramat David from 11:58 to 12:04.

Take-Off 11:25 – BANIAS to Luxor

Battle damage and serviceability issues started to impact the squadron, and only three BANIAS Vautours departed Ramat David to Luxor: Shlomo Keren in 21, Herzle Bodinger in 14, and Jacob Rafaeli with Shabtai Ben-Shoa in 33. Luxor was not attacked during the first wave. The AIR3 Mission Order to attack Luxor was issued in the wake of an AIR4 report of a section of Tu-16 bombers that took-off on a training mission that same morning, before the start of the Israeli attack, and was diverted to land at Luxor. Therefore, the intelligence overview presented to BANIAS during briefing asserted that the attackers should expect four aircraft at Luxor.

While planning this mission, it was noted that the Vautours would fly only 20 km north of Hurghada, an Egyptian fighter base that was not attacked. The worried Vautour air crews contacted AIR3 and were reassured that planning was at hand to raid Hurghada; four Squadron 119 Mirages departed Tel Nof to Hurghada at 11:35 and would initiate their attack just prior to the planned over flight of the Vautours.

Surprise awaited BANIAS, as Luxor was full of aircraft; the initial estimate of the BANIAS crews was that 20 An-12 airlifters and Tu-16 bombers were present at the Upper Egypt airfield. In line with preflight planning, the Vautour B joined the attackers during their third – and supposedly final – strafing pass for a BDA photographic pass, during which Shabtai Ben-Shoa counted 16 aircraft parked all over the airfield. At this point BANIAS 1 decided to press on with the strafing passes until all resident aircraft were destroyed. Herzle Bodinger recalled:

> We were briefed to attack Luxor because Egyptian bombers sought refuge in this distant air base that was supposedly beyond the limit of our radius of action. Many aircraft returned from the first wave with battle damage so only three Vautours could be spared to fly the mission to Luxor. One of the three aircraft was a Vautour B that could only lift two 500 kg bombs, rather than eight smaller bombs on Quadruple Bomb Racks, due to center of gravity issues. The Vautour B was also tasked to take BDA images so each time we strafed, the Vautour B flew a photographic pass.
>
> We flew an economical climb cruise along a straight line from Ramat David to Luxor. Hurghada was nearby that line. We were high so they could see us on their radar screens; all they had to do in order to frustrate our mission was to scramble their Hurghada based fighters to engage us. They need not even shot us down; a few acts of air combat maneuvers and we would not be able to make it to Luxor or if we would have been forced to

Luxor under BANIAS attack, a view to the northwest along runway 30.

A burning Tu-16 bomber at Luxor during the BANIAS attack. (PC)

> jettison our bombs then we would no longer be able to press ahead to Luxor. We knew that Squadron 119 was tasked to attack Hurghada and to intercept enemy aircraft, if there would be any over Hurghada, so from the moment we saw Hurghada we looked and waited. Suddenly we saw explosions and we were relieved, Hurghada was attacked and we could pass through.
>
> We descended and achieved absolute surprise. At first, anti-aircraft artillery did not open fire and when they did, after a long time, it was scattered and ineffective. The aircraft were dispersed all over the airfield so we flew our two bombing runs and started strafing. Each strafing pass we destroyed two aircraft with the Vautour B trailing behind us and photographing the progress of the destruction process.
>
> After three strafing passes there were still targets to be attacked so Shlomo Keren decided that we would continue until destruction of all targets. We flew six strafing passes and destroyed all targets. Keren's decision was correct but the additional passes consumed our fuel reserves. Just as we disengaged I was hit and soon afterwards I noticed that the left fuel gauge was dropping fast. There were two types of fuel gauges, total fuel and feeding fuel. Each engine had its own 400 liters feeding fuel tank and the troublesome gauge was of the left feeding tank. I reported to my leader and he replied 'you have a fuel leak, I can see.' Using the cross feed valve I tried to transfer fuel from the 'bad' side to the 'good' side but the right fuel gauge also began to drop; presumably the leak was from a critical point. I stopped crossover transfer and could only hope for the best. We were already cruising at 40,000 feet and had already covered about half the distance [back to base] when the left engine flamed out. I could not maintain altitude on one engine so I slowly descended until steady state was achieved at 25,000 feet. The two Vautours escorted me and I began to think that I might reach Eilat [airfield]. There was a discussion whether I should try to land at Eilat – a short runway at the time – and I decided to try if I would ever get there.
>
> The right engine flamed out as I flew over Eilat port and I glided to touch down exactly on the threshold of the runway. I was surprised that the landing run was so short and only then I realized that I never landed a net weight Vautour without bombs and fuel. I opened the canopy, walked to the wingtip and prepared to jump but by the time I got there, hundreds of citizens surrounded the aircraft. I then saw the firemen climbing on their ladder to the empty cockpit and I had to force my way to warn them that the ejection seat had not been secured. I am not a big guy so they looked at me and asked 'who are you?', 'I am the pilot' and I could see that they were puzzled how such a small pilot flew such a big aircraft!

Bodinger landed at Eilat at 13:23; BANIAS 1 and 3 continued to Ramat David, where they landed at 13:41 and 13:40, respectively. The excited Eilat citizens transported Bodinger to the City Mayor, where he waited until Squadron 110 reserve pilot Zeev Tavor flew a Vautour master maintainer from Ramat David to Eilat in the back seat of a Meteor trainer; Tavor departed Ramat David at 15:56 and returned to Ramat David at 16:30. The maintainer, Yossi Levin, made a temporary fix and Bodinger ferried – gear down – the Vautour back to Ramat David, thus ending a unique long range mission during which the two strafing Vautours claimed destruction of eight Tu-16 bombers and eight An-12 airlifters; all the aircraft that were present at Luxor!

FOCUS Second Wave Summary

The 31 IAF formations that took-off from 09:33 until 11:55 raided 14 Egyptian air bases (including six that were not attacked during the first wave), suffered no losses, and faced almost nonexistent opposition. The IAF claimed destruction of 94 enemy aircraft on the ground during the second wave, and Squadron 110 claimed 19 of these: three MiG-21s at Cairo West and eight Tu-16s and eight An-12s at Luxor.

The Vautours flew 17 sorties, including four air base attack missions, during the first wave, and this level of activity dropped to 11 sorties and three air base attack missions during the second wave. Average turnaround, the time frame from landing to take-off of a Vautour, was 70 minutes; ten times fold the popular figure.

More than four hours after H-Hour, the IAF had achieved air superiority against Egypt. Squadron 110 flew seven air base attack missions and claimed destruction of 42 enemy aircraft. Prewar estimates foresaw losses of 20%, but not a single Vautour was lost, yet.

FOCUS Third Wave

Jordanian artillery started to shell Ramat David at 11:40, and Syrian aircraft took-off to attack Israel from 11:50. The third FOCUS wave therefore targeted Jordanian and Syria air bases. Meanwhile, the WART Vautours were planned to fly another mission. Micha Gil was assigned to fly Vautour 65, but a malfunction delayed their departure until finally only Yoash Tsiddon with Menahem Eini flew this mission in Vautour 70 from 11:46 to 13:36. Interestingly, IAF reports indicated that Micha Gil did fly Vautour 65 from 13:31 to 14:14, but it is most likely this was a reporting error.

Take-Off 12:55 – JORDAN to Dmer

AIR4 evaluated that 40 MiG-21s were based at Dmer on the morning of 5 June 1967. Dmer was the main Syrian fighter base. Squadron 110 was tasked to launch the attack against Dmer; six formations of Mirages, Super Mysteres, and Mysteres would follow JORDAN and TIGRIS.

JORDAN was Levy Zur with Achikar Eyal in Vautour 31, Abraham Vilan in 07, Adam Zivoni in 09, and Uri Margalit in 67. The JORDAN Vautours flew north over Lebanon, turned east across the Lebanese-Syrian border, and headed south to arrive over Dmer from the north in order to accomplish tactical surprise. Levy Zur recalled:

> We flew to Dmer at low altitude but the Syrians expected us. When we pulled up to bomb, the local anti-aircraft artillery opened fire and a formation of four MiG-21s was taking-off, the first three MiG-21s already departed Dmer and the fourth was running along the runway.

In response to this development, JORDAN 1 ordered a change of plan; an immediate 360 degrees turn for the second bombing run. Levy Zur recalled:

> I looked back over my shoulder [upon completion of the second bombing run] and saw an air-to-air missile homing right onto my Vautour and a MiG-21 flying behind the missile. I broke to the left and both, the missile and the MiG-21, flashed by on

Vautour 07 – photographed landing at Ramat David during the winter of 1963/1964 – was lost over Dmer on 5 June 1967; the first Squadron 110 combat loss during the Six Day War. (PC)

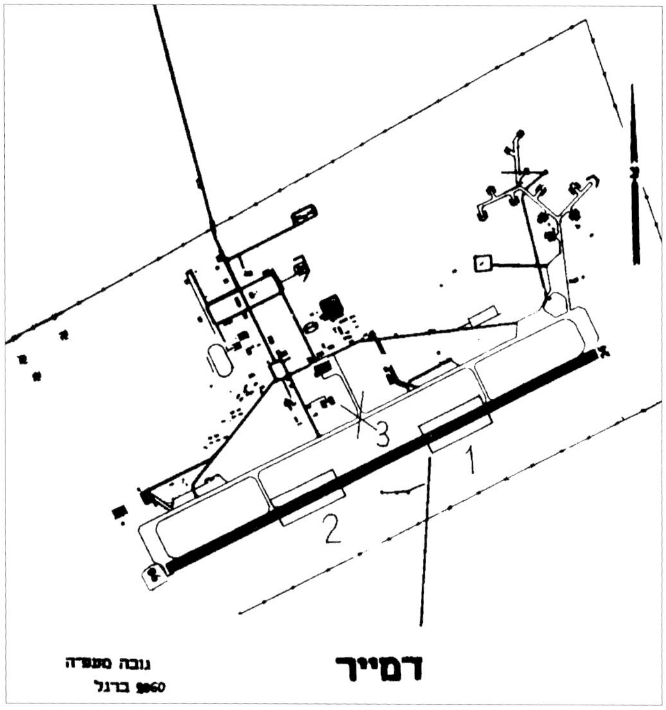

Dmer's single runway was 06/24, and AIR4 estimated that 40 MiG-21s were based there. (IDFA)

> our right. I continued to turn left, looking for my wingman, and I saw a Vautour spinning down but I did not see a parachute.

JORDAN 3 and 4 did not see a parachute either. Levy Zur recalled:

> I ordered to regroup in formation and we all headed west, flying at low altitude towards the Lebanon Valley en route back to base. Over the Lebanon Valley I turned left, heading south, and looked around. A MiG-17 was flying right next to me, to my left. Apparently, he also saw me at exactly the same moment and the two of us broke towards each other. We entered a 'scissors' maneuver right over the Lebanon Valley, flying low over the bottom of the valley with mountain ridges either side. This action was rather pointless from my point of view because the Vautour B was not armed with guns and the only weapon in hand was my personal pistol!
>
> Thus, during one of the 'scissors' acts, instead of reversing my turn, I turned on towards a canyon and exited the Lebanon Valley. The MiG-17 remained in the Lebanon Valley and I reported its presence there to the trailing pair.

Then the trailing pair, JORDAN 3 and 4 also engaged a MiG-17 over the Lebanon Valley. Adam Zivoni recalled:

> We attacked [Dmer] in the face of anti-aircraft fire and MiG-21s that were taking-off. The anti-aircraft fire hit [JORDAN] 2 and he was shot down; I only saw his aircraft crash but I did not see a parachute. Ignoring the MiGs, we flew two bombing runs and then two strafing passes.
>
> We disengaged and flew west towards Lebanon. I flew very low so I climbed a little and then I saw a MiG-17, apparently trying to attack me. I performed a 'high speed yo-yo' so instead of him being behind me. I was behind him. The MiG-17 pilot activated its afterburner, range between us opened up and the gun-sight of my aircraft was not serviceable – of all my Six Day War's sorties that was the one during which the gyro sight malfunctioned – but I opened fire. My aiming was only an estimate, the MiG-17 was too far away and my ammunition level was low due to the strafing passes. I missed and the MiG-17 disengaged.

JORDAN 2 pilot Abraham Vilan did eject from Vautour 07. He landed just outside the perimeter of Dmer air base and was taken a prisoner of war. Vilan did not see what actually hit his Vautour, nor did the other JORDAN air crews. The IAF attributed the loss of Vautour 07 to "anti-aircraft fire", but it is possible that a Syrian MiG-21 shot down the Vautour, perhaps even the same MiG-21 that attacked JORDAN 1, immediately after JORDAN 2 was shot down. David Nicolle and Tom Cooper relate – in their book *Arab MiG-19 and MiG-21 Units in Combat* (Osprey Publishing 2004) – the loss of JORDAN 2 to the claim made by SAF Squadron 5 MiG-21PF pilot Adeeb El Gar of a Vautour air-to-air kill; a kill that was claimed after an air combat during which the Syrian pilot launched both of his AA-2 Atoll air-to-air missiles.

Take-Off 13:31 – TIGRIS to Dmer

The second air base attack mission to Dmer took-off 36 minutes after the first, surely not in line with Focus B's vision of a second formation attacking the premiere Syrian MiG-21 air base only three minutes after the first. Originally, Focus B tasked six formations to raid Dmer from H-Hour until H+36. Actual third wave attacks against Dmer were not in line with the Focus B task assignment program but the result of real-time planning.

TIGRIS was manned by Simi Saar with Uri Talmor in Vautour 62, Michael Dvir in 08, Isaac Nir in 17, and Ran Zur in

Simi Saar with Uri Talmor flew Vautour 62 – photographed prior to war, jacked up during A-Level maintenance – on a FOCUS third wave mission to Dmer. (PC)

12; their time-on-target was 13:50. To achieve tactical surprise TIGRIS had to evade Syrian radar detection; they flew over geographically low areas that were shielded from preying radar beams of known Syrian radar stations. The final leg to Dmer was exposed to known Syrian radar coverage, but TIGRIS hoped that a late detection would still preserve tactical surprise. Simi Saar recalled:

> When I took-off, Levy Zur told me that he lost Abraham Vilan over Dmer; he spoke openly over the radio, 'Simi, pay attention, I lost one [aircraft], very intense anti-aircraft fire'. We were planned to fly two bombing runs and three strafing passes; we were programmed to fly such a pattern. We flew to Lebanon so that we would approach Dmer from the north; we flew at very low altitude. En route, I discussed the mission with my navigator Uri Talmor; he said 'Simi, I suggest to fly a single bombing run, to drop all bombs in a single bombing run and to evaluate during this bombing run how many, if any, strafing passes to fly'. I decided to get closer and to see what would happen. North of Dmer the terrain is hilly and when we pulled up a hell of anti-aircraft fire engulfed us. I immediately ordered 'change switches, single bombing run and [exit] north'. We flew that single bombing run from west to east, dropped all bombs, exited to the north and regrouped. All four of us were hit by the anti-aircraft fire, damage was light and we all returned safely but thanks to Levy Zur we did not lose another Vautour over Dmer.

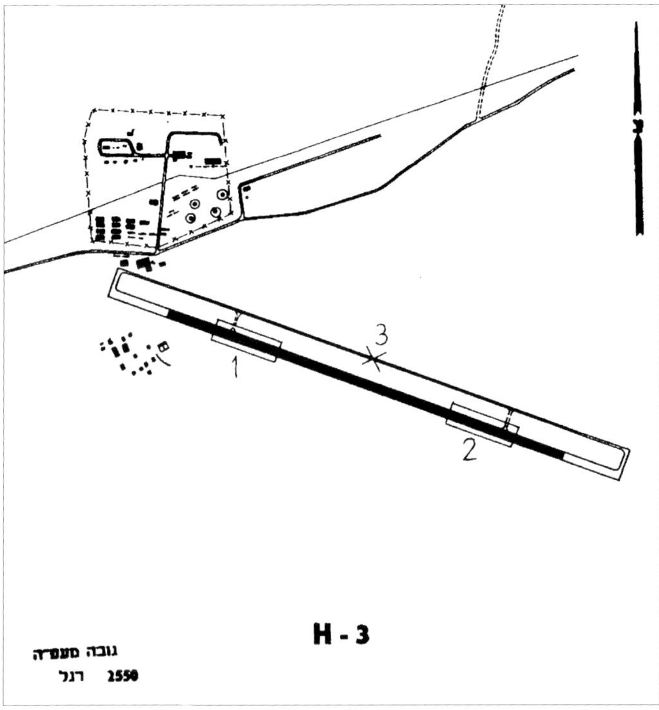

The only Squadron 110 Vautour to take-off from 13:31 to 16:50 were the VISTULA Vautours tasked to strike H-3 in Iraq, a forward air base not normally hosting resident squadrons. (IDFA)

From Third Wave to Fourth Wave

A total of 25 formations tasked to fly air base attack missions took-off from 12:15 to 14:10. During this time frame the IAF raided two Egyptian air bases, two Jordanian air bases, and four Syrian air bases. Air superiority over Egypt was maintained, the RJAF was wiped out, and the SAF lost some 50% of its combat aircraft. Seven IAF aircraft were lost, including a Vautour. Squadron 110 flew two air base attack missions during the third wave, down from four in the first wave and three in the second wave; the sorties count was 17, 11, and 9 for the first, second and third wave, respectively. This slow down was not unique to Squadron 110; the count of actual IAF air base attack missions declined from 45 in the first wave to 31 in the second wave, and to 24 in the third wave. Success of the air campaign, combat losses, serviceability issues, a gradual shift to support missions, and the transformation from pre-planned operations to real-time planning all contributed to the gradual slow down.

By 14:00 the IAF had ruled the clear summer Middle East skies from Cairo West, Bani Suef, and El Minaya in the west to Mafraq in the east, and from Luxor in the south to Dmer and Saikal in the north. FOCUS losses were lower than expected and FOCUS claims exceeded expectations. Therefore only ten formations tasked to fly air base attack missions took-off from 14:15 to 15:40 during the time frame of the fourth wave.

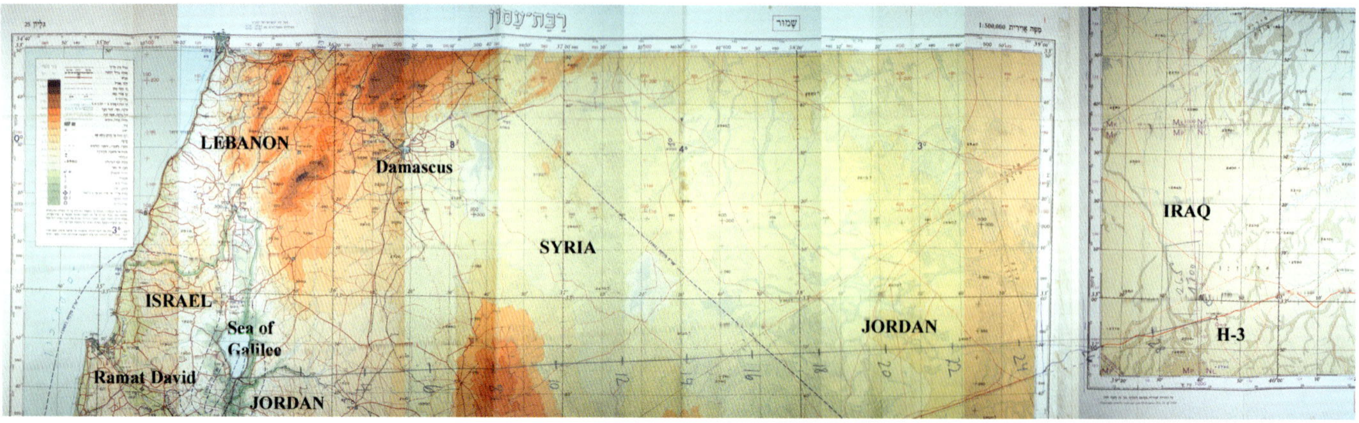

An authentic navigation map from Ramat David to H-3; a map that Squadron 117 Mirage pilot Judah Koren used on 6 June 1967. (PC)

The VISTULA H-3 pilots – Ran Goren, Gideon Magen, and Jacob Tal – in front of a Vautour – possibly Vautour 05 - loaded with bombs and adorned with mission markings on the vertical stabilizer just below the squadron badge. (PC)

The VISTULA pilots on the mission to H-3 as photographed after the war, from left to right: reserve pilot Ran Goren, reserve pilot Gideon Magen, and emergency posting pilot Jacob Tal. Magen and Goren returned to regular service after the war. (PC)

Take-Off 14:15 – VISTULA to H-3

The first fourth wave mission was thus flown by four Vautours that departed Ramat David in line with Mission Order 15/110 on an air base attack mission to H-3, Iraq. The expansion of the Six Day War's cycle of violence against a "second circle" state* has been one of the war's puzzles. Why did the IAF launch an attack against H-3, Iraq? Plan FOCUS clearly indicated that an Israeli attack against Iraqi air bases was wholly dependent upon Iraqi acts of war or aggression.

Did Iraq attack Israel prior to 14:15 on 5 June 1967? Official IAF research often states that the order to raid H-3 was issued after Iraqi Hunters raided Ramat David and/or after Iraqi Tu-16s bombed Israel. Yet the Tu-16 bombing mission was only flown on the morning of 6 June 1967, and was most likely an Iraqi reaction to the Israeli attack on H-3 the previous day. The alleged Hunters raid on Ramat David took place at 14:30 on 5 June 1967, 15 minutes after the Vautours departed the same air base. Moreover, proof that the marauding Hunters were indeed Iraqi had not surfaced yet. It is possible that the Israeli conclusion was based on circumstantial evidence; by then all Jordanian Hunters were presumably destroyed, so the only operational Hunters in the Middle East were by default Iraqi. A postwar claim that the IAF viewed H-3 as a Jordanian air base and for this reason it was attacked can be dismissed, as there is no indication of such an approach in Plan FOCUS.

Yet signs of potential Iraqi aggression were clearly in evidence. A sizeable Iraqi expeditionary force was already well prepared to reinforce Jordan in case of hostilities. More precisely, at 12:50 Israel intercepted an Egyptian message to Iraq that was essentially a plea for help, a request to launch Iraqi air operations against Israel. This interception probably tipped the scales. Israel had already launched a successful preemptive strike against Egypt and had responded to Jordanian and Syrian aggression. If the Iraqis were planning to attack Israel, the most logical path in wartime would be to attack first. The bottom line was crystal clear: Egypt asked Iraq to attack Israel, and the Israelis had the available force to strike first.

VISTULA manning was Gideon Magen with Alex Meltzer in Vautour 66 and Daniel Ilan in 18, and Ran Goren in 26 and Jacob Tal in 05. They had their target changed several times be-

* The Israeli "circles" theory emerged a long time after the Six Day War: the first circle is Arab states sharing a border with Israel (Egypt, Jordan, Lebanon, and Syria); the second circle is Arab states with a single buffer state in between, such as Iraq, in which case the buffer state is Jordan; and the third circle is hostile states with two buffer states in between, like Iran, in which case the buffer states are Jordan and Iraq.

fore they were tasked to fly an air base attack mission to H-3, with VISTULA 1 also tasked to photograph the Iraqi air base. There was no precise intelligence information concerning H-3: no details about based squadrons, no deployment layout of anti-aircraft artillery emplacement, nothing but coordinates of the Iraqi air base, the headings of the single runway, and inaccurate 1:1,000,000 scale maps. Gideon Magen recalled:

> The Fighting Executive asked me 'do you know where H-3 is?' and I retorted 'no, maybe in Jordan?' and he said 'no, it is in Syria'... Our only intelligence information was that H-3 was a NATO patterned airfield with one runway and a parallel taxiway, a 1:500,000 scale map and coordinates.

A low mission profile – in and out, all the way – was planned because the SAF was not completely wiped out. Due to the poor quality of the maps the route was in a straight line from the southernmost point of the Sea of Galilee, heading almost exactly east for some 28 minutes until rendezvous with a desert road leading to H-3 after some two more minutes of flying. Immediately after take-off Ilan reported a landing gear malfunction; the front wheel doors did not close. Magen ordered his wingman to return to base and VISTULA pressed ahead as a 3-ship. Gideon Magen recounted:

> We rendezvoused with the road and pulled up on time. We did not identify the target when Tal reported 'bogey, six o'clock'. A break to the right and I saw them, two MiG-21s right behind Goren and Tal. I ordered to keep looking for the target, followed the higher MiG and opened fire from a range of 800 yards; short burst, no hits. The lower MiG was in front of Tal; Tal opened fire, also from long range and also without apparent results, but he was in a 'sandwich' with the other MiG behind him. I tried to warn him but my external communication system ceased to function as a result of the guns burst; only the intercom still functioned.

Goren then warned Tal and Tal broke away. The inconclusive engagement continued for four or five minutes while the Vautours were striving to fly east, towards H-3, and Meltzer was scanning the scenery, looking out for H-3. When they finally spotted the Iraqi air base they were unable to report to their wingmen. Magen pulled up to bomb; he planned one bombing run from west to east – because he was flying a Vautour N with less fuel – and hoped that Goren and Tal would follow. While pulling up Magen saw two Hunters and a MiG-21 taxiing to take-off so he changed his aiming and dropped all eight bombs on the nearest quarter runway point, thus preventing the take-off of these aircraft and probably saving the day, as these Hunters were unable to take-off and intercept the Vautours. As VISTULA 1 pulled out of this dive bombing, Magen noted on their left a line of four or five aircraft, midway along the taxiway. Magen kept flying east to provide the pair, hopefully right behind him, enough space and time to bomb, turned 180 degrees and returned for a first strafing pass, aiming at the line that he had observed during bombing. Ran Goren recounted:

> Communication with Magen ceased after he opened fire. Because of the intensive evasive maneuvers the pair I led has been separated, so each aircraft flew independently. Yet, I had communication with Tal, but not with Magen. Suddenly we saw H-3, an airfield full of aircraft: silver MiGs, green Hunters and a large aircraft at the western end of the taxiway. I turned right to bomb from east to west, as planned, but because of the MiG threat [we were engaging in air combat], I decided to cut the long climb for the bombing dive position and to strafe first. I turned left and dived northwards, perpendicular to the runway towards the large aircraft that looked like a Tu-16; a burst of guns fire and that aircraft caught fire.

The large aircraft was not a Tu-16, but a transport aircraft. While Goren improvised that roughly south-north strafing pass, Tal improvised a bombing run against the single runway, generally flying west-east, and Magen returned to strafe, generally flying east-west and aiming at the line of aircraft along the taxiway. Tal's four bombs hit the runway, and Magen set on fire all aircraft at that line and also managed to strafe two aircraft parked in the western line; one of these caught fire before Magen squeezed his trigger, and it was the large aircraft that Goren strafed. A shadow of an aircraft, Goren's Vautour 26, flashed by underneath VISTULA 1's Vautour 66. Ran Goren recounted:

> Exactly at that moment Tal ordered me to break. He bombed from west to east and saw that a MiG launched an air-to-air missile. I broke, the missile missed and the MiG flashed on my right and zoomed upwards. I turned left and planned to bomb perpendicular to the runway but then I saw the two MiGs flying higher and Tal reported that none was following me so I changed my plan to bomb along the runway's axis from east to west. I saw two Hunters preparing for take off on the western end of the runway so I aimed to a point right in front of them to block their take off.

Goren also dropped all eight bombs in a single salvo. Magen flew a second strafing pass, aiming at two parked Hunters in the face of ineffective anti-aircraft fire – the two MiG-21s may have waited for the Vautours' attack to end in order to resume the engagement. Having exhausted the gun ammunition of Vautour 66, Magen noted that his fuel state was low and decided to disengage. Ran Goren recounted:

> I turned left 270 degrees and strafed the two Hunters that were still holding in take-off position. I flew a strafing pass from south to north and set them both on fire. Then I turned left and headed back home. Another Vautour also flew west. I assumed that this was Tal's aircraft and I ordered him to regroup with me; low altitude, 450 knots. Then I saw the third Vautour flying north of us; a brief discussion over the radio and I realized that I was flying alongside Magen – unable to communicate – with Tal north of us. Tal regrouped with us and then I notice a MiG-21, flying a kilometer or so right behind Magen. I ordered Magen to break but he did nothing, he could not hear me. I turned towards him and then I saw a trail of fire in pursuit of Magen, the missile hit the ground some 50 meters behind the Vautour; it was a miracle.

VISTULA flew west with Magen in the middle, Tal to his right, and Goren on his left side. At this point two MiG-21s closed behind VISTULA 1 and VISTULA 3. Goren saw the MiG-21 behind Magen, but did not see the MiG-21 behind his

Vautour. Magen saw the MiG-21 behind Goren, but did not see the MiG-21 that followed Vautour 66. They were unable to warn each other because of the communication failure of VISTULA 1. Gideon Magen recounted:

> Shortly after we regrouped, we saw a MiG-21 approaching Goren. Initially the MiG flew some three kilometers behind Goren. We were unable to warn him so I turned towards him and then I saw a missile exploding on the ground behind his Vautour. [The MiG-21s disengaged] and I turned back and headed west again. Tal was no longer where we expected him to be and we saw an explosion. We feared the worst but then we saw him trailing behind us. Apparently he dropped excessive bombs and this was the explosion that we saw.

Despite the interception and despite the improvised attack, the mission to H-3 was judged successful. The three Vautours evaded MiG-21s and still managed to attack. The leader reported a total claim of nine to 12 destroyed aircraft (VISTULA 1 alone strafed a line and claimed to have set on fire four or five fighters in a single pass), while the sub-leader's reported total claim was six MiG-21s destroyed, five Hunters destroyed, and one "large transport aircraft" destroyed. Ran Goren summed up:

> Three Vautours flew to attack an air base about which very little was known; they engaged two MiGs en route to the target; they pressed ahead and launched an attack while fighting those MiGs; their attack was improvised but successful, the runway was hit, some 12 aircraft were destroyed; they again engaged MiG-21s and evaded air-to-air missiles; they returned home safely; NOT BAD AT ALL.

The turning point of this mission was Gideon Magen's order not to jettison bombs, and to press ahead regardless of the MiG-21s. Ran Goren concluded:

> Gideon Magen displayed remarkable tenacity of purpose. Sometimes a single sentence can change the whole perspective. His order not to jettison bombs and tanks made it clear that we must press ahead regardless of risks.

FOCUS Fifth Wave

The attack against H-3 was the only mission that Squadron 110 flew during the fourth wave. The fifth and final wave included 18 formations that departed IAF air bases from 16:36 to 18:00, and were tasked to fly air base attack missions. Prior to the start of the fifth wave, Meteor 23 departed Ramat David at 15:56; the pilot was Zeev Tavor, and the passenger in the back seat was master maintainer Yossi Levin. Tavor flew Levin to Eilat and returned to Ramat David at 16:30. At Eilat, Levin patched up Vautour 14 and prepared the battle damaged aircraft for a flight back home to Ramat David.

Take-Off 16:50 – KISHON to Ras Banas

A possible sanctuary for residual Egyptian air power was Ras Banas, the air base that Squadron 110 photographed back in March 1963. It may have been expected that a Vautour N with a navigator in the back seat would fly lead, but the prime consideration was fuel, and the A had more fuel than the N;

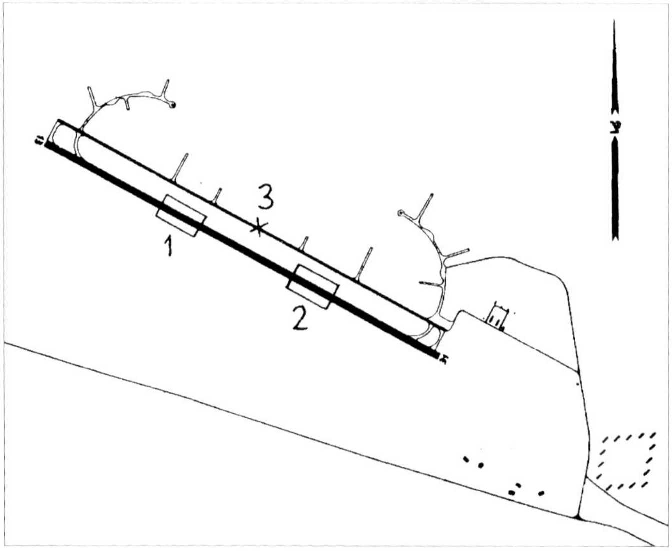

Ras Banas air base had a single runway (13/31), a parallel taxiway, dispersal positions, and barracks, but no aircraft were present at the time of the KISHON attack. (IDFA)

Zeev Tavor flew Yossi Levin from Ramat David to Eilat in Meteor 23; this photo of Meteor 23 was taken in 1964. (PC)

Pilots flying Vautour 18 had to abort Day 1 missions to H-3 and Cairo West due to technical malfunctions; this photo of Vautour 18 was taken from a HAS cell some time prior to war. (PC)

as a consequence four Vautour A aircraft flew the mission to Ras Banas: Isaac Golan in 21, Absalom Friedman in 09, Ovadia Nachman in 03, and Simon Zimon in 17. Isaac Golan recalled:

> The target was on the limit of our high altitude radius of action. We flew a long time, more than an hour in each direction, the most economical flight possible, climbing higher as fuel was consumed and weight went down. We arrived over Ras Banas and nothing was there, nothing at all, not a single aircraft to strafe. We bombed the runway and returned to base [at 19:20 after 2:30 flight hours].

Take-Off 17:15 – BANIAS to Cairo West
A fifth wave objective was to disrupt nighttime battle damage repair through attack with delay fuse bombs. BANIAS was Shlomo Keren with Elhanan Raz in Vautour 67, Ran Zur in 18, Beni Zohar with Shabtai Ben-Shoa in 62, and Uri Margalit in 26; the only Six Day War Squadron 110 air base attack mission with two navigators in its manning.

Vautour 18 again aborted an air base attack mission due to a malfunction, and Ran Zur returned to Ramat David at 18:15. En route to Cairo West the three Vautours strafed a radar station near Port Said airfield. A single bombing run was to be flown in order to minimize friction with the defending anti-aircraft artillery. The latter was indeed ready and damaged BANIAS 4 during the bombing run, but Uri Margalit pressed ahead regardless of damage.

The BANIAS secondary target was a SA-2 battery, but this was not tracked down and the Vautours strafed a P-30 radar station. On their way back they again strafed the radar station near Port Said and landed at Ramat David from 19:00 to 19:08.

Take-Off 17:31 – NIGER to Sinai
NIGER – Micha Gil with Menahem Eini in Vautour 31, Michael Dvir in 05, and Jacob Tal in 65 – flew the lights out mission to Bir Tamada. The Vautours were briefed to fly two bombing runs over the Bir Tamada runway and then split. The leader was

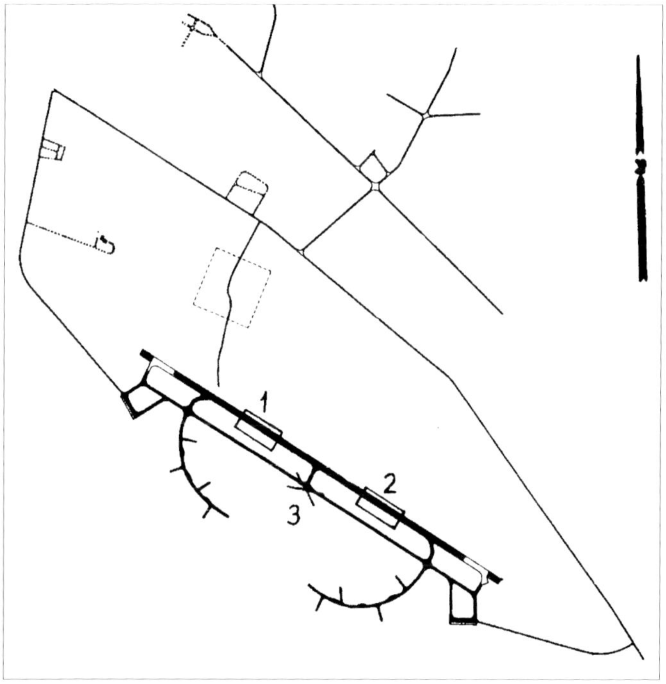

Prior to war, an EAF MiG-17 squadron deployed forward to Bir Tamada, AIR4 evaluated that 14 MiG-17s were at Bir Tamada on the morning of 5 June 1967; none remained there by the time NIGER attacked. (IDFA)

tasked to photograph, and the pair was tasked to fly battlefield air interdiction, strafing vehicles along roads in the Jabel Libni sector.

NIGER 1 dropped all eight 70 kg bombs during the first bombing run, while NIGER 2 and 3 dropped four bombs each. Due to crosswind they all missed. The two wingmen corrected aiming during their second bombing run and placed their remaining bombs right on the runway. No aircraft were observed at Bir Tamada, but anti-aircraft fire was active throughout the NIGER attack. As planned the Vautours then split. Micha Gil recalled:

NIGER 1 photographed this Il-14 in the Jabel Libni sector; a Squadron 101 Mirage that departed Hatzor at 08:08 that morning intercepted this Il-14, gunning the right engine of the transport aircraft and forcing its pilot to land in the desert. (PC)

> We were planned to bomb Bir Tamada and then photograph roads from Bir Tamada to Bir Gafgafa, Jabel Libni and El Arish. The runway at Bir Tamada was already holed; we missed it due to the strong crosswind and the fact that the sight of the B was less advanced than the sight of the A. We then flew our photography mission from low altitude, 200 feet. No one opened fire at us except, perhaps, once. Nearby Jabel Libni I saw an aircraft that landed in the desert. We took photos of this aircraft and called for the two attack aircraft. They arrived but had already expended all of their ammunition so they could not strafe this aircraft.

The strafing Vautours claimed the destruction of circa ten vehicles, and the NIGER trio landed at Ramat David from 18:43 to 18:47.

End of Day 1

Seven Vautours were on their way back home from fifth wave missions when the sun set at 18:40. Squadron 110 launched 16 missions from 07:17 to 17:31; those 16 missions involved 52 Vautour sorties, including two mission abortions: Vautour 18, which had to abort the missions to H-3 and to Cairo West. The 16 missions included 13 air base attack missions (48 sorties), two electronic warfare missions (three sorties), and a single reconnaissance mission (one sortie). The final two air base attack missions were actually composite operations: the mission to Cairo West also included suppression of enemy air defense, while the mission to Bir Tamada also included reconnaissance and support.

Squadron 110 raided eight EAF air bases (Bani Suef, Abu Sueir, Inchas, Cairo West, Luxor, Ras Banas, and Bir Tamada), a Syrian air base (Dmer), and an Iraqi air base (H-3). The Vautours claimed destruction of 52 enemy aircraft: 24 Tu-16 bombers,

Herzle Bodinger returned Vautour 14 from Eilat to Ramat David after Yossi Levin prepared the aircraft for the ferry flight; this photo of Vautour 14 taking-off from Ramat David was taken in 1964. (PC)

five Il-28 bombers, ten MiG-21 fighters, one MiG-17 fighter, three Hunters, eight An-12 airlifters, and one unidentified large transport aircraft.

The Squadron 110 contribution to IAF victory was therefore significant. The IAF launched 132 formations to attack 26 enemy air bases; a total of circa 474 air base attack sorties were flown; and destruction on the ground of almost 350 enemy aircraft was claimed for a total IAF loss of 24 aircraft. The relevant data for Squadron 110 was 13 formations that attacked eight enemy air bases in 48 sorties (including two abortions) and claimed destruction of 52 enemy aircraft on the ground for the loss of a single Vautour.

From Day 1 to Day 2

The success of FOCUS enabled the IAF to switch from an air superiority offensive to air support operations. As the heavy attack element of the IAF, Squadron 110 started to fly air support missions after all other attack squadrons initiated the shift from air superiority to air support. The first Squadron 110 formation to fly the air support mission was VISTULA, which departed Ramat David as a pair – Gideon Magen in Vautour 66 and Jacob Gal in 08 – at 18:45 to fly Mission Order 21/110. The VISTULA Vautours were loaded with four napalm bombs each, and were tasked to attack an Egyptian armored column in Sinai; this was presumably the elite Shazly Group. A standard nighttime attack formation was a 3-ship: an illumination leader loaded with flares and an attack pair. VISTULA took-off as a daytime attack pair, but actually attacked at nighttime. To the surprise of the VISTULA pilots, the Egyptian column was on the move with their lights on. They started their attack and flew several bombing runs, during which the lights were gradually switched off and anti-aircraft fire was opened. The pair claimed destruction of four vehicles, but Vautour 66 was hit by anti-aircraft fire, so the attack was called off and the pair returned to Ramat David at 19:45.

Next to take-off was TIGRIS, which departed Ramat David at 20:03 as a standard 3-ship nighttime attack formation: Simi Saar with Achikar Eyal in Vautour 31, Jacob Rafaeli in 05, and Adam Zivoni in 67. The bombing pair was loaded with eight 70 kg bombs each. TIGRIS was tasked to bomb Jordanian artillery near Qalqilyah that had shelled Israeli targets under the cover of

Vautour 21 flies over IDF Centurion tanks in the Negev Desert during the build up period that preceded the Six Day War; Squadron 110 initiated air support operations from the Day 1 VISTULA take-off at 18:45. Vautour 21 flew three Day 1 air superiority missions, then a nighttime air support mission from 21:15 and a Day 2 daytime air support mission from 04:45, but was then grounded until the late afternoon hours of 9 June 1967. (PC)

darkness. The Vautours saw nothing over Qalqilyah so they flew to Jerusalem, from where flashes of fire were clearly visible. However, TIGRIS was not authorized to attack in the Jerusalem sector so they returned to Qalqilyah, bombed the Jordanian artillery there, and returned to Ramat David at 20:56.

The second 3-ship to take-off was JORDAN: Levy Zur with Elhanan Raz in 33, Ran Goren in 21, and Giora Goren in 17; again, the bombers were armed with eight 70 kg bombs each. JORDAN took-off at 21:15 and landed at 22:15. Ran Goren recounted:

> I was ordered to go to sleep so that I would be on nighttime readiness but although I did not sleep the previous night, and flew two tough missions, I could not get myself to sleep.
>
> We were scrambled to attack Jordanian artillery nearby Qalqilyah, there were rumors that these guns shelled Tel Aviv. We set in the cockpits for an hour; waiting for Simi, Rafaeli and Zivoni to finish their attack. At last, we were authorized to start our engines. The whole base was blacked out and taxiing was very difficult. Black out impact was even worse in the air; Israel was completely blacked out and it was a moonless night. Orientation was very difficult, almost impossible. Flying in formation was also a challenge. All along the border with Jordan we saw flashes of fire; a particularly heavy exchange of fire was nearby Jenin where an armored battle was raging.
>
> I set course from Tel Aviv to Qalqilyah but saw nothing over there. Flashes were indeed noted there but it was impossible to pinpoint the exact location of the artillery emplacements. After orbiting over Qalqilyah for a long time, a flare illuminated and we spotted a flash from a hill to the east of Qalqilyah. Zur ordered to bomb that hill and I went in first, dropping all eight 70 kg bombs right on the center of that hill.
>
> Our bombing was not particularly effective; shelling continued. On the way back home we split to burn fuel; finding base in the darkness was also difficult.

As a lesson from these two missions, a lone Vautour armed with two 500 kg bombs departed Ramat David at 22:45. Call sign RIVER 292 was Daniel Ilan with Uri Talmor in Vautour 65. The new tactic was IDF illumination of the Jordanian artillery battery near Qalqilyah. It took some 30 minutes to refine the coordination of this joint operation. Finally, IDF illumination was right on spot in time for the Vautour to attack, and the two 500 kg bombs were dropped in this single attack; reportedly the Jordanian artillery was silenced in the wake of this bombing. RIVER 292 returned to Ramat David at 23:42.

The next air support mission again involved a 3-ship formation. VOLGA – Jacob Rafaeli with Achikar Eyal in 31, Adam Zivoni in 18, and Giora Goren in 09 – took-off at 01:16. They were tasked to attack Jordanian armor advancing along the road from Jericho to Jerusalem; no IDF units could illuminate this force for the Vautours. The Vautours spotted the dimmed lights of the Jordanian armored vehicles, illuminated the convoy, and attacked with 70 kg bombs; they landed at Ramat David from 02:10 to 02:11.

The 13th and last Vautour sortie on the night of 5 to 6 June 1967 departed Ramat David at 02:26: Ran Goren with Uri Talmor, call sign RIVER 286, in Vautour 62. Ran Goren recounted:

> It was already past midnight when I was woken up for another mission; illumination in support of our forces that were attacking the Om Kattef fortifications in Sinai. A single Vautour was tasked to fly that mission but I flew with navigator Uri Talmor and it was good to have someone to talk to while flying in such a dark night. We flew south at 10,000 feet and I was worried that we would not see Om Kattef but my worries evaporated as we got closer; the battlefield was clearly visible: tracers, explosions and fires. East of Om Kattef we saw three huge 'snakes', IDF convoys.
>
> We tried to contact the IDF forces but were unable to do so. We heard them but they did not hear us. We tried again

> and again to establish contact, without success. IAF ground control did not authorize our request to illuminate as planned without radio contact with the ground forces. This was really frustrating; a malfunction precluded us from supporting our forces. We departed Om Kattef after some 30 minutes and flew back north but asked ground control for an alternative target. We were vectored to Jerusalem and tasked to illuminate for a formation of Mysteres. Their target was a Jordanian convoy along the road from Jericho to Jerusalem.
>
> It was difficult to see Jerusalem except for flashes of battle. We illuminated east of Jerusalem but did not see a convoy. The next day I learnt that the objective of our action was to deter the convoy from moving until next morning when it was attacked and destroyed.

RIVER 286 landed at 03:54. Thus ended the first night of the war. During Day 2 (6 June 1967), the Vautours would mostly exploit the air superiority that was achieved the previous day, bombing Syrian forces in the Golan Heights and supporting IDF forces advancing into Sinai and the West Bank.

Take-Off 04:45 – VISTULA to the Golan Heights

In line with prewar plans, Israeli Navy commandos raided Egyptian and Syrian ports with Vautours on readiness to provide air support. Mission Order 3/110 tasked VISTULA – Gideon Magen in 18 and Simon Zimon in 21 – to support the Israeli Navy; the pair flew north, off the Syrian port of Baniyas, only to be notified that no air support was needed. Loaded with four napalm bombs, VISTULA asked for an alternative target and was ordered to contact a forward air controller that tasked the Vautours to attack a target northeast of the Sea of Galilee. They attacked as tasked and landed at 06:13 (Magen) and 06:00 (Zimon).

Take-Off 04:52 – JORDAN to the Golan Heights

The first JORDAN pair (Levy Zur with Menahem Eini in 31 and Uri Margalit in 03) waited for the second pair (Micha Gil in 14 and Jacob Tal in 12), which took-off at 05:00. The JORDAN mission was to attack Syrian forces along the road from the Upper Customs House to Quneitra. The Vautours approached at 15,000 feet, well above the 10,000 feet limit of Syrian anti-aircraft fire. Regardless of intense anti-aircraft fire, the Vautours dive bombed with four 500 kg bombs each and then strafed. Zur landed at 05:12, and the three other JORDAN Vautours landed at 05:34.

Take-Off 05:00 – RAPID to Gaza

RAPID – Ovadia Nachman with Elhanan Raz in 33 and Absalom Friedman in 08 – was also initially tasked to support the Israeli Navy, but took-off on a close air support mission. Loaded with four napalm bombs, the forward air controller directed RAPID to attack an anti-tank gun emplacement. They then flew three strafing passes; most likely only RAPID 2 actually strafed, and it was during this action that anti-aircraft fire damaged Vautour 08. Friedman was forced to land at Tel Nof at 06:12, while RAPID 1 returned to Ramat David at 06:13.

Take-Off 05:45 – WART

Yoash Tsiddon flew his third WART mission; Vautour 70 returned to Ramat David at 07:15.

Take-Off 06:31 – TIGRIS to H-3

Squadron 110 raided H-3 during Day 1, and the IDF evaluated that the Iraqis would counterattack on the morning of Day 2. In light of this evaluation, AIR3 planned a second strike against H-3. The Israeli evaluation was correct, and Iraq indeed launched an aerial counterattack. Reportedly four Tu-16 bombers were committed to this task. The Israeli narrative is that three bombers aborted the mission en route, so that only the commander of the Iraqi Tu-16 force pressed ahead. That Tu-16 bombed targets in midland Israel and then retreated back east via the Jezreel Valley, a flight path that would see the Iraqi bomber flying right over Ramat David air base.

A lesson from the Day 1 strike was assignment of an escort; a pair of Squadron 117 Mirages – the pilots were Judah Koren and Pessach Shraga – trailed TIGRIS: Simi Saar with Alex Meltzer in 67, Ran Zur in 05, Beni Zohar in 17, and Herzle Bodinger in 62. Each Vautour was loaded with eight 70 kg bombs. Simi Saar recalled:

> Excitement level was so high that I was unable to sleep during the night of 5 to 6 June, my second sleepless night. The following morning I was assigned to lead the mission to H-3. I was already strapped inside the cockpit when we heard a loud noise; a Tu-16 flying over Ramat David! The maintainers that strapped me in said 'look, a Boeing' and I retorted 'this is not a Boeing; it is a Tu-16!' The next moment I was all alone, everyone ran for cover. I activated the communication system, put on the helmet and heard the Ramat David control

Micha Gil flew two air support missions over the Syrian front in Vautour 14 during the morning of 6 June 1967; Vautour 14 was photographed flying over Ramat David with 1,300 liter drop tanks back in 1964. (IDFA)

A tail gun from the Iraqi Tu-16 that was shot down after flying over Ramat David on the morning of 6 June 1967 is on display at the IAF Museum. The IAF credited the Tu-16 kill to the combined action of a Squadron 117 Mirage that launched a Shafrir missile and Air Defense Battalion 881, which engaged the Iraqi bomber with anti-aircraft fire. (PC)

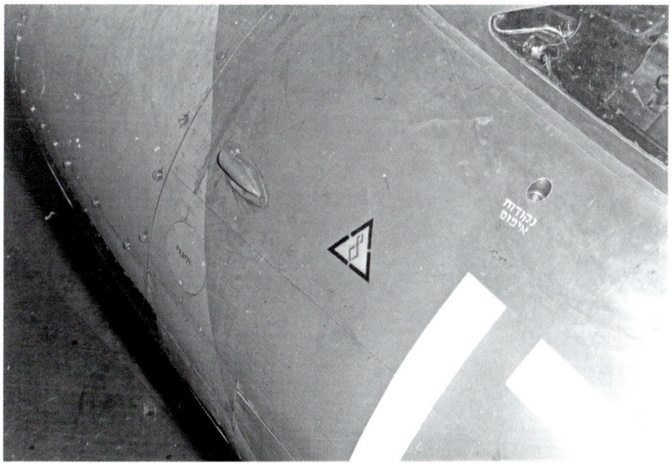

The kill marking painted on the left nose section of Vautour 17 after Beni Zohar was credited with an Iraqi Hunter kill over H-3 on 6 June 1967, the only IAF Vautour pilot and aircraft to be credited with an air-to-air kill. (PC)

Beni Zohar flew Vautour 17 to H-3 as TIGRIS 3; Vautour 17 was photographed landing at Ramat David some time before 1966. (PC)

> tower vectoring a pair of Mirages to intercept that Tu-16 and it was duly shot down.
>
> The Tu-16 did not bomb Ramat David; it only strafed [from its tail gun] but still caused a lot of fuss. Once this episode was over, we resumed our preparations and took-off. Our maps were not good, the terrain short of H-3 was featureless desert and the only navigation aids were compass and watch. Our plan was to head east towards the H road and from there along the road to H-3. The moment we pulled up, to initiate our bombing pattern, I saw a Hunter landing, already running on the runway, and another Hunter flying right in front of me, it was on final to land. I could have shot it down but our objective was to attack so I ignored that Hunter and pressed ahead with bombing.

The Hunter in front of TIGRIS 1 retracted its undercarriage, aborted its landing and flew away. Beni Zohar recalled:

> I flew the first bombing run, dropped half of the bombs and then I heard [TIGRIS] 2 warning [TIGRIS] 1 that a Hunter was right behind him. I looked around and I saw them turning left at a very low altitude, almost scratching the ground. I followed [TIGRIS] 2, a little bit curious; the situation did not seem to be desperate. Anyway I reported [to TIGRIS 1] 'I am following you' and this enabled [TIGRIS 2] to level its wings, climb away and fly a second bombing run. Thus I actually followed the Hunter. My intention was to tell [TIGRIS] 1 how to avoid being shot down; [TIGRIS] 1 asked me 'what is going on?' and I said 'he is still following you'. Then, it all took seconds, the Hunter broke away; maybe the pilot of the Hunter saw me or maybe someone warned him of my presence. I reported to [TIGRIS] 1 'he left you' and [TIGRIS] 1 replied 'good'. Meanwhile, because of that break, the Hunter flew right in front of me, really big, huge and I was closing fast, very fast. My sight was still in bombing mode, not in air-to-air mode, so I aimed in front of the Hunter and opened fire, a very short burst. At that moment, for a split second, I saw the Hunter like a model; I could see every single screw. At the time we both turned left; then he reversed his turn to the right and I admired its ability to maneuver in such a way. I pulled up and waited to see what he would do; I expected that he would turn after me and I planned to dive but I saw that he continued in a shallow turn so I followed him and then he crashed. There was no ejection, just a shallow turn; a kind of maneuver that was not an air combat any more and I did not have any trouble following him from above, as I was, still loaded with half of the bombs, and he simply descended and crashed.

Postwar Beni Zohar was credited with a kill; the only air-to-air kill credited to an IAF pilot flying a Vautour in combat. The escorting Mirages also engaged Hunters; the Mirage leader opened fire at a Hunter but did not see it crash. During debrief, a Vautour pilot reported that he saw a pilot ejecting from a Hunter that then crashed. Based on this information Judah Koren was credited with an air-to-air kill. As for the Vautours, they pressed ahead with the attack, flying a second bombing run and then strafing. Simi Saar recalled:

> A Hunter flew right behind me but I did jot jettison the fuel tanks or the bombs. We turned at very low altitude and I saw that he was closing on me. I also saw Zohar following that Hunter and then he shot down the Hunter and I returned for another bombing run. Next we flew three strafing passes and comfortably regrouped for the flight back home, it was then that MiG-21s intercepted us.

Separately, both leaders decided that it was time to go home and ordered a retreat. The first TIGRIS pair was already heading west when the Mirage leader saw a MiG-21 trailing the second pair that had just completed a strafing pass. Beni Zohar recalled:

> I exited the strafing pass and Simi ordered to disengage. Exactly at that moment I saw a MiG-21 diving at me in such a beautiful dive so I reported 'I can not [disengage], I am in scissors with a MiG-21!' I actually reported after the first scissors act; I broke, he broke, he pulled up high to dive at me again and then Koren, the Mirage pilot, said 'leave him, leave him, he is mine!' This was really funny because I was not the aggressor, I was merely defending myself. Anyway, Koren really took advantage of him [the MiG-21] pulling up and it was the only time that I saw an air-to-air kill. It was all very close; very, very close. I saw the guns flash when the Mirage opened fire and almost instantly the MiG-21 froze in the air, caught fire and spun down. Koren's action was so skillful, swift and simple.

The Mirages then zoomed upwards, returning to Ramat David at high altitude, while the Vautours disengaged at low altitude. Herzle Bodinger recalled:

> The mission to H-3 was considered so important that the wing commander attended our briefing. We then exited the squadron building and walked to the aircraft, my aircraft was nearby the squadron building. I did the usual 'walk around', signed the aircraft logbook and was about to climb to the cockpit when I saw a large aircraft approaching from south-east. It was the size of a Boeing 707 and it flew slowly, maybe a little more than 200 knots, at an altitude of 1,000 to 2,000 feet. This was not a Boeing but a Tu-16 with closed bomb-bay doors that did not see Ramat David; it flew over Ramat David but did not see it. Ramat David was camouflaged in such a way that we, the local pilots, could hardly see it from the air. The previous day, a pair of Hunters arrived to attack Ramat David; they were unable to see the air base so they attacked a nearby village. The runway was painted brown, all markings were removed and it looked like the surrounding fields. The base was remarkably camouflaged.
>
> I asked the maintainers 'do you recognize this aircraft?' They answered that it was a Boeing 707 so I told them it was a Tu-16. Exactly at this moment the siren was sounded all over the base, the maintainers ran for cover and I was standing there, all alone next to my aircraft. It is amazing how much happened at such a short time. Simultaneously, the anti-aircraft artillery emplacement beside the intersection of the runways, quite close to my aircraft, opened fire; the wing commander exited our squadron building, rushed to a nearby 0.5" machine guns – many of these weapons were dispersed in the base – and also opened fire. Then a pair of Mirages attacked the Tu-16 that within a short while started to smoke, turned right, initiated a shallow descent in the direction hence it came and crashed. As a matter of fact, the Tu-16 only saw Ramat David after the local defenses opened fire, at which point the rear gunner of the bomber returned fire, hitting buildings in the base but causing no casualties. The bomber did cause casualties when it crashed into an IDF base [to the east of Ramat David, killing 16 IDF service persons].
>
> The Tu-16 action was over and we took-off from runway 15, heading east towards H-3. We were planned to fly to H-3 at low altitude and to return from H-3 at low altitude; the Mirages had to return at high altitude [because fuel consumption is higher at lower altitude]. Our maps were old and we did not know where exactly H-3 was; we set course from the southern tip of the Sea of Galilee and flew east to a desert road that was to lead us to H-3. We indeed hit that road and right there was the Iraqi expeditionary force, a lot of Iraqi armor. We remembered this and pressed ahead to H-3.
>
> We flew along that road when suddenly we saw dust rising upwards, aircraft took-off from H-3 right in front of us. The plan was that the Mirages would engage enemy aircraft while we would attack from east to west so if anyone would get hit, he would continue flying west towards Israel. We pulled up and saw Hunters landing, the first was already running on the runway. Our leader bombed the runway; this Hunter ran into the craters and crashed. The second Hunter was on final so he aborted landing and followed our leader. The Hunter was a very dangerous opponent, armed with four 30 mm guns, robust and maneuverable. We warned him, he broke, the Hunter followed, [TIGRIS] 3 shot down that Hunter and I was right behind all of this action. I looked down, saw the shadow of my aircraft on the ground and did not see another shadow behind me, nothing threatened me and I decided to fly the second bombing run. As I pulled up I saw, right in front of me, a Hunter with flaps down and gear down, also preparing to land. I switched sight to air-to-air mode and input the Hunter's span, 36 feet, but I was flying at 450 knots and he flew at 180 knots or so. I overtook him so fast that I only managed to fire a very, very brief burst. Maybe the H-3 control tower warned that pilot because at that moment he rolled over and broke down towards the ground. I decided not to follow him and resumed my bombing run. I was the only one flying over there and the local anti-aircraft artillery opened fire at me. I pointed my nose at the anti-aircraft artillery emplacements, returned fire that silenced those guns but I was no longer aligned with the runway axis and my bombs exploded on the apron where several aircraft were parked.
>
> As I pulled out, I heard orders to disengage and saw Judah Koren shooting down a MiG-21. We regrouped, the Mirages parted and climbed to return at high altitude while we still had ammunition left so we spread in line abreast and strafed the Iraqi armored force that we saw on our way to H-3.

The TIGRIS Vautours returned to Ramat David at 08:10; they claimed the destruction of seven Hunters (six on the ground and one in the air), two MiG-21s (on the ground), and seven vehicles in addition to the damage that their bombs caused to the Iraqi infrastructure at H-3.

Take-Off 07:45 – JORDAN to the Golan Heights
Levy Zur with Elhanan Raz in 33, Michael Dvir in 08, and Jacob Tal in 26 were tasked to attack an artillery battery north of the Upper Customs house with four 500 kg bombs. In the first pass JORDAN 1 and 3 hit the target, but the bombs of JORDAN 2 did not drop. Zur ordered JORDAN 3 to return to base and orbited overhead while JORDAN 2 flew a second bombing attack; this time the bombs were released and the target was hit. They landed from 08:00 to 08:03.

Take-Off 08:05 – BANIAS to the Golan Heights
The Syrians launched a local offensive in the Dan sector, and BANIAS was tasked to attack Syrian tanks. Shlomo Keren with Shabtai Ben-Shoa in 31, Uri Margalit in 03, and Micha Gil in 14 attacked with four 500 kg bombs each and claimed three tanks set on fire. They returned to Ramat David at 08:35.

Take-Off 09:01 – VISTULA to the Golan Heights
Gideon Magen with Menahem Eini in 66 and Simon Zimon in 12 were tasked to attack artillery emplacements in the south sector of the Golan Heights with four 500 kg bombs. They bombed twice, and were then ordered to strafe fortifications in the central sector of the Golan Heights. Anti-aircraft fire damaged Vautour 66 during the second strafing pass. Eini was lightly injured and his map was destroyed; Magen landed 66 at 09:52 and Zimon followed at 09:53.

Take-Off 09:43 – RAPID to Sinai
Ovadia Nachman with Elhanan Raz in 33, Michael Dvir in 08, Jacob Gal in 26, and Jacob Tal in 62 were tasked to attack armor with eight 70 kg bombs each. They orbited on a holding track for quite a while until the previous formation finished its attack. As a safeguard from anti-aircraft fire, RAPID was ordered to bomb from medium altitude and was not authorized to strafe. The Vautours bombed as briefed and landed from 10:50 to 10:57.

Take-Off 12:33 – JORDAN to the Golan Heights
Levy Zur with Shabtai Ben-Shoa in 65, Adam Zivoni in 03, Beni Zohar in 09, and Zeev Tavor in 05 were tasked to attack mortars with four 500 kg bombs. En route to their target, Zur spotted a convoy and asked permission to attack but the answer was negative; another formation was already tasked to attack that convoy. Pinpointing the mortars was a challenge; the JORDAN Vautours orbited for quite some time over the place where the target was supposed to be until Zohar saw the mortars. No anti-aircraft fire was noted, so they bombed and strafed prior to return to base from 13:06 to 13:09.

Take-Off 13:23 – BANIAS to the Golan Heights
Shlomo Keren with Alex Meltzer in 62, Herzle Bodinger in 12, Jacob Rafaeli in 67, and Micha Gil in 18 were tasked to attack a convoy with eight 70 kg bombs. The Vautours bombed and strafed. It was during a strafing pass that anti-aircraft fire damaged Vautour 67. The ailerons did not function, handling was difficult, the wing was on fire, and the aircraft crawled along at 250 knots. Rafaeli headed west with Keren as escort. Keren recommended ejection due to the fire, but as Rafaeli prepared to

Vautour 08 flew as JORDAN 2 on a Day 2 mission that departed Ramat David at 07:45, this photo of Vautour 08 was taken in 1964. (PC)

A Vautour lands with 1,300 liter drop tanks and empty Quadruple Bomb Racks; the shift from longer range air superiority offensive to shorter range air support operations included a change of typical weapons load from eight 70 kg bombs to four 500 kg bombs. (PC)

Reserve pilot Jacob Rafaeli belly landed the battle damaged Vautour 67 on 6 June 1967; the Vautour was then transported to IAI and was repaired. Rafaeli himself returned the repaired and repainted Vautour 67 from IAI to 110 sometime before 1 October 1967. This photo was taken in the HAS immediately after this ferry flight from Lod to Ramat David. (PC)

eject he noted that the fire was less intense; presumably, the fire had consumed all of the hydraulic fluid. Levy Zur then suggested to Rafaeli to continue flying for as long as the aircraft would remain controllable. Vautour 67 limped all the way to Ramat David but the undercarriage did not detract, so the aircraft was belly landed on the runway at 14:03. The three other BANIAS Vautours landed from 14:08 to 14:11.

Take-Off 13:40 – KISHON to the Golan Heights

Isaac Golan in 14, Simon Zimon in 26, Gideon Magen with Achikar Eyal in 31, and Giora Goren in 08 were tasked to attack with four 500 kg bombs the same convoy that BANIAS had attacked. En route to that convoy the target was changed to an artillery battery. By then the whole Golan Heights was covered in smoke and pinpointing a target was a challenge, more so if the target was not learnt on the ground but assigned in the air. Nevertheless, Golan pinpointed the artillery battery and the Vautours bombed their target. BANIAS 3 returned to base and landed at 14:20, while the other three Vautours headed north, looking for targets to strafe. Eventually they joined a formation of Mysteres that strafed armored vehicles. It was during these strafing passes that anti-aircraft fire damaged Vautour 08. Goren flew west and ejected over Israeli territory, while Zimon landed at 14:33 and Golan returned at 15:00.

Take-Off 16:21 – TIGRIS to Sinai

Simi Saar with Achikar Eyal in 33, Herzle Bodinger in 65, Ran Goren in 03, and Zeev Tavor in 17 were tasked to attack along the road to Bir Tamada with four 500 kg bombs. Ran Goren recounted:

> I went to sleep at 04:00 and gunfire woke me up at 06:00; it was the Tu-16 that flew over Ramat David and was shot down. When this episode was over, I returned to sleep – waking up from time to time to listen to the news – until noon when I reported to the squadron.
>
> Our mission was to attack armor nearby Bir Tamada. We flew to Sinai at high altitude, sensing security that no enemy aircraft would be able to endanger us; air superiority is great! Many other aircraft were also in the air, every formation was assigned to attack a segment of a road. Nearby Bir Tamada we noted a few fortifications and some scattered tanks heading west in retreat; the air base was neutralized but the local anti-aircraft artillery was still active and opened fire at us.

Vautour 08's final departure from Ramat David was at 13:40 on 6 June 1967; Giora Goren had to eject from Vautour 08 after Syrian anti-aircraft fire damaged the aircraft. (PC)

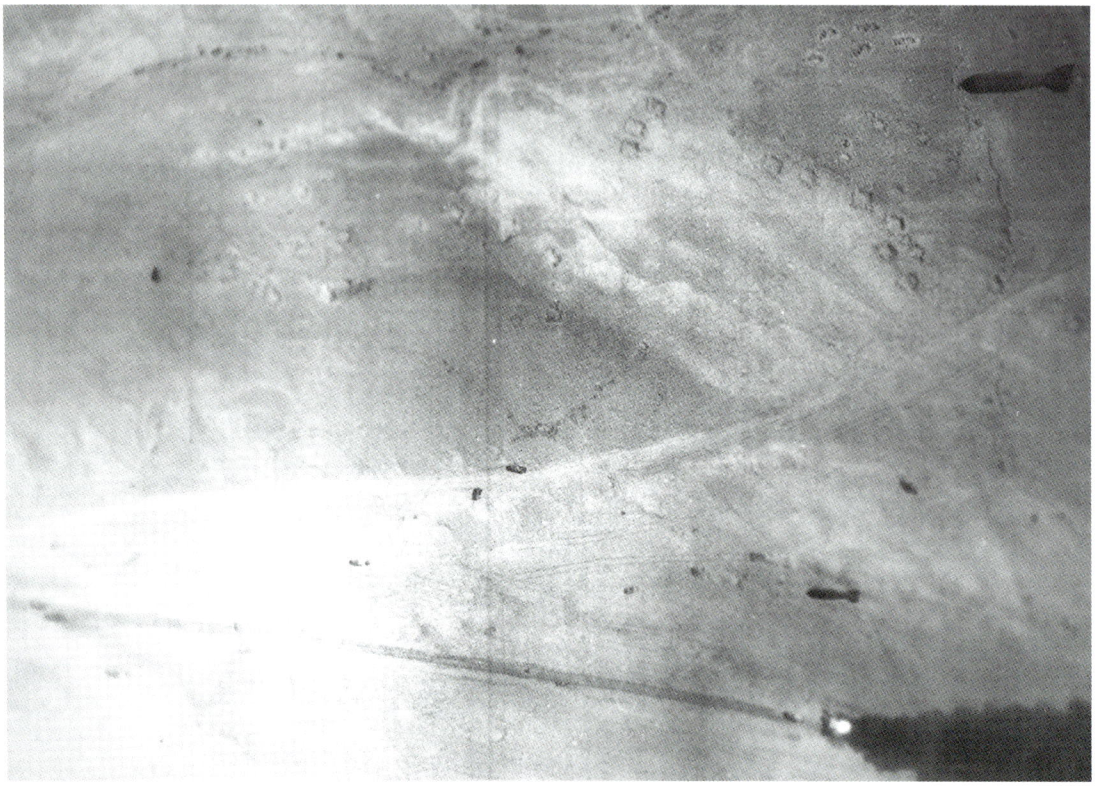

Two bombs drop from a Vautour B during air support operations over Sinai. (PC)

> We bombed the fortification in two runs. My bombs did not drop during the second run so the first pair returned to base while I tried to bomb a tank on the move. Aiming was good and the bombs seemed to have exploded right on the tank but a few seconds later the tank drove out of the smoke, seemingly intact.

TIGRIS 1 landed at 17:32, TIGRIS 2 at 17:25, and TIGRIS 3 and 4 at 17:27.

Take-Off 16:34 – KISHON to Sinai
Isaac Golan with Shabtai Ben-Shoa in 31, Daniel Ilan in 09, Isaac Nir in 05, and Absalom Friedman in 18 were also tasked to attack in the sector of Bir Tamada with four 500 kg bombs. They noted a significant deployment of armored forces east of the Mitla Pass and flew four bombing runs from medium altitude, dropping one bomb at a time. Strafing was not authorized. KISHON reported hits on nine tanks and an anti-aircraft artillery emplacement, and they landed from 17:43 to 17:45.

Take-Off 16:49 – NILE to Sinai
Beni Zohar with Alex Meltzer in 62, Uri Margalit in 26, and Simon Zimon in 12 were tasked to interdict in the Bir Gafgafa sector with four 500 kg bombs. NILE reported hits on two tanks, plus several trucks, and returned to Ramat David at 18:14.

Take-Off 18:00 – BANIAS to Jerusalem
Shlomo Keren with Achikar Eyal in 65 and Jacob Gal in 17 were tasked to fly close air support in the Jerusalem sector armed with four napalm bombs. They were assigned to interdict along the Jerusalem to Jericho road, bombing and strafing in the face of intense anti-aircraft fire that damaged both aircraft. BANIAS 1 landed at 18:37 and BANIAS 2 followed at 18:38.

Take-Off 18:19 – JORDAN to the West Bank
Levy Zur with Shabtai Ben-Shoa in 33 and Zeev Tavor in 03 were scrambled to fly air support over the West Bank, armed with four napalm bombs. They flew to Jenin but were not needed there; it was already past sunset at 18:41 so they flew north, attacked targets of opportunity along the road from Nablus to Jenin, and returned to Ramat David. JORDAN 1 landed at 19:00 and JORDAN 2 followed at 19:13.

From Day 2 to Day 3
52 daylight sorties were flown on 6 June 1967; the 17 missions included a single air base attack, a single electronic warfare mission, and 15 air support missions (eight over the Syrian front, two over the Jordanian front, and five over the Egyptian front). Vautour 08 was lost and Vautour 67 was out of action; both were hit during strafing over the Golan Heights.

The first mission on the night of 6 to 7 June 1967 was flown in a daytime 4-ship formation that took-off at 18:56. KISHON – Isaac Golan with Alex Meltzer in 31, Isaac Nir in 09, Ran Goren in 05, and Daniel Ilan in 12 – was tasked to bomb the Syrian Upper Customs House fortification with four 500 kg bombs. It was already past sunset and visibility was poor, so Golan ordered to fly a single bombing dive. KISHON 1 to 3 hit the target and the four Vautours landed from 19:12 to 19:24.

The second nighttime mission departed Ramat David at 23:41. VISTULA – Gideon Magen with Uri Talmor in 65, Adam Zivoni in 03, and Ran Zur in 18 – was tasked to interdict, with four 500 kg bombs, an Iraqi brigade reportedly advancing from Jericho to Jerusalem. VISTULA 1 illuminated the Jericho to Jerusalem road but no vehicles were seen. The Vautours bombed the road and returned to land from 00:22 to 00:25.

NIGER – Micha Gil with Elhanan Raz in 31, Ovadia Nachman in 05, and Michael Dvir in 09 – followed VISTULA at 00:05, and were tasked to fly the same mission, also armed with four 500 kg bombs. NIGER 1 illuminated the road from Jerusalem to Jericho and again, no convoy was seen; the two bombers attacked tanks that were observed scattered around the road east of Jerusalem. The NIGER Vautours landed at 00:55.

The fourth and final nighttime mission departed Ramat David at 00:30. RIVER 283 – Beni Zohar with Achikar Eyal in Vautour 33 – was tasked to illuminate in support of IDF forces in Sinai. The Vautour was loaded with 16 flares, but communication with the forward air controller was far from perfect. It took quite a while until the IDF force marked the required illumination point. The Vautour then dropped three flares, but was asked to stop illuminating because the ground forces were not ready yet. A few minutes later they were asked to resume illumination, maintaining continuous illumination for the next 25 minutes until asked to stop illuminating. They returned to Ramat David at 02:05.

Take-Off 05:38 – JORDAN to Sinai
Levy Zur with Menahem Eini in 31, Uri Margalit in 26, and Jacob Rafaeli in 03 were tasked to interdict in the Bir Gafgafa sector. JORDAN 1 was armed with four 250 kg bombs; JORDAN 2 and 3 were loaded with 16 70 kg bombs. Bir Gafgafa and other bases all seemed damaged and deserted. All vehicles observed were heading west, retreating from Sinai; the Vautours attacked those vehicles in the face of no opposition at all, except for a single anti-aircraft gun emplacement at Bir Gafgafa that was still active. JORDAN 1 landed at 06:37, and JORDAN 2 and 3 followed at 06:53.

Fast turnaround was no longer an issue during Day 2 operations. Vautour 09 flew two Day 2 daylight missions: the first departed Ramat David at 12:33 and the second took-off at 16:34. (PC)

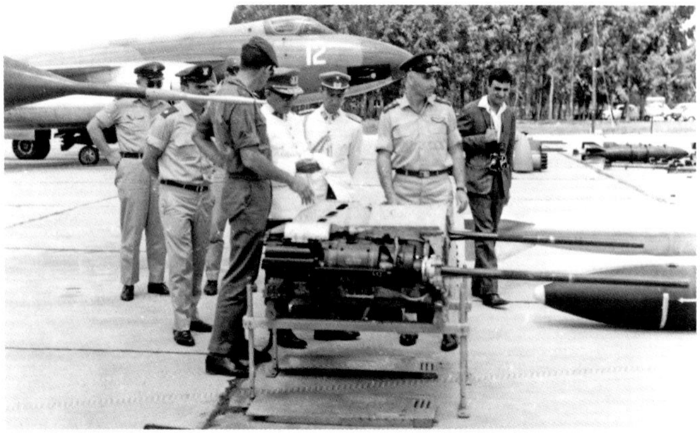

Vautour 12 was photographed in May 1967 during a visit to Ramat David of the Chilean Chief of Staff. Ramat David Commander Ezekiel Somech is second from right. (PC)

Take-Off 06:47 – BANIAS to Sinai
Shlomo Keren with Alex Meltzer in 65, Absalom Friedman in 12, and Simon Zimon in 14 were also tasked to fly interdiction over Bir Gafgafa. All three Vautours were armed with four 500 kg bombs. They actually attacked along the road from Jabel Libni, where hundreds of vehicles were driving west. Each Vautour flew four bombing attacks, dropping one bomb per attack, and an estimated eight to nine strafing passes, firing two guns only to save ammunition and to enable so many attacks. Overall, it was estimated more than 100 vehicles were destroyed. The BANIAS Vautours returned to base at 08:04.

Take-Off 06:57 – TIGRIS to the Golan Heights
Simi Saar with Shabtai Ben-Shoa in 33, Isaac Nir in 17, Ran Goren in 66, and Zeev Tavor in 09 were tasked to bomb Aleika barracks with 16 70 kg bombs each. The target was large; anti-aircraft fire was not noted, but there is no such thing as an event less mission. The bombs of TIGRIS 4 did not drop during the first attack, so he dropped all bombs during the second attack. The bombs of TIGRIS 3 did not drop during the second attack, so he dived several more times, but only two bombs were released and Ran Goren had to land with six bombs still attached to the Quadruple Bomb Racks. Thus, TIGRIS 1 and 2 landed at 07:24 with TIGRIS 3 and 4 following at 07:25.

Take-Off 07:10 – KISHON to the Golan Heights
Isaac Golan with Elhanan Raz in 62 and Jacob Tal in 18 were tasked to bomb a brigade headquarters near Masadeh with four 500 kg bombs. They hit that target and still had bombs left, so ground control authorized them to bomb along the Quneitra to Damascus road; it was then that a bomb that Kishon 1 dropped from medium altitude scored a direct hit on a moving vehicle. KISHON 2 landed at 07:48 and KISHON 1 at 07:50.

Take-Off 08:29 – WART
Yoash Tsiddon flew his fourth WART mission, returning to Ramat David in Vautour 70 at 10:14.

Take-Off 08:31 – RAPID to Sinai
Ovadia Nachman with Menahem Eini in 31, Jacob Gal in 05, and Herzle Bodinger in 26 attacked stranded vehicles in the Mitla Pass with 250 and 500 kg bombs. RAPID 1 returned to base and landed at 09:40, while Gal and Bodinger flew west to strafe the SA-2 battery at Ayon Mossa. Apparently the battery was being evacuated and only an antenna remained. The Vautours strafed that antenna and landed at 09:53.

Take-Off 08:52 – VISTULA to Sinai
Gideon Magen with Shabtai Ben-Shoa in 66, Daniel Ilan in 18, Levy Zur in 09, and Isaac Nir in 62 were also tasked to fly inter-

A Vautour over the Mitla Pass; a killing ground for hundreds of retreating Egyptian vehicles. (PC)

diction over the Mitla Pass armed with four 500 kg bombs. En route to the target, VISTULA 1 suffered a fuel system malfunction. Magen aborted the mission and landed at 09:31; Zur took over as leader. They bombed the vehicles at the Mitla Pass and strafed the SA-2 battery at Ayon Mossa in the face of intense anti-aircraft fire. Not positively sure that this was indeed their SA-2 target, they flew south along the Sinai coast of the Gulf of Suez. They did not see another SA-2 battery (there were none) and turned back as they ran out of fuel. On their way back they saw a convoy along the Bir Tamada to Kabrit road, so they strafed and returned to Ramat David with minimal fuel at 10:06.

Take-Off 09:40 – NILE to Sinai
Beni Zohar with Achikar Eyal in 33 and Zeev Tavor in 12 were tasked to attack the vehicles at the Mitla Pass with 500 kg bombs. The Vautours bombed, photographed, and landed at 10:50.

Take-Off 10:23 – BANIAS to H-3
Shlomo Keren with Alex Meltzer in 65, Ezekiel Somech in 17, Isaac Golan in 14, and Absalom Friedman in 03 were tasked to attack H-3 with two 500 kg bombs each. The escort was bolstered from two to four Mirages, but these were also loaded with two bombs; the rationale was to exploit the escort to the fullest, so the Mirages would bomb first, then engage enemy aircraft to defend the Vautours. This concept proved to be a gross mistake that would cost the attack force dearly. Isaac Golan recalled:

> I was 'fighting executive' and Gideon Magen was assigned to fly to H-3 as [BANIAS] 3. Magen flew to H-3 during the first day attack so I said 'Gideon, you were already there. Take over as 'fighting executive' and I will fly to H-3 as [BANIAS] 3' and so it was. Previously, wing commander Ezekiel Somech called the squadron on the phone and asked to fly that mission so I assigned him as [BANIAS] 2. For Meltzer it was the third mission to H-3 and he protested against that manning.

Part of a Vautour – 14 or 65 – is handed over to IAF Commander Eliezer Shkedy on 12 August 2004 after retrieval from Iraq. (IDF)

Shlomo Keren examines the Aircraft Servicing Form prior to a flight in Vautour 09; Keren was killed on 7 June 1967 while flying his eighth Six Day War mission. (PC)

Vautour 65 was lost as BANIAS 1 during the 7 June 1967 mission to H-3, pilot Shlomo Keren and navigator Alex Meltzer were killed. This photo of 65 was taken in April 1966. (PC)

Vautour 14 landing at Ramat David in January 1967. Isaac Golan ejected from this aircraft over H-3 on 7 June 1967. (PC)

We flew along the 'usual' route to H-3; a straight line east to the desert road and along the desert road to H-3. Some ten minutes into the flight, the sight of my Vautour malfunctioned but I pressed ahead. En route we flew right over a radar station that may have warned H-3 about our approach. The Vautours flew lead and the Mirages trailed us. The plan was that upon pull up to attack, the Mirages would split to pairs, the pairs would overtake us on both sides, they would bomb first and climb to fly a defensive combat air patrol over H-3. The pull up was a bit shallow and as we turned in to attack we saw [enemy] aircraft diving after us from above.

The Mirages bombed and immediately entered air combat. We flew two bombing runs; I and my wingman also flew three strafing passes. When we split into pairs, possibly after the second bombing run but maybe after the first, a Hunter followed Somech. Instead of splitting, I closed on that Hunter and I opened fire to scare him away, I was too far to really threaten him and in any case I could not aim properly, the sight was unserviceable. But the Hunter turned away; maybe they learnt a lesson from the previous day action when a Hunter was shot down in similar scenario.

Then Somech ordered to disengage. We heard over the radio the air combat chatter of the Mirage pilots, and roughly from this moment on nobody saw Keren and Meltzer, but nothing threatened me yet and we were still flying in an organized pair so I decided to press ahead with the attack and to fly the strafing passes. We flew three strafing passes and only when we completed the third and last strafing pass I saw a Hunter right behind me. I turned left, my wingman turned right and I was all alone with that Hunter. We started 'scissors' at low altitude, north of H-3, and the Hunter was a very maneuverable aircraft, much more maneuverable than a Vautour. He opened fire each time we crossed and at a certain phase he probably hit my aircraft, I lost control and ejected.

The previous day, after the second strike against H-3, the Iraqis decided to stop basing aircraft at H-3 and to only use H-3 as a staging airfield. For this reason, the retreating RJAF Hunter pilots that reached H-3 on 6 June 1967 (in a convoy that witnessed the TIGRIS strafing on the return flight from the 6 June 1967 mission to H-3) were ordered to continue their journey to Habbaniya, from where the Iraqis operated three squadrons of Hunters. Among that group of Jordanian pilots was Saif ul Azam, a Pakistan Air Force pilot on a tour of service with the RJAF since November 1966. This group of RJAF pilots reached Habbaniya at 21:00 on the evening of 6 June. The next morning, a formation of four Hunters was ordered to fly a combat air patrol over Habbaniya. The leader was the Pakistani Saif ul Azam, his wingman was the Jordanian Ihsan Shurdom (RJAF Commander 1983 to 1993), and two Iraqi pilots flew in the trailing pair. The Hunters were five miles short of H-3 at 25,000 feet when the IAF aircraft pulled up to attack. Saif ul Azam dived after the Vautours but was engaged by the Mirages, so he ordered to split his formation into two pairs. In the ensuing air combat, Saif ul Azam claimed a Mirage and a two-seater Vautour. The narrative of Saif ul Azam did not include a "scissors" maneuver with a Vautour, so he did not engage Golan. Both Ihsan Shurdom and an Iraqi pilot named Samir reported shooting down a Vautour, while the other Iraqi pilot, named Muhammad, was shot down and killed.

The IAF attack force lost a Mirage (pilot Gideon Dror ejected) and two Vautours: Keren and Meltzer were killed, and Golan ejected and was taken prisoner of war. BANIAS 2 and 4 returned to Ramat David at 11:55.

Squadron 110 lost its Acting Deputy A and Acting Deputy B. Wing 1 Operations Officer Jacob Gal was assigned to Squadron 110 as Deputy B.

Take-Off 11:45 – Reconnaissance Over Sinai

RIVER 273 – Adam Zivoni with Elhanan Raz in 31 – was tasked to photograph the road from Bir Tamada to the killing ground at

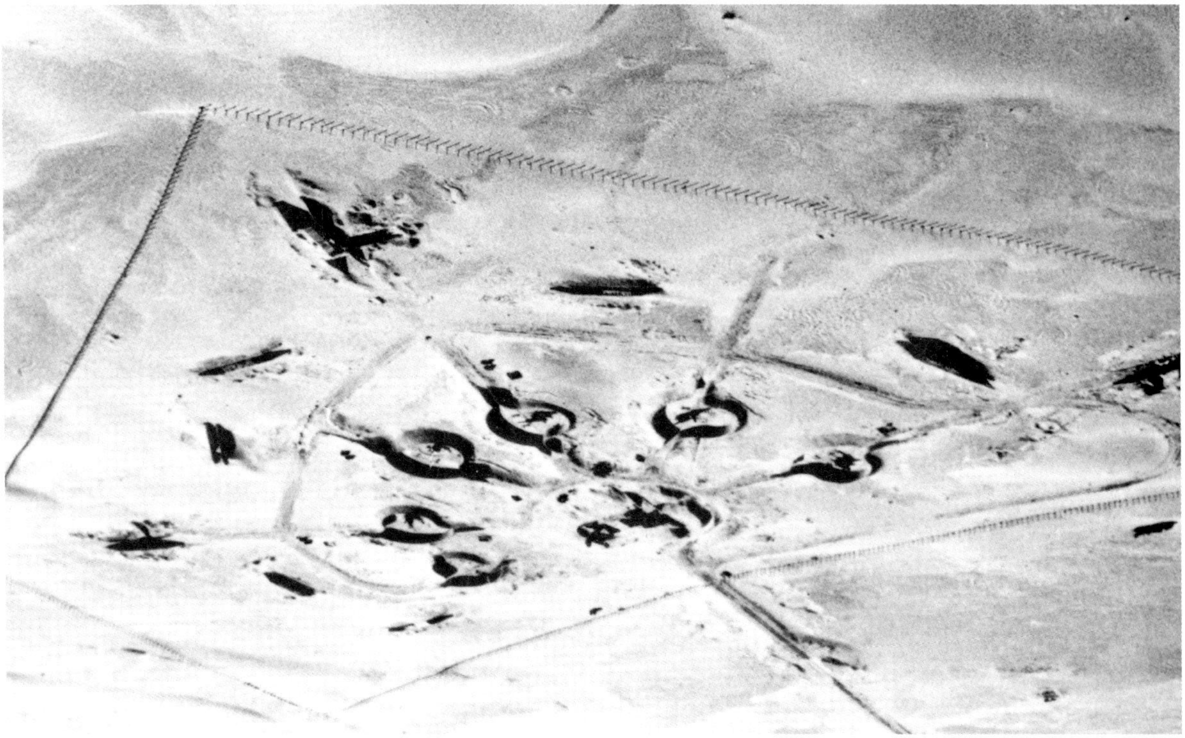

The Egyptian SA-2 battery west of the Mitla Pass that YARMOUK attacked on 7 June 1967; the IDF captured this battery almost intact, and the Egyptians evacuated the other SA-2 battery in Sinai that was based at Ayon Mossa. (PC)

A Vautour photographing Egyptian vehicles attempting to retreat from Sinai. (PC)

the Mitla Pass. Having completed their reconnaissance mission, RIVER 273 was asked to fly a visual reconnaissance from the Mitla Pass to Ras Sudar; the only noteworthy observations were Vautours attacking a SA-2 battery west of the Mitla Pass – this was the YARMOUK section – and a burning oil tank at Ras Sudar. RIVER 273 landed at 13:01.

Take-Off 11:50 – YARMOUK to Sinai

Ran Goren with Shabtai Ben-Shoa in 62 and Michael Dvir in 12 were tasked to fly interdiction over the Mitla Pass with four napalm bombs. En route to the Mitla Pass, ground control notified YARMOUK that IDF forces were approaching the Mitla Pass; to avoid potential "friendly fire", the Vautours were ordered to attack west of the Mitla Pass. The pair saw thousands of vehicles stranded in the killing ground of the Mitla Pass. Many of these were burnt out, but only a few vehicles were observed west of the Mitla Pass. They claimed hits on five tanks and two trucks, and then attacked the SA-2 battery west of the Mitla Pass that was actually inactive. While at it, YARMOUK 2 reported an electrical malfunction so the pair returned to base, landing at 13:05.

Take-Off 14:33 – Reconnaissance Over Sinai

RIVER 277 – Jacob Rafaeli with Menahem Eini in 33 – was tasked to photograph roads in the central sector of Sinai. The Vautour flew, undisturbed, over the battlefields that raged along the desert roads. An armored battle was noted near Bir Tamada, so they took photos as planned and landed at 16:00.

Take-Off 16:07 – TIGRIS to the Golan Heights

Simi Saar with Uri Talmor in 66 and Ran Zur in 09 were armed with four 500 kg bombs; they were tasked to bomb an artillery battery and radar station nearby Quneitra. The pair bombed the artillery battery but did not find the radar station; instead, they flew interdiction along a road in this sector, including strafing, and landed at 16:57.

Take-Off 17:22 – VISTULA to the Golan Heights

Gideon Magen in 03, Ovadia Nachman in 17, and Simon Zimon in 12 were armed with 500 kg bombs and tasked to bomb some sort of Syrian army installation, which they did with ten bombs. They were then assigned to attack a radar station east of Sheikh Miskin; they attacked this radar station with their two remaining bombs, strafed, and returned to base at 18:21.

Take-Off 17:27 – YARKON to the Golan Heights

Isaac Nir in 62, Levy Zur in 05, and Zeev Tavor in 18 were armed with four 500 kg bombs. Zur was flying as tutor during Nir's first wartime mission as leader. YARKON was tasked to bomb in the north sector of the Golan Heights, but many other IAF aircraft were already attacking Syrian targets, and the Vautours were ordered to wait for quite some time. When finally authorized to attack, the weather – clouds from 5,000 to 12,000 feet – precluded bombing over the north sector, so Nir asked for an alternative target in the south sector. The new target was mortar emplacements. Weather over the south sector was also poor, cloudy and windy, while the setting sun hampered visibility. They flew two bombing dives and returned to base at 18:30.

Take-Off 17:29 – WART

Yoash Tsiddon flew his fifth WART mission. His navigator was Oded Erez – who returned to Israel from his assignment in France – on his first Six Day War mission. They returned to Ramat David in Vautour 70 at 18:15.

Oded Erez, sitting, flew the first of his nine Six Day War missions in the late afternoon hours of 7 June 1967; standing behind him are, from left to right: Levy Zur, Adam Zivoni, Giora Goren, and Elhanan Raz. (PC)

Take-Off 17:54 – VOLGA to the Golan Heights

Jacob Rafaeli with Achikar Eyal in 31 and Michael Dvir in 26 were armed with four 500 kg bombs and tasked to bomb a target near Masadeh. They too had to orbit on hold for quite a while until they were authorized to cross the border, and they too were unable to attack because of the poor weather over the north sector of the Golan Heights. VOLGA was not assigned to attack an alternative target, but instead ordered to dump bombs over the Mediterranean Sea and return to base, landing at 19:00.

From Day 3 to Day 4

The Vautours flew 17 daylight missions on 7 June 1967: one air base attack mission, two reconnaissance missions, two electronic warfare missions, and 12 air support missions (six over Sinai and six over the Golan Heights). Those 17 missions included 41 sorties: four air base attack sorties, two reconnaissance sorties, two electronic warfare sorties, and 33 air support sorties. Two Vautours were lost to Hunters during the air base attack mission to H-3: Shlomo Keren and Alex Meltzer were killed when Vautour 65 was lost, and Isaac Golan ejected from Vautour 14 and would be a prisoner of war in Iraq until 26 June 1967.

By the end of Day 3, the IDF had completed the occupation of the West Bank from Jordan and IDF forces were advancing west to the Suez Canal as occupation of Sinai from Egypt was in progress, while the Syrian front remained static. Israel was clearly winning the war, and for Squadron 110 the meaning was a well deserved night off; not a single Vautour departed Ramat David from 17:54 on 7 June to 04:59 on 8 June 1967, Day 4 of the Six Day War.

Take-Off 04:59 – WART

The IDF advanced west towards the Suez Canal and the IAF launched an air superiority offensive against SA-2 batteries west of the Suez Canal. Only Mirages were tasked to attack the SA-2 batteries, and Squadron 110's contribution was WART; the first Day 4 WART mission was flown by Yoash Tsiddon with Oded Erez in Vautour 70. They flew on station off Port Said, reported sighting of a ship that was actually the *Liberty*, and landed at 06:53.

Take-Off 07:40 to 07:46 – Golan Heights

ARNON – Adam Zivoni with Achikar Eyal in 62 and Ran Goren in 18 took-off at 07:40 armed with four 500 kg bombs; they were tasked to silence an anti-aircraft gun emplacement near Masadeh. The emplacement was well camouflaged and the pair missed during their first bombing dive. Return fire revealed the exact position of the emplacement, and ARNON accomplished the mission in a second bombing dive. They landed at 08:09.

DANUBE took-off at 07:46 with Deputy B Jacob Gal in 26, flying his first Six Day War sortie as mission leader with Ovadia Nachman as wingman in 12. They returned to Ramat David at 08:20.

Take-Off 08:11 – Reconnaissance Over Syria

Beni Zohar with Shabtai Ben-Shoa in 31 and Micha Gil with Menahem Eini in 33 photographed the rear sector of the Syrian front, flying from north to south. Weather was perfect, and the Vautours, totally undisturbed during their mission, returned to base at 08:41.

Vautour 17 flew three missions on 7 June 1967: Isaac Nir flew as TIGRIS 2 an air support mission over the Golan Heights, Ezekiel Somech flew as BANIAS 2 to H-3, and Ovadia Nachman flew as VISTULA 3 an air support mission over the Golan Heights. This photo of 4317 was taken in April 1966. (PC)

Herzle Bodinger flew Vautour 17 on 8 June 1967 during his first Six Day War mission as leader. (PC)

Take-Off 08:20 – WART

Yoash Tsiddon with Oded Erez again departed Ramat David in Vautour 70 for an electronic warfare mission in support of Mirage attacks against Suez Canal based SA-2 batteries; they flew over Bir Gafgafa and Kabrit, landing at 10:15.

Take-Off 08:54 to 09:30 – Golan Heights

Herzle Bodinger flew Vautour 17 during his first Six Day War mission as leader; his wingman was Jacob Tal in 03. TIBER took-off at 08:54 armed with four 500 kg bombs; the pair was tasked to attack an anti-aircraft gun emplacement. Just like ARNON, TIBER misidentified the emplacement in the first attack and bombed the emplacement during the second bombing dive. They landed at 09:20.

VISTULA – Gideon Magen in 62 and Zeev Tavor in 26 – took-off at 09:20 armed with four 500 kg bombs. They dive bombed fortifications near the Upper Customs House. By then the Vautours operated from medium altitude to minimize exposure to Syrian anti-aircraft fire.

VOLGA – Jacob Rafaeli with Elhanan Raz in 66 and Absalom Friedman in 18 – took-off at 09:28 after a scramble armed with four 500 kg bombs. Assigned with target in flight, they bombed some sort of trenches.

TIGRIS – Simi Saar in 18 and Ran Zur in 09 – took-off at 09:30 and attacked an anti-aircraft artillery emplacement south of Quneitra with four 500 kg bombs each.

Take-Off 11:21 – WART

Zeev Tavor flew the third Day 4 WART mission in Vautour 70; he returned to Ramat David at 13:18.

Take-Off 13:32 – YARKON to the Golan Heights

Isaac Nir in 09 and Simon Zimon in 17 were tasked to bomb a fortification west of Mound Azaziat with four 500 kg bombs each. They flew four bombing dives, dropping one bomb in every dive, and reported six hits, while one bomb missed and one bomb failed to separate from its rack. No anti-aircraft fire was noted and they landed at 14:10.

Take-Off 13:41 – Reconnaissance Over Sinai

Simi Saar with Achikar Eyal in 33 and Micha Gil with Elhanan Raz in 31 were tasked to photograph the Mitla Pass. For the first time during the Six Day War, a Vautour reconnaissance mission was assigned with an escort: a pair of Squadron 117 Mirages that departed Ramat David at 13:46. Right over the Mitla Pass, the Israeli pilots saw two MiG-17s attacking IDF forces with rockets. The Mirages intercepted those MiG-17s – Mirage pilot Ehud Henkin was credited with a kill – and the Vautours aborted their reconnaissance mission because they were briefed not to photograph without an escort. The TIGRIS Vautours landed at 14:42.

Take-Off 14:15 – VISTULA to the Golan Heights

Gideon Magen in 26 and Uri Margalit in 18 were tasked to bomb a radar station near Daraa with four 500 kg bombs each. The location of the radar station was not as briefed – it was north of a road rather than south of a road – but VISTULA bombed the radar station and strafed nearby trucks that were assumed to be an organic element of that radar station. They landed at 14:49.

Take-Off 15:48 to 17:39 – Golan Heights

Six pairs departed Ramat David from 15:48 to 17:39 to hammer Syrian targets in the Golan Heights. VOLGA – Jacob Rafaeli with Uri Talmor in 62 and Ran Zur in 09 – took-off at 15:48, dive bombed a fortification with four 500 kg bombs each, and landed at 16:18 (VOLGA 2) and 16:22 (VOLGA 1).

THAMES – Michael Dvir with Menahem Eini in 66 and Adam Zivoni in 05 – took-off at 16:04. Dvir was flying his first Six Day War mission as leader, and THAMES was scrambled to a holding pattern until assigned with a target. They dive bombed the Mound Azaziat fortification from medium altitude; each Vautour dropped four 500 kg bombs. Anti-aircraft fire was light, but a request to strafe was denied and THAMES landed at 16:32.

JORDAN – Levy Zur in 18 and Absalom Friedman in 03 – took-off at 16:28 with four 500 kg bombs each. Visibility deteriorated due to dust, smoke and the setting sun; the pair bombed fortifications, did not strafe, and landed at 16:58.

NILE – Beni Zohar in 26 and Uri Margalit in 17 – departed at 17:02 armed with four 500 kg bombs each. They dive bombed an anti-aircraft artillery emplacement in the north sector in the face of intense anti-aircraft fire and returned to land at 17:31.

YARMOUK – Ran Goren with Oded Erez in 62 and Zeev Tavor in 09 – took-off at 17:15 with four 500 kg bombs each. Their mission was to bomb an artillery battery northwest of Quneitra. The emplacements of this battery seemed empty, but YARMOUK was ordered to attack anyway. They dive bombed twice in the face of anti-aircraft fire from an emplacement to the north of their target.

A Vautour flying along a road in Sinai and photographing burning Egyptian vehicles; such was the impact of the FOCUS air superiority offensive on subsequent IAF air support operations, especially over Sinai, where the Egyptian armed forces were fully exposed to Israeli air strikes. (PC)

VOLGA 1 pilot Jacob Rafaeli (on left ladder) and navigator Uri Talmor (on right ladder) with 110 maintainers, and Vautour 62 armed with four 500 kg bombs, prior to take-off at 15:48 on 8 June 1967. (PC)

Vautour 62 – already named LILITH – prior to the VOLGA mission that departed Ramat David at 15:48 on 8 June 1967. (PC)

DANUBE – Jacob Gal in 18 and Simon Zimon in 05 – were scrambled at 17:39 with four 500 kg bombs each. They flew a racetrack pattern until ground control assigned a target, the anti-aircraft artillery emplacement that engaged YARMOUK. Poor visibility made identification difficult for DANUBE, but the wingman managed to spot the target and the Vautours attacked. Strong winds impacted the safer medium altitude dive bombing pattern more than during vulnerable low level attack, and only two bombs seemed to have hit their mark. The DANUBE Vautours landed at 18:02.

From Day 4 to Day 5

The Vautours flew 19 daylight missions on 8 June 1967: 14 attack missions (all over Syria), three electronic warfare missions, and two reconnaissance missions (one over Syria and one over Egypt); the respective sorties count was 28, three, and four for a total of 35 sorties. Obviously, the center of gravity, for the IAF in general and for Squadron 110 in particular, had shifted north to the Golan Heights.

IDF Command North prepared plans for an offensive against Syria; the objectives were to occupy the Golan Heights, and to push back the Syrian forces some 15 kilometers to the east and thus remove the threat of Syrian artillery fire. The Israeli government objected to full scale fighting in the Golan Heights for as long as fighting raged in Sinai and the West Bank. During Day 4 fighting in the West Bank practically ceased, and combat in Sinai was scaled down to mostly mop up operations. Previously, on the evening of 7 June 1967, the Israeli government had authorized a limited offensive against Syria. The IDF opinion was that a limited offensive against Syria was rather pointless, so government level discussions concerning Israeli policy continued throughout 8 June 1967. That evening, from 19:00, the Israeli Cabinet discussed options against Syria. Minister of Defense Moshe Dayan objected the proposition to launch an offensive, stating that he "would rather move ten villages 15 kilometers to the west" than launch an offensive to move the Syrian artillery 15 kilometers to the east. The Cabinet decided to wait.

Under such uncertainty, nighttime action was strictly limited and only two 3-ship formations of Vautours took-off on the night of 8 to 9 June 1967; both were tasked to attack at the Golan Heights. ARNON – Adam Zivoni with Shabtai Ben-Shoa in 62, Beni Zohar in 05, and Jacob Tal in 03 – departed at 18:30 and landed at 19:12. VISTULA – Gideon Magen with Elhanan Raz in 33, Ran Goren in 05, and Ran Zur in 18 – was scrambled at 20:35. They were scrambled to silence an artillery battery, but were ordered to hold until Squadron 109 Mysteres finished their attack on the same target. Flying in formation over a blacked out nation, in a pitch black night, was a tough challenge. This time waiting did not pay off; the Mysteres silenced the artillery battery and the Vautours were ordered to return to base, landing at 21:07 (VISTULA 1) and 21:16 (VISTULA 2 and 3).

As noted, the Israeli government decided to delay a decision concerning an offensive against Syria. Prime Minister Levy Eshkol and IDF Chief of Staff Isaac Rabin supported an offensive against Syria; Minister of Defense Moshe Dayan opposed. Then, at 06:00 on the morning of 9 June 1967, Minister of Defense Dayan arrived for a briefing at the IDF GHQ. Previously, reports from IAF pilots operating over the Golan Heights indicated that Syrian forces were showing signs of fatigue, Egypt had agreed to a cease fire at 22:30 on 8 June 1967, Damascus Radio announced Syrian willingness to accept a cease fire at 03:20 on 9 June 1967, and IDF Command South reported a firm foothold on the east bank of the Suez Canal at 05:54. Sometime before 07:00, Minister of Defense Dayan ordered IDF Command North to launch an offensive against Syria.

IDF Command North – already fully prepared to launch an offensive against Syria – initiated a battle procedure with H-Hour set for 11:30. Meanwhile, the IAF actually ceased flying air support missions over the Golan Heights; there was no apparent need to strike Syria from the air, as fighting in the West Bank and in Sinai had already ceased, or was drawing to a close, and Syria had stated a willingness to cease fire. Only after the Command North battle procedure started to gather momentum were IAF strikes resumed. The first IAF attack formation of this preparation fire phase against Syrian targets in the Golan Heights took-off at 09:25; it was Squadron 110's NILE section.

Day 5 Take-Off 09:25 to 18:25 – Golan Heights

Only a few attack formations could operate simultaneously over the relatively small terrain compartment of the Golan Heights. A dozen pairs departed IAF air bases from 09:25 to 09:52, tasked to attack Syrian targets: fortifications, artillery batteries, and anti-aircraft artillery emplacements. Rather than take-off, bomb, and land, many formations had to fly prolonged nerve-racking racetracks patterns while waiting for ground control to authorize their attacks.

NILE – Beni Zohar in 09 and Isaac Nir in 18 – took-off at 09:25, armed with four 500 kg bombs and tasked to attack mortars near the village of Banias. The first formation to take-off on a Day 5 attack mission to the Golan Heights did not have to wait for ground control clearance. The Vautours bombed, noted five hits and three near misses, and landed at 09:44.

Vautour 05 – in this photo with a prefix digit "5" and lacking the red nose cone – flew two daytime missions on 8 June 1967; both were air support missions over the Golan Heights flown during the afternoon hours, the first as THAMES 2 and the second as DANUBE 2. Vautour 05 then flew two nighttime missions on the evening of 8 June 1967; the pilots were Beni Zohar (from 18:30 to 19:12) and Ran Goren (from 20:35 to 21:16). (PC)

Isaac Nir in the cockpit of Vautour 18 – named DEMON – prior to take-off on an air support mission to the Golan Heights as NILE 2 at 09:25 on 9 June 1967. (PC)

The only Six Day War WART mission to be flown against Syria departed Ramat David at 09:49. Yoash Tsiddon with Oded Erez flew Vautour 66, opposite the Golan Heights, and landed at 10:32.

YARMOUK – Ran Goren in 17 and Absalom Friedman in 03 – took-off at 09:50 armed with four 500 kg bombs each. This time the Vautours flew a typical Golan Heights attack mission: waiting for a target while flying racetrack over Safed. The assigned target was fortifications that YARMOUK bombed circa 10:10 prior to landing at 10:27.

JORDAN – Levy Zur with Uri Talmor in 62 and Uri Margalit in 05 – took-off at 10:10. They were assigned, in flight, to attack, with four 500 kg bombs each, an ammunition dump that exploded; they landed at 10:50.

TIGRIS – Simi Saar in 09 and Micha Gil in 18 – took-off at 11:14 armed with four 500 kg bombs each. In flight they were assigned to attack targets in the Mound Zaatar sector. They landed at 11:43.

RAPID – Ovadia Nachman in 26 and Adam Zivoni in 03 – took-off at 11:40, armed with four 500 kg bombs each, tasked to attack an artillery battery south of Quneitra. The pair flew several passes at medium altitude until they positively identified their target. They dive bombed twice, missing in the first attack and scoring hits in the second. RAPID returned to Ramat David at 12:14.

DANUBE – Jacob Gal with Achikar Eyal in 62 and Michael Dvir in 17 – took-off at 12:46 armed with four napalm bombs and was assigned a target in flight. The assigned target (mortars) did not match their weapons, so DANUBE asked for an alternative target. This second target seemed to be deserted, so DANUBE again asked for an alternative target; the second alternative target was deployed forces at Vasset Junction. The Vautours bombed in the face of intense anti-aircraft fire from an emplacement which defended this vital junction. DANUBE then strafed that anti-aircraft artillery emplacement until it was silenced. The Vautours landed at 13:24.

ELBE – Daniel Ilan in 03 and Jacob Rafaeli in 26 – took-off at 13:22; a first Six Day War mission lead for Ilan. Armed with four 400 kg bombs, the pair was scrambled from cockpit readiness and assigned to fly close air support against a fortification in the north sector under attack by IDF forces. ELBE landed at 14:02.

TIBER – Herzle Bodinger in 05 and Zeev Tavor in 12 – took-off at 14:25. Armed with four napalm bombs each, TIBER flew close air support: four napalm delivery runs and a single strafing pass, after which the supported IDF force was observed to advance. TIBER landed at 15:12.

VISTULA – Gideon Magen with Shabtai Ben-Shoa in 62 and Jacob Tal in 09 – took-off from 16:05. Armed with four 400 kg bombs each, the VISTULA pair bombed Vasset and landed at 16:44.

JORDAN – Levy Zur in 17 and Yoash Tsiddon with Oded Erez in 66 – took-off at 16:48; this was the first offensive mission for Tsiddon. JORDAN attacked artillery in two dive bombings and landed at 17:20 (JORDAN 1) and 17:33. Anti-aircraft fire damaged Vautour 66 and JORDAN 2 landed, safely, with inoperative brakes.

YARKON – Isaac Nir in 26 and Ran Zur in 18 – took-off at 17:16. Armed with four 400 kg bombs, YARKON was tasked to attack a fortification south of Pik, in the south sector. No anti-aircraft fire was noted so the Vautours flew four dive bombing attacks, dropping one bomb per dive, and returned to base at 17:35.

TIGRIS – Simi Saar in 09 and Ovadia Nachman in 03 – took-off at 17:53. They bombed an anti-aircraft artillery emplacement with four 400 kg bombs each in the face of extremely intense anti-aircraft fire and returned to Ramat David at 18:38.

TIBER – Herzle Bodinger with Menahem Eini in 62 and Simon Zimon in 05 – took-off at 18:25. Loaded with four 400 kg bombs and tasked to attack artillery east of Nafah. With sunset at 18:42, visibility was an issue and pinpointing the target proved difficult; it was the navigator who was the first to spot the target. TIBER attacked, also in the face of extremely intense anti-aircraft fire, and landed at 18:59.

From Day 5 to Day 6

The squadron flew 14 daytime missions on 9 June 1967; 13 of these were air support missions flown in pairs over the Golan Heights. The exception was a single WART mission; sorties count was therefore 27. In the air support mission, the Vautours bombed from medium altitude, trading accuracy for safety, and mostly avoided strafing. Such policy paid off nicely, as not a single Vautour suffered severe battle damage and none was shot down.

A single mission was flown during the night of 9 to 10 June 1967. NILE – Beni Zohar in 17, Micha Gil in 26, and Michael Dvir in 05 – was scrambled to silence artillery near Pik. They

Zeev Tavor returned the repaired Vautour 21, landing at Ramat David at 18:38 on 8 June 1967. Vautour 21 did not fly missions from 06:00 on 6 June 1967 (previously it flew five missions) to 06:45 on 10 June 1967 (subsequently flying three missions); this photo of Vautour 21 was taken in April 1966. (PC)

took-off at 23:33. The illuminating Vautour was loaded with 16 flares and the attackers were armed with four 500 kg bombs each. The Vautours attacked at 23:55: NILE 1 landed at 00:07, NILE 3 at 00:16, and NILE 2 at 00:26.

Day 6 Take-Off 04:50 to 13:40 – Golan Heights

DANUBE – Jacob Gal in 03 and Jacob Tal in 18 – was honored to initiate the Squadron 110 offensive on the sixth and final day of the war. They took-off at 04:50, armed with four 400 kg bombs, and attacked an anti-aircraft artillery emplacement west of Quneitra. DANUBE reported destruction of this emplacement, but return fire from other emplacements damaged Vautour 18. They landed at 05:18.

VOLGA – Jacob Rafaeli in 09 and Micha Gil in 05 – took-off at 05:09. They were scrambled with four 400 kg bombs each and were tasked, in the air, to attack mortars southeast of the Upper Customs House. They landed at 05:51.

TIBER – Herzle Bodinger in 66 and Ovadia Nachman in 17 – took-off at 05:58. Also armed with four 400 kg bombs, TIBER was tasked to fly interdiction along the road from Quneitra to Damascus. They bombed an anti-aircraft artillery emplacement outside Quneitra and strafed vehicles along the road, reporting destruction of three armored vehicles. They returned to base at 06:33.

SEINE – Uri Margalit in 21, Ran Zur in 12, and Absalom Friedman in 26 – took-off at 06:45; the first Six Day War mission lead for Margalit. SEINE was also planned to fly as a pair, while Friedman was planned to fly as wingman in another pair, but the Vautour of Friedman's leader was unserviceable, so Friedman joined SEINE. They were tasked to attack fortifications with 400 kg bombs and flew three dive bombings, reporting heavy damage to the fortifications, but also indicating that these were seemingly deserted. Friedman landed at 07:15, while SEINE 1 and 2 landed at 07:26 because the airspeed indicator of Vautour 21 malfunctioned. Vautour 21 suffered battle damage in the early morning hours of 6 June; Zeev Tavor departed Ramat David in 21 at 07:45 on 6 June and flew the aircraft to IAI for repairs. Tavor returned 21 from IAI on 9 June, landing at 18:38, so it was assumed that the airspeed malfunction was related to battle damage or repair work. The technique for landing without an air speed indicator was to fly beside a "shepherd" aircraft that approached the runway at the correct airspeed.

JORDAN – Levy Zur in 03 and Daniel Ilan in 09 – took-off at 07:06. In the air they were assigned to attack the Upper Customs House. They bombed with four 400 kg bombs each in the face of anti-aircraft fire; request to strafe was denied and they landed at 07:26.

TIGRIS – Simi Saar with Achikar Eyal in 33 and Adam Zivoni with Menahem Eini in 31 – took-off at 07:32 and flew a high altitude reconnaissance mission from Sheikh Miskin to Damascus. They landed at 08:15.

VISTULA – Gideon Magen with Uri Talmor in 62 and Zeev Tavor with Oded Erez in 66 – took-off at 07:51 armed with four 400 kg bombs. They were tasked to fly interdiction along the road from Quneitra to Damascus, but traffic was sparse; they bombed two bridges and strafed vehicles, reporting the destruction of three trucks.

NILE – Beni Zohar in 18 and Ran Goren in 26 – took-off at 08:17, also armed with four 400 kg bombs. The target was a fortification near Masadeh. There was no anti-aircraft fire so they flew four dive bombing patterns, dropping one bomb per dive, and returned to base at 08:47.

DANUBE – Jacob Gal in 21 and Nachman Ovadia in 12 – took-off at 08:53. The lead Vautour was loaded with ten 400 kg bombs, six in the bomb-bay and four under the wings; the wingman's load was four 400 kg bombs. They were tasked to fly close air support of an IDF force in the south sector. The friendly force marked its position with white smoke and the Vautours flew five dive bombings, dropping two bombs per dive. They landed at 09:28.

JORDAN – Levy Zur in 18 and Absalom Friedman in 03 – took-off at 09:38. Their bomb load was the same as that of DANUBE. They were scrambled, assigned a target, bombed and landed at 10:00.

YARMOUK – Ran Goren with Elhanan Raz in 66 and Uri Margalit in 17 – took-off at 10:01. Armed with four 400 kg bombs, they were tasked to look for targets of opportunity. They bombed and strafed vehicles west of Quneitra, and there was no anti-aircraft fire. YARMOUK 2 landed at 10:42 and YAR-

A Squadron 110 end of war photo, standing left to right: Herzle Bodinger, Shabtai Ben-Shoa, Ran Goren, Uri Margalit, Uri Talmor, Menahem Eini, Jacob Rafaeli, Oded Erez, Beni Zohar, Achikar Eyal, Yoash Tsiddon, Adam Zivoni, Ovadia Nachman, Levy Zur, and Zeev Tavor; sitting on fuselage and wing left to right: Daniel Ilan, Absalom Friedman, Gideon Magen, Micha Gil, Michael Dvir, Simi Saar, Jacob Gal, Elhanan Raz, and Ran Zur; standing on fuselage and wing left to right: Giora Goren, Jacob Tal, Simon Zimon, and Isaac Nir. (PC)

MOUK 1 returned to base at 10:45; they reported hits on ten vehicles, including three that caught fire.

VISTULA – Gideon Magen with Shabtai Ben-Shoa in 62 and Michael Dvir in 12 – took-off at 10:30. Both Vautours were armed with ten 400 kg bombs and they were tasked, in the air, to attack vehicles near the village of Hoshniya. They bombed and expended all the ammunition of their guns in four strafing passes, claiming the destruction of some 20 vehicles and landing at 11:11.

TIGRIS – Simi Saar in 21 and Ran Zur in 18 – was tasked to attack enemy forces near the village of Salokiya, but Vautour 18 malfunctioned prior to take-off. Instead of scrapping the mission altogether Simi took-off alone, at 11:06, and attacked with ten 400 kg bombs. He did not notice anti-aircraft fire and returned to base at 11:34.

YARKON – Isaac Nir in 05, Zeev Tavor in 09, and Ran Zur in 18 – took-off at 11:17; Vautour 18 was quickly repaired, so Ran Zur joined the next pair, YARKON. Armed with ten 400 kg bombs each, YARKON was tasked to attack vehicles near Nafah, but had to wait until a formation of Ouragans finished their attack against the same target. There was no anti-aircraft fire so they dropped one bomb at a time. Upon completion of bombing Ran Zur returned to base – his fuel state was lower due to its long gestation prior to take-off; he landed at 11:58. The other two Vautours pressed ahead with strafing passes: Tavor landed at 12:05 and Nir at 12:08. YARKON reported the destruction of one tank and 13 trucks.

DANUBE – Jacob Gal in 26 and Absalom Friedman in 03 – took-off at 11:57. They attacked vehicles outside Hoshniya with 14 400 kg bombs and two strafing passes, claiming the destruction of 14 vehicles and landing at 12:34.

JORDAN – Levy Zur with Oded Erez in 62, Jacob Tal in 12, Herzle Bodinger with Uri Talmor in 66, and Jacob Rafaeli in 17 – took-off at 12:27. They bombed Taufik, in the south sector of the Golan Heights, prior to an IDF offensive in this direction. The four JORDAN Vautours landed at 12:57.

YARMOUK – Ran Goren in 03 and Uri Margalit in 18 – took-off at 13:40 armed with ten 400 kg bombs each. They were originally tasked to fly interdiction along the Quneitra to Damascus road, but YARMOUK 2 reported engine trouble and the pair was diverted to bomb Pik, in the south sector of the Golan Heights. Uri Margalit recalled:

> An engine malfunctioned immediately after take-off so squadron operations ordered me to jettison my bombs over the sea [and to return to base]. I resented this suggestion as I was already in the air and fully armed so I asked for a closer alternative target. We were re-tasked to fly close air support in the sector of Pik. IAF ground control directed us to attack west of the village from were our forces were under continuous fire that stopped their advance. I flew on one engine so I decided to fly a shallow attack dive and to drop all bombs – both from the bomb-bay and from the wings – in a single run from about 10,000 feet. Once I did so the aircraft seemed to have jumped in the air, relieved at once of such a heavy load. Then the forward air controller reported that the source of fire we had just attacked had been silenced and our forces were able to resume their advance.

They landed at 14:04, with YARMOUK 2 actually performing a single engine landing.

While Squadron 110 was busy bombing Syrian targets, IDF forces in Sinai started to clear roads and collect war booty. (PC)

Then, at 15:03, Vautour 70 departed for yet another WART mission. Zeev Tavor with Oded Erez flew their electronic warfare patrol off Port Said and returned to land at 16:01; the last Vautour landing on the sixth and final day of the Six Day War.

From 04:50 to 15:03 on 10 June 1967, Squadron 110 launched 18 missions that accounted for 38 sorties: 16 air support missions over Syria (35 sorties), a reconnaissance mission to Syria (two sorties), and a WART mission against Egypt (one sortie).

110 in the Six Day War

The squadron flew four types of missions during the Six Day War: air superiority, air support, reconnaissance, and electronic warfare. A few missions were integrated efforts that included more than one type of mission – air superiority and reconnaissance or air superiority and air support – but only the main mission is referred to in the following discussion.

Air superiority totaled 15 missions (56 sorties). The Vautours flew ten missions (36 sorties) against Egyptian air bases, two missions (eight sorties) against Syrian air bases, and three missions (12 sorties) against Iraqi air bases. The attacking formations reported the destruction of 60 aircraft on the ground and one Hunter in the air. Three Vautours were lost in this air superiority campaign: Vautour 07 over Dmer, Syria (pilot prisoner of war); Vautour 65 over H-3, Iraq (pilot and navigator killed); and Vautour 14 also over H-3 (pilot prisoner of war).

Air support accounted for 83 missions (202 sorties), including 14 missions (38 sorties) over Sinai (Egypt), eight missions (20 sorties) over the West Bank (Jordan), and 61 missions (144 sorties) over the Golan Heights (Syria). A single aircraft was lost during these air support missions, which benefited immensely from the success of the earlier air superiority campaign; this was Vautour 08 in an air support mission over the Golan Heights (pilot ejected over Israeli territory).

Reconnaissance covered six missions (nine sorties): four (five sorties) over Egypt and two (four sorties) over Syria.

Electronic warfare was the tasking of ten missions (11 sorties): nine (ten) over Egypt and one (one) over Syria.

The 25 pilots, seven navigators, and 18 Vautours flew 114 missions that accounted for 278 sorties; an average of 15 sorties per aircraft and 11 sorties per pilot. Ten years after arrival in Israel and nine years after IAF service introduction, the Vautour went to war and proved its worth as a true "heavy attack aircraft".

Isaac Golan returned from Iraq, via Jordan, on 26 June 1967; Ramat David Wing Commander Ezekiel Somech greeted Golan in the presence of Squadron 110 Commander Levy Zur (holding his chin) and Ramat David Flying Squadron Commander Shabtai Gilboa (left). (PC)

6

ATTRITION TO RETIREMENT

1967 - 1971

SIX DAYS OF WAR CHANGED THE MIDDLE EAST. The IAF air superiority campaign was so devastating and so successful that the IDF was able to occupy the Sinai Peninsula from Egypt, the West Bank from Jordan, and the Golan Heights from Syria within such a short time frame. The Israeli victory was not followed by a diplomatic campaign to end the Arab-Israeli conflict. Egypt President Gamal Abdel Nasser stamped the slogan "what had been taken by force will be returned by force", while in September 1967 Arab leaders adopted the "three no" policy at the Khartoum Summit: no recognition of Israel, no negotiation with Israel, and no peace. Hostilities in the Middle East continued.

Postwar Reorganization

Four Vautours were lost, a pilot (Shlomo Keren) and a navigator (Alex Meltzer) were killed, a Vautour pilot that was "slipped back" to Squadron 116 prior to war (Jonathan Zorea) was killed in a Mystere, and two pilots were prisoners of war (Abraham Vilan in Syria and Isaac Golan in Iraq).

The "slipped back" pilots Ilan Lior and Moshe Teman returned to resume their training towards qualification as operational Vautour pilots. Emergency posting service persons returned to their duties. Reserves returned to their civil occupations.

The active IAF Vautour fleet was down to 15 aircraft: eight A, two B, and five N. In line with this depletion, and because other squadrons had suffered heavier losses, no new young pilots were assigned to 110 at the start of IAF Year 1967 Term 2 and Term 3. Few pilots were assigned from the squadron; Ran Goren and Isaac Nir started their Mirage Conversion Course in July 1967.

More Punch for the Vautour

MAT5 suggested on 25 June 1967 to qualify the Vautour to lift 65 kg bombs, 120 kg bombs, or anti-runway bombs on the Light Bomb Rack or Quadruple Bomb Rack. The suggestion to qualify the anti-runway bomb on the Vautour was rejected. AIR2 ordered, on 6 August, to test 65 and 120 kg bombs loaded on the Light Bomb Rack or Quadruple Bomb Rack on Vautour 62. Levy Zur flew the first test in Vautour 62 on 12 September; Abraham Vilan with Jacob Shmueli flew the following four test flights from 18 to 20 September. AIR2 qualified, on 10 April 1968, two configurations: four Quadruple Bomb Racks with 16 120 kg bombs and four Quadruple Bomb Racks with 14 100 kg bombs (no bombs on the inner rear points on the inner Quadruple Bomb Rack due to risk that the released bomb might impact the engine).

Only three Vautours (65, 66 and 70) were modified for WART prior to the Six Day War, and one of those (65) was lost. Thus, it was decided to wire for WART all N aircraft and two A aircraft, as indicated in a MAT2 paper from 11 July 1967.

Another lesson from the war was a requirement that AIR2 issued on 12 July 1967 for standardization of Vautour bombing systems through installation of the Vautour A bombing system, including intervalometer, in the N aircraft. A prototype intervalometer installation was fitted in Vautour 67 and AIR2 qualified this modification on 9 February 1968.

Jacob Agassi took over from Ezekiel Somech as Ramat David Wing Commander on 30 June 1967 with Vautour 26 – with prefix digit "1" – in the center of the parade flanked by two Squadron 117 Mirages, two Squadron 109 Mysteres, and two Squadron 107 Ouragans. (PC)

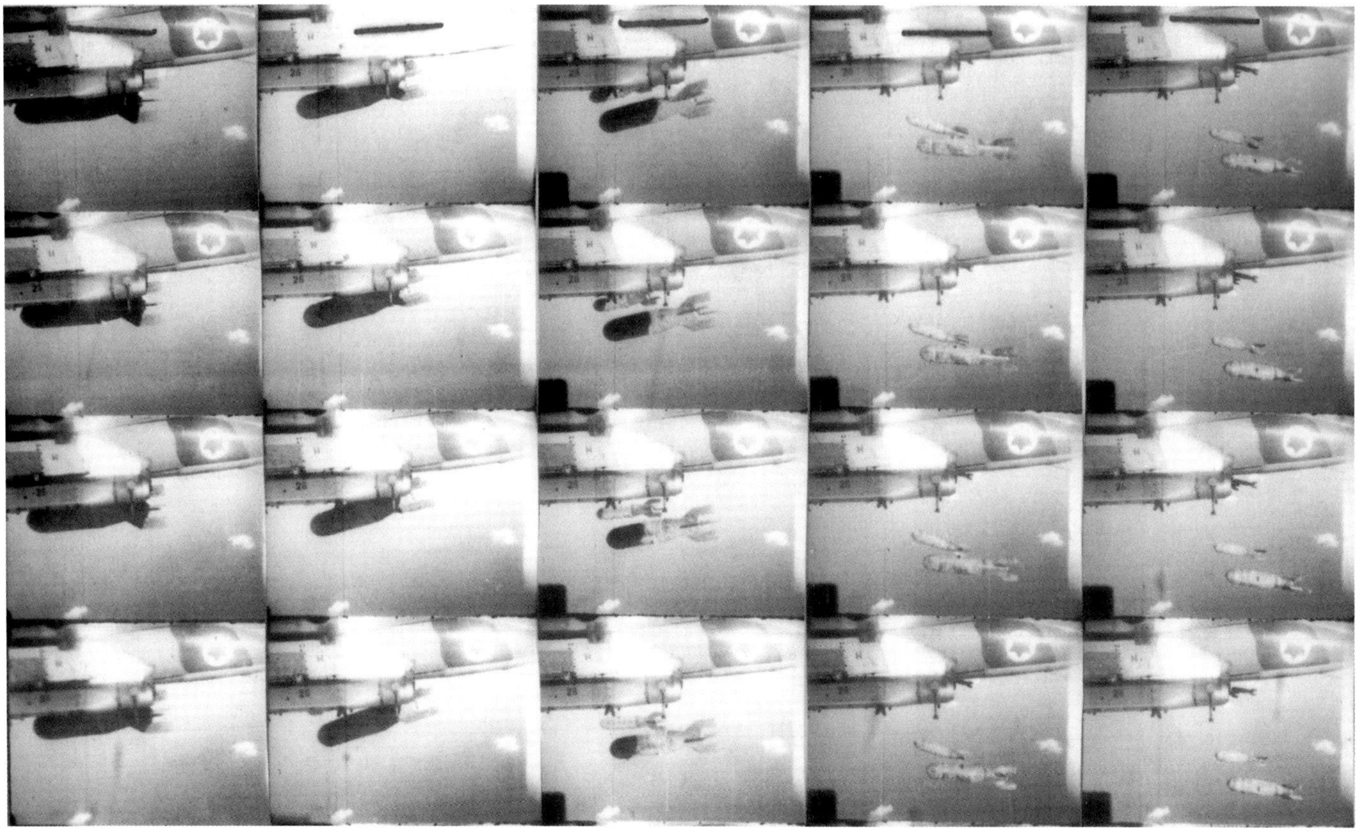

Bombs are released from Vautour 62 during the September 1967 series of Quadruple Bomb Rack test flights. (IDFA)

Only eight Vautours flew in formation during IAF Day on 6 July 1967; all with 1,250 liter fuel tanks, except for Vautour 70, with 1,300 liter drop tanks. Micha Gil flew Vautour 33 with Menahem Eini. (PC)

COMBINE Test Resumption

Vautour 70 reached Israel to test COMBINE. The radar produced poor ground mapping and promising maritime strike capabilities during its first test. IAI ELTA improved the radar and MAT8 issued, on 23 July 1967, a request to test the improved COMBINE. Vautour 70 returned to IAI to be fitted with COMBINE; the second round of COMBINE tests started on 10 September 1967 when Isaac Golan with Isaac Erez flew Vautour 70.

In Action Again

Armed clashes along the Suez Canal, the new border between Egypt and Israel, continued past cease fire. IAF air support was occasionally called; Isaac Golan, in Vautour 03, bombed Egyptian targets in the Port Ibrahim sector on 14 July 1967.

Vautour 28 and Vautour 30

The loss of four Vautours in the Six Day War generated interest in rebuilding the Technical School's instructional airframe – the Vau-

From 1967, the IAF initiated a switch from the darker two tone pre-Six Day War camouflage scheme (as 18) to the lighter three tone post Six Day War camouflage scheme (as 12). Vautour 18 is adorned with mission and kill markings on its vertical stabilizer. (PC)

tour pre-production 07 aircraft – to IAF service standard. First step was allocation, on 26 September 1967, of IAF Number 4328.

Meanwhile, Vautour A production number 7 was purchased from France. AIR3 issued, on 29 November 1967, Mission Order 67/110 to cover the ferry flight from France to Israel. Preparations lasted longer than planned. IAF Chief Test Pilot Daniel Shapira flew the Vautour (call sign 4X-FCB GREENERY 104) from Chateaudun to Brindisi, and hence to Israel on 21 December 1967. Placed in the hands of IAI for modification into IAF standard, this Vautour became IAF Number 4330.

Terner Takes Over

Levy Zur handed over command of the squadron to Jacob Terner on 1 October 1967. Aki Arzi was assigned to the squadron as Deputy A. By then the Vautours were named: 03 was THE SILENCER, 05 was THE PERSUADER, 09 was THE DESTROYER, 12 was TORNADO, 17 was THE PANTHER, 18 was DEMON, 21 was COBRA, 26 was PYTHON, 31 was THE LITTLE BROTHER, 33 was THE BIG BROTHER, 61 was VAMPIRE, 62 was LILITH, 66 was DRAGON, 67 was CYCLONE, and 70 was FANTOMAS. The look of the Vautours also gradually changed from the darker pre-Six Day War camouflage scheme to a lighter post Six Day War scheme.

INS Eilat

The Vautour was the most capable IAF illumination platform, qualified to operate with up to 18 flares in the bomb-bay and/or up to 16 flares under the wings. Back on 10 November 1966, Squadron 110 Commander Levy Zur proposed to increase the internal capacity to 27 flares. This proposal may have been a lesson from an operational illumination mission that Isaac Golan flew in Vautour 17 on the night of 9 to 10 November 1966.

The requested modification was implemented, and MAT5 issued on 25 September 1967 a requirement to test the new capacity of 27 illumination bombs in the bomb-bay. AIR2 issued the corresponding test order on 15 October 1967.

Six days later, Egyptian missile boats sunk the Israeli Navy Ship Eilat, off Port Said, on the evening of 21 October 1967. Squadron 110 was scrambled to illuminate the rescue operation; the first to pinpoint the location of the survivors was Jacob Gal with Shabtai Ben-Shoa, and the pilots and navigators who followed up this illumination action were Aki Arzi, Micha Gil, Simi Saar, and Adam Zivoni with Achikar Eyal in 33. The crew of the destroyer numbered 199; 47 of them were killed.

The 27 flares in the bomb-bay modification was tested on 7 November 1967 in Vautour 21 piloted by Micha Gil. In the

The Squadron 110 command hand-over parade on 1 October 1967 with Flying Flight on the left and Technical Flight on the right, from left to right: a trailer tanker, Vautour 26 still in pre-Six Day War scheme, a Meteor trainer and Vautour 67 in post Six Day War scheme. (PC)

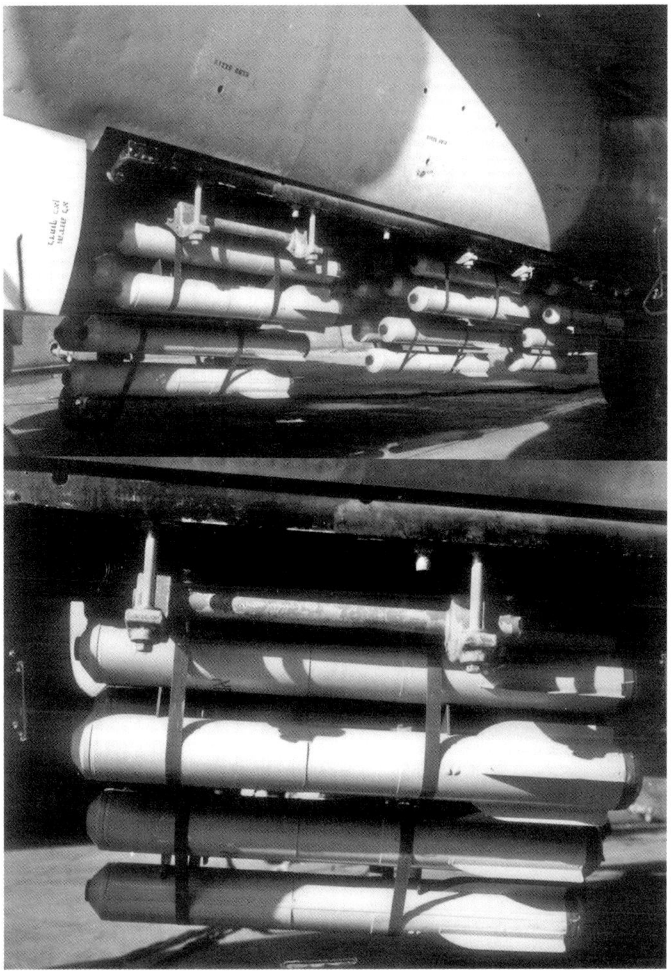

The bomb-bay configuration of 27 flares as tested on Vautour 21 in November 1967. (IDFA)

wake of this flight MAT5 recommended, on 23 November, to qualify this modification, and AIR2 qualification was issued on 25 February 1968.

Additional External Stations

MAT4 reported on 16 November 1967 that the design of an additional weapon station under the wing, outboard of the existing stations, had been completed. The new station was tested on a spare wing and was suggested for Shafrir 1, Shafrir 2, or the Light Bomb Rack loaded with two flares or two 77 kg bombs. A proposition to qualify the new station for 100 kg bombs was also presented.

MAT5 responded on 5 December 1967, stating that the development of the additional external station was an anachronism. The requirement was issued in 1965; the Shafrir 1 was in short supply and due to be withdrawn from service during IAF Year 1969, while the Shafrir 2 was not expected to be available to attack aircraft before 1971. The Vautour was already modified to lift 27 flares, so the addition of four flares seemed negligible; IAF stock of 77 kg bombs was limited, and the physical dimensions of the 100 kg bombs precluded their use on the Light Bomb Rack. MAT5 recommended to suspend development of an additional wing station and to re-evaluate the necessity of the operational requirement for such a modification. The IAF Vautours were never to be equipped with an additional wing station.

A bomb is loaded to the outer right wing station of Vautour 26; the 1965 requirement to fit a third under wing station was considered anachronistic by December 1967. (PC)

The Demise of THE LITTLE BROTHER

Deputy A Aki Arzi with Elhanan Raz in Vautour 31 and Deputy B Jacob Gal with Menahem Eini in Vautour 66 departed Ramat David at 13:30 on 1 December 1967 to fly a low altitude reconnaissance mission over the Egyptian coast, west of the Gulf of Suez; the objective of this mission was to photograph a site that was reportedly prepared for an SA-2 battery. While flying over Egyptian territory, circa 14:00, Gal and Eini noted some kind of leak or smoke emanating from the left engine of Vautour 31. They reported to their leader but received no answer. Vautour 31 then turned east. Vautour 66 followed and got closer, at which point Gal and Eini saw that the left engine of Vautour 31 was on fire. Jacob Gal recalled:

> The Vautour of Arzi and Raz was hit and I radioed to them to pull up in the direction of the Gulf of Suez. The aircraft was on fire. Suddenly, the right wing was torn away and the aircraft dived to the sea. I yelled over the radio 'eject' but I only saw one ejection and a parachute.

Vautour 31 crashed into the sea roughly halfway over the Gulf of Suez. Arzi and Raz were declared missing in action.

Deputy B Jacob Gal took over as Deputy A. Squadron 110 Commander Jacob Terner travelled to Kibbutz Kfar Giladi to persuade the assembly of members to approve the return from reserve service to active duty of Naphtali Porat. The Kibbutz assembly approved and Naphtali Porat – previously a Squadron 117 Mirage reserve pilot – returned to active duty as Squadron 110 Deputy B.

Naphtali Porat still in the cockpit of Vautour 18 after his solo flight on 29 December 1967. (PC)

Vautour 70 Reconnaissance

The loss of Vautour 31 scaled down the number of reconnaissance Vautours to two. To make up for this loss, it was decided to modify Vautour 70 as a reconnaissance aircraft. MAT3 Chief Haim Yaron reported on 29 December 1967 that modification priorities were: three K-38 cameras for high altitude photography plus guns pack for self defense, GOVERNOR as installed in Vautour 66, and a PACK camera as fitted to Mirage reconnaissance aircraft.

First test flights were flown in May 1968; Beni Zohar with Achikar Eyal flew Mission 51/110 on 9 May. The tested cameras included K-38 (with 24" and 36" lenses) and Zeiss (with 6" and 12" lenses) for high altitude photography and GOVERNOR for low altitude reconnaissance. The AIR2 report from 26 May concluded that the Vautour 70 reconnaissance products were not yet up to the standard of Vautour 33, so it was recommended that the process of "refine and test" would continue.

AIR2 qualified Vautour 70 for high (K-38) and low (Zeiss) altitude reconnaissance on 29 August 1968. Planned installation of guns pack and PACK and GOVERNOR cameras was to be accomplished at a later date.

From EAR to INFERNO

Clashes along the Israeli-Jordanian border also raged, mostly involving artillery duels with the Iraqi expeditionary force in north Jordan and incursions of PLO squads. IAF aircraft, including the Vautours, shared readiness to support IDF patrols that encountered PLO squads and were on readiness to counter Iraqi artillery. Squadron 110 action over the Jordanian border started on 11 February 1968. That day emergency posting navigator Achikar Eyal flew three missions over Jordan, all in 33: reconnaissance with Jacob Gal, reconnaissance with Micha Gil, and illumination with Herzle Bodinger.

AIR2 issued in May 1967 an operational requirement for conversion of Vautour 70 as a reconnaissance aircraft and test flights started in May 1968. (BIAF)

A 4-ship formation en route to attack targets in Jordan during Operation INFERNO; the lead Vautour 67 is armed with four bombs under its wings, and the wingmen Vautours (09, 05 and 21) are loaded with two 1,250 liter drop tanks and two bombs under their wings, plus six bombs inside the bomb-bay; 67 and 21 are already in the post Six Day War lighter scheme; 09 and 05 are still in the darker pre-Six Day War scheme. (PC)

Operation EAR, on 15 February 1968, was an IAF offensive against Iraqi and Jordanian targets in the north sector of the border; IAF strikes lasted from 15:00 to 23:00, and 110 launched two formations that attacked a Jordanian barracks.

Operation INFERNO was launched on 21 March 1968; the IAF supported an IDF offensive to mop up PLO nests in the central sector, and the main IDF effort was towards the town Karameh. The Vautours flew air support missions throughout the day as well as reconnaissance, and an attack against a Jordanian radar station nearby Mazar. Levy Zur flew three missions (reconnaissance with Achikar Eyal in 33, attack with Achikar Eyal in 66, and attack with Shabtai Ben-Shoa in 66), and Achikar Eyal flew three missions (the two with Levy Zur, plus an attack with Jacob Terner); Jonathan Shahar flew two missions (attack in 05 and the radar station attack in 18); and Zeev Tavor flew two attack missions in 17 and 61. Strikes ceased that evening, but the following day Jacob Gal with Achikar Eyal flew two reconnaissance missions over Jordan in 33.

INFERNO highlighted several operational issues that Squadron 110 Commander Jacob Terner addressed on 31 March 1968: a second UHF transceiver instead of the VHF radio-2 set, replacement of the old radio compass with a modern instrument, installation of a gunnery camera (the sight camera was useless when the guns were fired, it was only fit for tracking and ranging practice) as was fitted to Vautour 07 (which was lost during the Six Day War), more fuel through qualification of 500 or 625 liter external fuel tanks loaded to the outer wing stations, or enlarged bomb-bay fuel tanks, and the modification of all N aircraft to operate, alternately, in the attack, reconnaissance, and electronic warfare missions.

AIR2 officer Ephraim Ashkenazi referred on 12 April 1968 to Squadron 110's list of issues: a second UHF was to be installed, the original radio compass was not to be replaced

Vautour 66 landing at Ramat David in January 1968; Jacob Terner with Achikar Eyal flew attack Mission 1/110 in 66 on 21 March 1968, and Levy Zur then flew two attack missions in 66 on 21 March 1968. (PC)

due to other priorities (the orders for modern radio compass instruments were assigned to equip the Mirage 5 interdictors with Fouga trainers and Bell 205 helicopters in line for follow up lots); the gunnery camera as fitted to Vautour 07 was indeed qualified on 6 April 1967, but since then the IAF had initiated transition from French to American hardware, so it was proposed to install an American camera with corresponding delay to this project; Material Department's engineers rejected the proposition to fit 500 or 625 liter fuel tanks to the external wing stations; production of more and/or improved bomb-bay fuel tanks was rejected because "by the time production will be completed, there will be enough Skyhawk aircraft to relieve the Vautour of long range missions", and AIR3 did not issue an operational requirement to modify more N aircraft for reconnaissance.

In the wake of this correspondence, MAT7 suggested on 15 April 1968 that Vautour 30 would be used to test the prototype installation of a second UHF radio because IAI was busy modifying this aircraft to IAF standard with a delivery date set for late April.

Wing to Wing

During an air base attack exercise over Ramat David on 14 March 1968, wingman Moshe Teman reported to his leader a red warning light in his cockpit, indicating that the doors of the front main wheels were not locked. His leader stopped the exercise, got closer, and flew underneath the wingman's Vautour to inspect the status of the front main wheels doors. This inspection took place while the aircraft were flying relatively low and heading in the direction of Nazareth. Apparently all seemed well with the front main wheel doors, so the leader – noting that they were approaching the hills of Nazareth and having no space to dive away – skidded to the left and climbed away. It was then that the right wing of the leader's Vautour 03 rubbed the left wing of the wingman's Vautour 62.

Squadron 109 Commander Ohad Shadmi, also in the air at the time, got closer in his Skyhawk to inspect the damage and both Vautours landed safely. The two Vautours were repaired, with Vautour 03 flying again by May 1968 and Vautour 62 returning to the air by October 1968.

SMASH and CRAB

Operations along the Jordanian border continued with evolving tactics aimed to calm down hostilities. Three such new tactics were heliborne special operations, armed reconnaissance along the border, and immediate reaction.

SMASH 1, on 8 April 1968, was a heliborne special operation; the objective was to mop up PLO nests north of Aqaba. In preparation to SMASH 1, Jacob Gal with Achikar Eyal photographed the sector of the operation in Vautour 33 on 4 April.

CRAB was the code for an armed reconnaissance along the Jordanian border; Herzle Bodinger with Achikar Eyal flew a CRAB mission in Vautour 61 on 21 April 1968.

Immediate reaction involved the scramble of combat aircraft to attack immediately after a border clash erupted. CONGO – Jacob Terner, Abraham Vilan, Micha Gil, and Herzle Bodinger – were thus scrambled on 4 June 1968 to attack a Jordanian artillery battery.

From Vautour 28 to Vautour 20

The Technical School instructional airframe was transported to IAI for refurbishment to IAF service standard. The pre-work survey concluded that such refurbishment was possible, but would consume a lot of effort and would never bring the pre-production Vautour to true IAF standard. For example, the dimensions of the Vautour 28's cockpit were smaller than the cockpits of production Vautour A aircraft. During this survey it was suggested to rebuild Vautour 20 with parts from Vautour 28 so that the product of the refurbishment would be to IAF standard.

The idea was not far fetched; this was the original scheme back in 1964, and before that plan foundered in 1965 due to financial issues no longer relevant. While cost mattered in 1965, the combat losses sustained in the Six Day War and

Damage to the right wing of Vautour 03 after the collision on 14 March 1968. (PC)

Damage to the left wing of Vautour 62 after the collision on 14 March 1968. (PC)

The Independence Day fly-past on 2 May 1968 included 14 Vautours in three formations: a leading 4-ship and two flanking 5-ship formations. Jacob Terner with Achikar Eyal in Vautour 33 flew lead (the 4-ship also included Zeev Tavor with Yair David, Jacob Gal with Jacob Shmueli, and Daniel Ilan). Naphtali Porat with Shabtai Ben-Shoa were in the lead of the right 5-ship with Jonathan Shahar in 05, Ovadia Nachman in 12, Beni Zohar, and Giora Goren. Micha Gil (in 30) flew lead in the left 5-ship formation, leading Gideon Magen with Menahem Eini, Abraham Vilan, Adam Zivoni, and Jacob Rafaeli. Also in this participants photo are Herzle Bodinger (right below the Vespa's front seat) with Uri Talmor (fifth from left in photo) and Technical Flight Commander Nathan Segall (walking left to right in front of the cypress tree). (PC)

Vautour 33 leads 12, 17, 18 and 67; the Vautours are in varied schemes (18 and 33 darker; 12, 17 and 67 lighter) and configurations (12, 17 and 33 with two 1,300 liter drop tanks; 67 with four Quadruple Bomb Racks and 18 with two 1,300 liter drop tanks and two Quadruple Bomb Racks). (PC)

the French embargo resulted in an attitude that every Vautour counted regardless of cost. Thus, MAT2 reported on 29 May 1968 a Material Department Chief order to initiate a similar survey concerning the feasibility to rebuild Vautour 20 instead of 28. Pending the results of this new survey, Material Department Chief Israel Rot ordered on 3 June to suspend work on Vautour 28.

Vautour 20 and Vautour 28 survey completion was reported on 4 September 1968. The conclusion was that the rebuild of Vautour 20, with parts from Vautour 28, was feasible. In line with this finding, Material Department Chief ordered to repair Vautour 20.

Saving Vautour 33

Vautour 33 was the sole survivor of the IAF's four Vautour B aircraft, and it too was almost lost on the morning of 20 June 1968 when flown by Micha Gil with Shabtai Ben-Shoa. For the past year Vautour 33 had been engaged in mapping missions over Sinai Peninsula and the Red Sea. Micha Gil recalled:

> We flew alone. Note once we heard over the radio that our fighters were scrambled [to engage enemy aircraft that endangered the reconnaissance Vautour] but I can not recall that I had to abort such a mapping mission.

Vautour 33 trapped in the net at Ramat David on 20 June 1968; the convex leading edge wing is apparent. (PC)

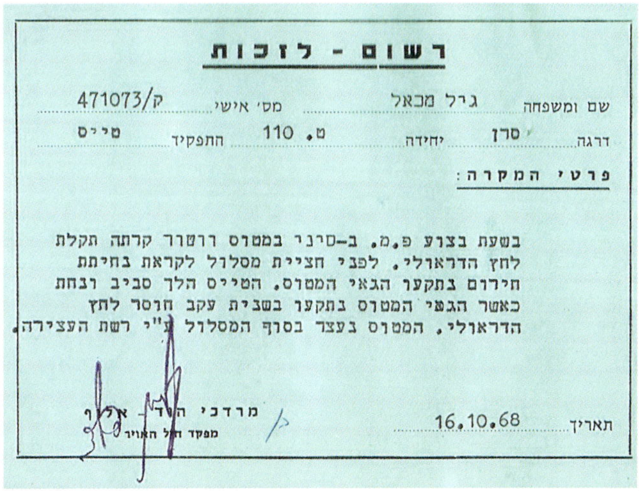

רשום - לזכות

שם ומשפחה גיל מכאל מסי אישי 471073/ק

דרגה סרן יחידה ט. 110 התפקיד טייס

פרטי המקרה:

בשעת בצוע פ.מ. ב-סיני במטוס ווטור קרתה תקלת לחץ הדראולי. לפני חציית מסלול לקראת בחיתת חירום בתקער הגאי המטוס. הטייס הלך סביב ונחת כאשר הגאי המטוס בתקער בשבית עקב חוסר לחץ הדראולי. המטוס בעצר בסוף המסלול ע"י רשת העצירה.

מרדכי הוד - אלוף
מפקד חיל האויר

תאריך 16.10.68

The "Green Note" to Micha Gil for saving Vautour 33 that IAF Commander Moti Hod signed on 16 October 1968. (PC)

Damage to Vautour 33 was light, and it was flying again by August 1968. (PC)

These missions lasted for a long time, up to two hours, and were flown at high altitude, 45,000 to 48,000 feet. It was very cold, no matter how much heating the aircraft generated, and in the end the windshield was covered with ice.

It was on our way back home when a hydraulic system warning light came on. This was not too alarming because there were two hydraulic systems; the regular system malfunctioned but the emergency system functioned. The aircraft was fully controllable until we approached to land. When we entered the landing pattern I noted two Skyhawks waiting to take-off; the pilots were [Squadron 109 Commander] Ohad Shadmi and [Wing 1 Commander] Jacob Agassi. They were holding beside the edge of the runway. When I entered final, I had to correct heading but the stick stuck and I could not steer the aircraft that was heading straight towards those two Skyhawks. I hastily opened the throttles, the Vautour used to lift its nose up when power was applied, and it was possible to control the aircraft directionally with the engines. I managed to abort landing and climbed away but by then both warning lights of the hydraulic systems, regular and emergency, were on. The aircraft manual was unequivocal, eject if warning lights of the two hydraulic systems were on. I reported my situation and said that I would fly towards Haifa Bay and eject over the Mediterranean Sea.

While flying to the sea I noticed that the emergency system's hydraulic pressure recovered and it was again possible to control the aircraft. Shabtai [Ben-Shoa] and I analyzed this issue and I also discussed options with Agassi [who was still holding on the ground in the cockpit of his Skyhawk]. Our analysis was that the action to pull up the nose, with the rather large [monoblock] elevator [during final approach], drained hydraulic pressure. The selected solution was to fly a long, shallow final approach, a little faster to improve safety margins. Our ejection seats were useless below 800 feet so I said to Shabtai that if anything would go wrong then I would climb and we would eject.

We entered a longer, shallower final approach and indeed, as expected, when I pulled up, the stick stuck. I stopped moving the stick backwards, waited for the hydraulic pressure to recover and then moved the stick a little bit more backwards, waiting again for the pressure to build up and so on. Because of this unusual approach we landed faster, the braking parachute detached and the tires exploded during braking. I switched off the engines and we ran into the net at 100 knots. Next thing I remember is absolute silence; the aircraft stopped, the net was wrapped over the aircraft and total silence surrounded us.

Vautour 33 was saved. The net stopped the reconnaissance Vautour at 09:20; damage to the aircraft was light (IAF Category 2; Cat 5 was total loss), and the problem was traced to a leak from the right engine. Vautour 33 was flying again by August 1968 and Micha Gil received a "Green Note" that IAF Commander Moti Hod signed on 16 October 1968, for saving the only surviving Israeli Vautour B.

Vautour 29

An "opportunity to purchase" from France another Vautour, production aircraft number 8, was reported on 24 June 1968. Vautour A8 was previously used to test air refueling both as receiver and tanker. The French priced the redundant Vautour at 2.595 million French Francs and the IMOD reported, on 9 August 1968, an agreed cost of 1.5 million FF. Allocation of IMOD budget was approved on 14 August 1968, and the Vautour was flown to Chateaudun for preparation for the ferry flight to Israel. By 25 August it was reported that the French authorities objected to a ferry flight, but approved dismantling of the aircraft for transport to Israel. Dismantling was completed by 13 September when packing the aircraft to shipment was already underway. The new IAF Vautour was already in Israel on 28 October when MAT2 stated that Vautour 29 "should be assembled and prepared for service introduction".

IAF Commander Moti Hod (right) guides visitors at Ramat David in December 1968 with the still silver Vautour 29 in the background; IAF airlifters transported Vautour 29 to Ramat David, where it was assembled. (PC)

Vautour 66 upon completion of D-Level maintenance at IAI in January 1969. The Six Day War ribbon is painted on the left side of the vertical stabilizer with mission and kill markings on the right side. (BIAF)

Reconnaissance Shortfall

AIR2 rejected Squadron 110's proposition to modify all Vautour N aircraft for reconnaissance, but the saving of Vautour 33 highlighted a reconnaissance shortfall. Vautour 66 was undergoing a lengthy periodic depot level maintenance at IAI, so if Vautour 33 would have been lost, only Vautour 70 would remain of the reconnaissance qualified Vautours.

AIR3 officer Isaac Golan issued on 2 July 1968 an operational requirement for an additional reconnaissance Vautour. The presented rational was that the IAF fielded only four reconnaissance aircraft: two Mirages and two Vautours. The Mirages were planned for operations over heavily defended sectors, and the maximum scale of Mirage photos was 1:8,000, while the Vautours produced photos at up to 1:5,000 scales, and the demand for imagery intelligence was ever increasing. For these reasons AIR3 suggested that Vautour 66 should be brought to the same standard as Vautour 70.

AIR2 officer Menahem Eini reported on 10 July 1968 that Air Department Chief approved the requested modification to Vautour 66, which was to be fitted with K-38 cameras for high altitude photography – as in Vautour 70 – in addition to its low altitude qualification with the GOVERNOR camera.

Missing in the Mont Blanc

Vautour pilot Ilan Lior departed Israel for Europe in July 1968 together with two friends; all three of them had graduated Flying School Class 49 back in March 1966. Two of them climbed to the top of Mont Blanc, but only one returned from this journey. Ilan Lior went missing on 30 July 1968.

This tragedy was the third death among the four pilots assigned to Squadron 110 in IAF Year 1966 Term 3 and Year 1967 Term 1. Nahum Roman was killed in a Meteor preparation flight prior to the Vautour Conversion Course on 28 November 1966. Jonathan Zorea was "slipped back" to Squadron 116 Mystere prior to the Six Day War and was killed over Syria on 5 June 1967. Ilan Lior went missing on Mont Blanc on 30 July 1968. The fourth tragedy struck on 24 February 1969 when Moshe Teman – by then a Flying School instructor and a Squadron 110 emergency posting pilot – was killed in a Fouga trainer accident.

Vulture Strike

Flying at low altitude exposed the Vautour crews to the danger of bird strikes; collisions with birds were not rare, and a number of IAF aircraft were lost over the years, but not Vautours, not even when a black vulture hit Vautour 67 during a low altitude navigation exercise on 6 August 1968. Micha Gil recalled:

> It was training for the navigator – [Shaya] Harsit – of navigating a reconnaissance profile. We flew over Sinai, very low, the fastest possible, perhaps as fast as 550 knots, when suddenly something flashed by. This something was a huge Black Vulture that was ingested into an engine. The impact was mighty, the aircraft swerved and we immediately lost power in this engine but I was in control, we were able to climb on one engine and headed to a successful single engine emergency landing at Hatzerim.

The IAF Day fly-past on 15 July 1968 featured a combined 16-ship formation of French attack aircraft: Squadron 113 Ouragans, Squadron 116 Mysteres, Squadron 105 Super Mysteres, and Squadron 110 Vautours. (IGPO)

Vautour 67 during a rehearsal that preceded the IAF Day fly-past on 15 July 1968, some three weeks before it suffered a vulture strike. (PC)

Vautour 62 tested American rocket pods from 21 May 1968; the two initially tested combinations were four 19 2.75" rockets pods or two such pods and two 1,300 liter drop tanks. (IDFA)

The safety of a combat aircraft with two engines again paid off. Vautour 67 – at the time assigned to test the GRAPEVINE electronic warfare system – returned to the air soon afterwards.

American Rockets

The IAF transition from French to American hardware impacted the Vautour on 28 August 1968, when AIR2 issued an order to test the Vautour with two or four made in the USA rocket pods, each containing 19 2.75" rockets. The objectives of the test program, to be flown in Vautour 62, were to explore the flight envelope with two rocket pods (and two external fuel tanks) and with four pods, to define launch limitations, and to yield fuel consumption for the tested configurations.

Test flights from 21 May to 15 September 1968 actually covered two types of American rocket pods: the 19 2.75" rocket pod and the four 5" rocket pod. Test analysis lasted longer than usual; MAT5 recommended on 30 June 1969 to qualify the Vautour American rocket pods combination at flight envelope limitations of 500 knots, Mach 0.95, 5 G, a roll rate of 120 degrees per second and 30,000 feet, while launch limitations were 500 knots, Mach 0.95, 0.8 to 1.2 G, and a dive angle of up to 45 degrees.

Vautour 30 during the Suez War Day turnaround competition on 29 October 1968. (PC)

GRAPEVINE Test

The importance of electronic warfare was constantly growing, and the Vautour was the first IAF combat aircraft to be so deeply involved in the development and deployment of electronic weapons. MAT9 requested on 30 July 1968 to test the new GRAPEVINE system fitted into a pod.

AIR2 issued on 6 September 1968 a GRAPEVINE test order, specifying two flights aimed to explore the flight envelope of the Vautour with the GRAPEVINE pod and to test the electronic features of the pod. The first test flight, flown on 16 September in Vautour 67, accomplished all of the AIR2 set objectives, so the second test flight was cancelled.

Fuel Matters

The squadron's suggestion to fit 500 or 625 liter fuel tanks to the outer wing station was technically not feasible, so an alternative proposition was to qualify the Vautour to operate with 1,700 liter fuel tanks in addition to the already qualified 1,300 and 1,250 liter tanks. AIR2 ordered in November 1968 to test the 1,700 liter fuel tanks; the assigned aircraft was Vautour 70. Jacob Gal with Shabtai Ben-Shoa flew the two test flights in 70 on 24 March 1969. The initial objective was to fit the larger fuel tanks only to reconnaissance Vautours, but Deputy A Gal proposed on 26 March 1969 that the 1,700 liter fuel tanks should be qualified for all Vautour aircraft. This was not possible, primarily because there were only 12 such tanks in IAF inventory. MAT5 concluded on 30 April 1969 that the Vautour 1,700 liter fuel tank combination was indeed proven but under certain limitations: maximum angle of flaps was 20 degrees, maximum 2 G, and no rolls. Due to the small number of available tanks it was suggested to avoid a jettison test. The MAT5 recommendation was to qualify the Vautour 1,700 liter fuel tank combination only for reconnaissance Vautours on special missions.

Meanwhile, Vautour 29, the former French tanker, was assembled and prepared for IAF service. By then the IAF was operating Skyhawks, which were capable of air refueling but lacked tankers. Therefore, AIR2 Chief Uri Talmor indicated that Vautour 29 should be brought to serviceable status with air refuel capability to be retained, both as receiver and tanker. This was not to be, and the air refueling equipment was duly removed from Vautour 29 by January 1969.

Vautour 70 was utilized to test 1,700 liter drop tanks in March 1969. (PC)

Vautours 66, 67 and 26 flew as TIGER 1, 2 and 4, respectively, during Operation BATTERY on 24 February 1969. (PC)

CRICKET and NINEVEH

The combat aircraft readiness to support IDF activity along the Jordanian border was coded CRICKET from November 1968.

Operation NINEVEH was staged on 4 December 1968; a major air offensive against the Iraqi expeditionary force in north Jordan. The squadron launched two formations: the first included Jacob Terner with Achikar Eyal, Herzle Bodinger in Vautour 05, and Gideon Magen and Beni Zohar. Vautour 05 returned with minor damage to a Quadruple Bomb Rack from small arms fire.

A 3-ship formation – Herzle Bodinger, Ran Zur, and Isaac Gat – was scrambled on 17 February 1969 to support an IDF patrol along the Jordanian border.

BATTERY

Operation BATTERY was launched on the morning of 24 February 1969; IAF aircraft attacked PLO targets in Syria from 07:00. Squadron 110's part included TIGER – Jacob Terner with Uri Talmor in 66, Jacob Gal in 67, Herzle Bodinger in 17, and Abraham Rodan in 26 – which bombed (TIGER 1 and 2 with four napalm bombs and TIGER 3 and 4 with six 250 kg bombs and 14 100 kg bombs) and strafed (each expending 135 to 400 rounds), and a pair – Gideon Magen with Shabtai Ben-Shoa in 33 and Micha Gil with Jacob Shmueli in 70 – that photographed.

The Syrians scrambled interceptors and the TIGER Vautours briefly engaged MiGs. TIGER had already attacked, so at the time of engagement their mission was to return home safely; the Vautours disengaged and returned to Ramat David while air combat patrol Mirages entered combat with the MiGs.

Vautour 03 Lost Over Jordan

A pair – Herzle Bodinger and Abraham Rodan – was tasked, on 17 March 1969, to strafe Jordanian vehicles. Anti-aircraft fire damaged both Vautours. A pair and a 4-ship formation raided, on 26 March, PLO targets near Salt, Jordan; the objective was PLO leader Yasser Arafat, but Arafat escaped unhurt.

The objective of Operation OAR 162 on 22 April 1969 was to attack a radar station near Mazar, Jordan. STERLING – Jacob Terner in 17, Micha Gil in 12, Naphtali Porat in 03, and Isaac Gat in 05 – departed Ramat David at 07:00. The Vautours were planned to bomb from medium altitude. Over the target Vautour 03 exploded, and Naphtali Porat was killed.

OAR 162 was the final Vautour flight for Micha Gil; having started conversion to the Skyhawk – flying his Skyhawk solo on 18 April 1969 – Gil was still listed in the Squadron 110 roster pending qualification as an operational Skyhawk pilot. Reserve pilot Gideon Magen returned to regular service in November 1968; his planned assignment was Squadron 69 Deputy B, and pending activation of the Ramat David based Phantom squadron, he attended a Squadron 117 Mirage Conversion Course and continued to fly missions in the Vautour. In the wake of OAR 162, Magen took over as Acting Deputy B from 4 May 1969.

Vautour 03 was lost over Jordan on 22 April 1969; this photo was taken on 29 December 1968 after David Brisk flew his Vautour solo in 4303. (PC)

Gideon Magen returns Vautour 67 to Ramat David on 18 June 1969 with four empty Quadruple Bomb Racks after an attack mission over Jordan. (PC)

Start of Attrition

The same day that Gideon Magen entered office as Acting Deputy B (4 May 1969) he flew with Shabtai Ben-Shoa a COPSE armed reconnaissance mission over south Jordan in Vautour 67. The next day, Magen and Ben-Shoa again flew 67 on a COPSE mission. Two days later, on 7 May, Magen, still in 67 but this time with Jacob Shmueli, flew Operation OAR 19, an attack against a PLO target in Jordan.

Down south, the Attrition War between Egypt and Israel, mostly fought along the Suez Canal, "officially" opened on 8 March 1969. IAF involvement was fairly limited during the first phase of that conflict from March to July 1969. A new mission for Squadron 110 was artillery spotting; Gideon Magen with Shabtai Ben-Shoa tested the feasibility of this concept in Vautour 61 on 13 May 1969 and, the following day, they flew an artillery spotting mission over the Port Said sector in 67 that lasted 2:05 flight hours.

Gideon Magen attacked Jordanian fortifications in 09 on 19 May 1969; flew two close air support missions during an IDF offensive against Jordan on 21 May (the first with Daniel Wittmann in 61 and the second in 29); and attacked a PLO target in Jordan, in Vautour 30, on 28 May. Thus, the Squadron 110 Acting Deputy B flew a total of 11 operational sorties in May 1968: the aforementioned seven missions over Jordan and artillery spotting, plus three reconnaissance missions over Jordan.

Operational activity intensified, and so were efforts to improve the operational viability of the Vautour; AIR2 issued test orders concerning a modification aimed to minimize the high oil consumption of the Vautour's engines (the test aircraft was Vautour 21) and IFF installation (proposed prototype fit in 33, while 70 was already fitted with a temporary installation) on 1 and 3 June 1969, respectively.

Gideon Magen, with Jonathan Erel, attacked Jordanian fortifications in 67 on 18 June 1969 and flew, with Jacob Shmueli in 66, a BDA mission to photograph with GOVERNOR. The following day a CRICKET readiness pair – Herzle Bodinger with Shabtai Ben-Shoa and David Brisk – was scrambled to support an IDF patrol in contact with a PLO squad near the Jordanian border. Magen, in Vautour 30, bombed a PLO target in Jordan on 1 July.

American Bombs

Vautour American rocket qualification completed, the next phase in the Vautour's Americanization process was qualification of American bombs. AIR2 issued on 2 July 1969 an order to test the Vautour Mark 82 bombs combination. The test was to be flown in Vautour 62 and to cover two phases: exploration of flight envelope with Mark 82 bombs loaded on the wings and definition of bomb release parameters.

BOXER 1 to 6

Moti Yeshurun took over from Jacob Terner as Squadron 110 Commander on 13 July 1969. A week later, the squadron was again involved in large scale operations against Egypt.

Israel escalated the Attrition War from 20 July 1969 with an IAF offensive against Egyptian artillery, fortifications, and installations west of the Suez Canal. Operation BOXER ran from 20 to 28 July in six daily phases; the squadron flew 58 combat sorties, an average of ten sorties per BOXER.

BOXER 1 – 20 July 1969 – sorties included Gideon Magen in 18 (bombing a fortification in the lead of a 4-ship formation), Ami Alkalai with Oded Erez in 67, and Jonathan Shahar with Eli Tovel in 70 (flying a WART mission).

BOXER 2 – 22 July 1969 – involved Gideon Magen in 29 (leading a 4-ship formation that bombed the Sweet Water Canal west of the Suez Canal), Ovadia Nachman in 12, Jonathan Shahar in 18, and Herzle Bodinger with Oded Erez in 62.

Vautour 61 was destroyed in an accident at Ramat David on 5 July 1969, pilot Abraham Rodan and Navigator Eli Tovel were OK. (PC)

Vautour 62 (left) and Vautour 26 flying low over Hatzerim during the IAF Day fly-past on 17 July 1969. (PC)

Moti Yeshurun took over from Jacob Terner as Squadron 110 Commander in this parade on 13 July 1969; Deputy A Herzle Bodinger is facing the parade. Flying Flight, left to right: Dani Haloutz, Abraham Vilan, Isaac Gat, Ami Alkalai, Daniel Wittmann, Moshe Goldwasser, Hagai Tamir, Abraham Rodan, Jacob Shmueli, and Eli Tovel. (PC)

BOXER 3 – 24 July 1969 – included Gideon Magen in 29 (leading a pair that bombed and strafed in cooperation with Skyhawks), Ovadia Nachman in 09, and Jonathan Shahar in 17.

BOXER 4 – 25 July 1969 – saw Gideon Magen, in 29, leading a pair that bombed a bridge and strafed an encampment of tents and Simon Zimon with Oded Erez in 66.

BOXER 5 – 26 July 1969 – sorties included Moti Yeshurun with Oded Erez in 62, Gideon Magen with Eli Tovel in 62 (bombing bunkers west of the canal), and Ovadia Nachman in 26.

From SUNRISE 1 to RAVIV

Towards the end of BOXER, Gideon Magen with Shabtai Ben-Shoa, in Vautour 62, were scrambled on the evening of 26 July 1969 to illuminate over Jordan after Katyusha rocket launchers attacked Israel.

The following evening – 27 July 1969 – Israel reacted with Operation SUNRISE 1. Squadron 110 Vautours illuminated a Syrian artillery battery on the move in north Jordan; Squadron 105 Super Mysteres and Squadron 109 Skyhawks attacked and destroyed that artillery battery.

Two 4-ship formations of Vautours bombed King Abdullah Canal in Jordan on 10 August 1969; the first formation included Moti Yeshurun, Jacob Agassi, Beni Zohar, and Isaac Gat, while the second formation was composed of Michael Dvir, David Brisk, Abraham Rodan, and Ami Alkalai.

Vautours – including Gideon Magen with Daniel Wittmann in 62 – attacked PLO targets in Lebanon on 5 September 1969. Two days later (on 7 September) Dani Haloutz flew his first operational sortie, flying as wingman and escorting a reconnaissance Vautour on a mission over Lebanon.

IDF Operation RAVIV on 9 September 1969 was an armored raid across the Gulf of Suez; IAF air support included Gideon Magen with Uri Talmor in 67 and Michael Dvir with Oded Erez in 62.

Vautour Modification Discussions

An AIR2 meeting on 9 November 1969 discussed modifications to the Vautour; a few of these had been in discussion since 1964. The old plans to fit Vautours with a second UHF radio and an improved radio compass and gunnery camera were not implemented yet. Even modifications that were already qualified, such as a gunnery camera, were not implemented due to, among other reasons, a shift from French hardware to American equipment. Thus, it was decided to fit a second UHF radio, but an American ARC-51 set instead of the French TRAP 21 originally planned for installation in the Vautour. Likewise, it was agreed to use the American KS-27 as a gunnery camera instead of the French camera fitted to Vautour 07, an installation that was qualified prior to the Six Day War.

Additionally, the IFF installation program was not yet finished. AIR2 recommended that MAT4 should finish design of an integrated modification – that would include all of these systems – by 1 March 1970, with actual modification work to start from May or June 1970. Shortly afterwards, on 24 November 1969, AIR2 indicated that a Mini Pan gunnery camera fitted underneath the nose section was preferred to the vertical stabilizer leading edge installation of the KS-27.

Vautour 61 Reconnaissance Request

Vautour 70 was due for D-Level maintenance. The Mirage was the principle IAF reconnaissance aircraft while six RF-4E Phantoms were on order; pending delivery of those Phantom reconnaissance aircraft, the IAF launched a stopgap program to modify two F-4E Phantoms for reconnaissance. Still, the Vautour was regarded as a valuable reconnaissance platform at a time when IDF appetite for imagery intelligence was ever increasing.

In line with these considerations, AIR3 recommended on 14 November 1969 to modify the damaged Vautour 61 for reconnaissance with a high altitude K-38 camera. It was suggested that if such a modification would be approved, then it should be implemented before Vautour 70 was grounded for D-

Vautour 62 under tow at Ramat David in August 1969. (PC)

Vautours 12 and 18 represented 110 in the Ramat David Suez War Day turnaround competition on 29 October 1969. (PC)

Level work. However, damage to Vautour 61 was too extensive to warrant repair (plus availability of Phantoms and Skyhawks was also a consideration). Material Department Chief ordered on 15 January 1970 that Vautour 61 would not be repaired, and the aircraft was officially struck off charge on 18 January 1970.

From Vautour to Phantom

The impact on Squadron 110 of the Phantom's introduction – the new IAF "heavy attack aircraft" – was not limited to reconnaissance. Ramat David transformed from an all French equipped air base to an American hardware dominated wing. Four squadrons operated from Ramat David by the end of the Six Day War: Squadron 107 Ouragan, Squadron 109 Mystere, Squadron 110 Vautour, and Squadron 117 Mirage. By November 1969 the Ramat David wing included Squadron 69 Phantom, Squadron 109 Skyhawk, Squadron 110 Vautour, and Squadron 117 Mirage.

The most profound impact on Squadron 110 was the conversion of Vautour pilots and navigators to the Phantom. The first four Phantom navigators formerly flew the Vautour: Achikar Eyal, Menahem Eini, Saul Levy, and Yair David converted to the Phantom in the USA from March 1969. Three Vautour pilots (Levy Zur, Gideon Magen, and Abraham Rodan) and three Vautour navigators (Shabtai Ben-Shoa, Jonathan Erel, and Moshe Goldwasser) started their Phantom Conversion Course in August 1969. Vautour pilots Michael Dvir and Dani Haloutz and Vautour navigator Daniel Wittmann were assigned to the IAF Year 1969 Term 3 Phantom Conversion Course.

Two Flying School Class 58 graduates (Ariel Cohen and Joel Feldsho) were assigned to the IAF Year 1969 Term 3 Vautour Conversion Course. (PC)

Jonathan Shahar flew an attack mission over the Suez Canal in Vautour 26 on 3 December 1969. (PC)

December 1969

Attrition War air operations intensified. The Vautours flew attack, reconnaissance, and electronic warfare over and across the Suez Canal, with at least one mission flown almost every day. Reserve pilots Ovadia Nachman and Jonathan Shahar flew 18 and 26, respectively, to attack Egyptian targets across the Suez Canal on 3 December 1969. It was a Vautour reconnaissance mission on 23 December that generated photos of an isolated P-12 radar station; a radar station that an IDF heliborne special operation targeted on the night of 26 to 27 December. The P-12 radar was captured and flown to Sinai.

BLOSSOM WART

The decline of the Vautour as the IAF's "heavy attack aircraft" was inevitable. The IAF order of battle for January 1970 included two Phantom squadrons, the new "heavy attack aircraft", and three squadrons of Skyhawks, the new "light attack aircraft" that had load and range comparable or superior to the load and range of the old "heavy attack aircraft", the Vautour.

A fine example of this state of affairs was the launch of Operation BLOSSOM on 7 January 1970; a series of long range bombings against military installations in Egypt that lasted until April 1970. The Phantoms flew the BLOSSOM raids, while Vautour participation was limited to WART electronic warfare patrol in support of the intruding Phantoms. MAT3 reported on 17 February 1970 that all four remaining Vautour N aircraft were modified for WART. Electronic warfare was in high demand, and it was decided to increase the number of Vautour WART aircraft from four to six. Since there were no more Vautour N aircraft, MAT4 was instructed to design a WART modification of a Vautour A aircraft.

Nighttime Flights

An Attrition War characteristic was that IAF squadrons continued to fly operational missions and training flights. On a "normal" day during that conflict, some of the pilots of 110 were tasked to fly "wartime" missions over the Suez Canal frontier while others were assigned to fly "peacetime" training flights. Such was the night of 19 to 20 February 1970. Jonathan Shahar recalled:

> I did not fly the Vautour from 1966 when I was assigned to command a Mystere squadron. In 1969 I retired from IAF service, joined El Al and resumed flying the Vautour as a reserve pilot; at first refreshment flights, then I was re-qualified as an operational pilot and finally I was re-qualified for nighttime missions. One evening I was assigned to fly a nighttime training mission [in Vautour 21], leading a 3-ship formation to the range at Jabel Libni. That night the squadron was also tasked to bomb with napalm targets in the El Ballah sector along the Suez Canal. Prior to take-off, my Number 3 reported a malfunction so we took-off as a pair. Number 2 [reserve pilot Ovadia Nachman in 29] reported, immediately after take-off, that the landing gear of his aircraft failed to retract; instead of emptying fuel tanks in the air [in order to lighten the aircraft to normal landing weight], he requested – and was granted – permission to practice GCA. I pressed ahead alone; illuminated for myself, bombed and returned to Ramat David.

That night, in parallel with training, the squadron also launched two missions. A total of 13 attack aircraft pairs – including two Vautour pairs – operated during the night over the west bank of the Suez Canal to harass Egyptian troops. Deputy A Herzle Bodinger was the leader of the first Vautour pair to attack. Herzle Bodinger recalled:

> We attacked with napalm bombs that were dropped from very low altitude without illumination, only in the light of the moon.

Emergency posting pilot Jacob Gal was the leader of the second pair; his wingman was Ami Alkalai in Vautour 05. Jacob Gal recalled:

> The mission was to drop four napalm bombs on top of Egyptian fortifications west of the Suez Canal. Our tactic was to attack in singles; one aircraft orbited east of the Suez Canal, monitored the attack of the other aircraft west of the Suez Canal and reported hazards such as anti-aircraft fire. So we did and I attacked first. I dropped my bomb right on target, saw tracers and Alkalai reported that I was under anti-aircraft fire. We then switched roles; Alkalai attacked and I watched over him, reporting tracers, anti-aircraft fire. This way each one of us attacked four times. In his fourth attack, anti-aircraft fire hit Alkalai's aircraft. He climbed and headed to Sinai but reported handling difficulties and that his aircraft was on fire. I reported his situation to ground control and, after I realized that he was no longer in control of his aircraft, I ordered him to eject.

Ami Alkalai ejected safely from Vautour 05 and returned from Sinai to Ramat David in a rescue helicopter.

Ovadia Nachman flew Vautour 29 on the evening of 19 February 1970; the same evening that Vautour 05 was lost. (PC)

The farewell ceremony from Meteors - left to right - 15, 21 and 17 in front of the Squadron 117 building on 26 February 1970. (PC)

Farewell Meteor

Squadron 110 had been the caretaker of the Meteor trainers since 1964. The Meteor trainers were utilized to prepare pilots for their Vautour Conversion Course, to tow targets for air-to-air gunnery, and even resumed operational missions. Gideon Magen flew two COPSE missions over Jordan during 1968: in Meteor 18 (with Squadron 117 pilot Itamar Neuner) on 24 March 1968 and in Meteor 15 (with Ami Alkalai) on 30 June 1968.

Three Meteors – 15, which was named HORNED VIPER, 17 was named VIPER, and 21 was named ADDER – remained in service until 26 February 1970, when the type was finally withdrawn from IAF service. Squadron 110 operated the Meteors from 1964 until 1970, but the farewell parade was staged in front of the building of Squadron 117 that operated the Meteors from 1953 until 1961. When the ceremony was over the three Meteors departed Ramat David bound for IAI at Lod with Moti Yeshurun flying Meteor 21 and Herzle Bodinger flying Meteor 15.

David Zait after his Vautour solo flight at the start of IAF Year 1970 Term 1. (PC)

Year 1970 Term 1

The IAF inventory of Vautour aircraft was down to 14 by 1 April 1970: nine A (09, 12, 17, 18, 20, 21, 26, 29 and 30), one B (33), and four N (62, 66, 67 and 70). The logistic support of such a small and diverse fleet of aircraft – especially in light of the French embargo and IAF transition to American equipment – was getting more and more difficult every passing day. Still, IAF policy was to avoid major order of

Vautour 18 (with prefix digit "3") and 26 over Hatzerim during the IAF Day fly-past on 16 July 1970. (PC)

battle changes during wartime and to wait for acquisition of additional Phantoms and/or Skyhawks prior to retirement of French aircraft.

During IAF Year 1970 Term 1 (April to July) the Vautours accumulated 450 combat sorties. Attack accounted for the lion's share of these missions – the Vautours dropped 700 tons of bombs – with the balance of missions tasked to fly electronic warfare, illumination, and reconnaissance. Reserve pilots Ovadia Nachman and Jonathan Shahar flew ten and seven missions, respectively, during this term. Nachman flew seven attack missions (six against Egypt and one against Syria), two electronic warfare WART missions, and one reconnaissance escort mission (over Lebanon). Shahar flew seven attack missions (six against Egypt and one against the PLO in Jordan).

Anti-aircraft fire damaged Vautour 66, which David Brisk with Shabtai Ben-Shoa flew on a reconnaissance mission over Port Said on 22 April 1970. Shrapnel smashed the canopy and they descended from 30,000 feet to an emergency landing at Refidim, in Sinai. This was the eighth mission over enemy territory for Brisk since 1 April 1970 (including two missions over Syria, two missions over Jordan, and four missions over Egypt).

A Soviet air defense division deployed to Egypt from the spring of 1970. The coverage of the Egyptian-Soviet air defense network slowly advanced east towards the Suez Canal. By late June the coverage of the Egyptian-Soviet air defense network had approached the central sector of the Suez Canal. Only the Phantoms attacked elements of the Egyptian-Soviet air defense network, mostly SAM batteries, while the less capable attack aircraft – including Vautour – were relegated to operations against targets still outside the coverage of the Egyptian-Soviet air defense network.

Two Phantoms were lost over Egypt on 30 June 1970. Both crews ejected; two pilots and a navigator (Achikar Eyal) were taken prisoners of war, while the other navigator escaped. A helicopter rescued the Phantom navigator that evening; Deputy A Herzle Bodinger with Eli Tovel, in Vautour 62, flew the WART mission in support of the rescue.

Three Vautours were tasked to fly stand off electronic warfare missions – two with WART against SA-2 and one against SA-3 – during Operation CHALLENGE on 18 July 1970. Eight pairs of Phantoms raided the vanguard of the Egyptian-Soviet air defense network west of Ismailia; the IAF claimed the destruction of four SAM batteries at the cost of one Phantom shot down.

Eldad Maayan flew his Vautour solo in 4329 during August 1970. (PC)

End of Attrition

The Attrition War between Egypt and Israel ended in a cease fire on 7 August 1970. Sporadic clashes along and over the Israeli-Jordanian, Israeli-Lebanese, and Israeli-Syrian borders continued.

Squadron 110 Commander Moti Yeshurun requested in September 1970 to replace the reconnaissance Vautours' self-defense Shafrir 1 missiles with the more capable Shafrir 2. However, the Attrition War was over and delivery of more Phantoms and Skyhawks was expected, so the days of the Vautour in IAF service were clearly numbered; it was highly unlikely that any significant modifications would be applied to the Vautours.

Herzle Bodinger returned Vautour 20 from IAI – where this photo was taken – to Ramat David on 14 August 1970. (PC)

The IAF Year 1970 Term 2 Vautour Conversion Course was the last; pilots Eldad Maayan and Isaac Peleg, and navigator Zvi Lipchitz, were the last IAF air crews assigned to fly the Vautour.

The Return of Vautour 20

Squadron 110 regular pilot Herzle Bodinger flew Vautour 20 when the aircraft was damaged in an accident on 28 June 1964. Squadron 110 Deputy A Herzle Bodinger was honored to fly for the first time the repaired Vautour 20. On the morning of 30 July 1970, Bodinger flew a mission across the Suez Canal, attacking mortars in the Deversoir sector; later that day – a typical Attrition War day that included a mix of "wartime" missions and "peacetime" flights – Bodinger tested the still unpainted Vautour 20; a first flight for the repaired aircraft after six years, one month, and two days.

Bodinger test flew Vautour 20 again on 4 and 5 August 1970. The repaired aircraft was then painted, named JAGUAR, and ferried from IAI to Ramat David on 14 August; the pilot was Squadron 110 Deputy A Herzle Bodinger.

Transition Terms

The IAF Year overlapped the Israeli Fiscal Year from 1 April to the end of the following March, and was divided into three Terms. Those three terms did not last from 1 April to the end of July, from 1 August to the end of November, and from 1 December to the end of March. Each term lasted for a number of weeks that ended in a Flying School Class graduation parade on the last Thursday of the term; then a week was reserved for large scale end of term exercises and the next term started the following Sunday. Thus, Flying School Class 64 graduation on Thursday, 18 March 1971 ended IAF Year 1970 Term 3. Following a week of end of term exercise, IAF Year 1971 Term 1 started on Sunday 28 March 1971, and it was on that day that David Zait ejected from Vautour 12.

Also at the start of IAF Year 1971 Term 1, Squadron 109 loaned four A-4H Skyhawks to sister Squadron 110. The 109 Skyhawks were taxied to the Squadron 110 complex and were to introduce 110 to Skyhawk operations pending planned deliveries of A-4E Skyhawks for 110. In order to preserve the operational status of the squadron during conversion from Vautour to Skyhawk Vautour flights continued, albeit at a reduced scale, until the end of this transition term.

Herzle Bodinger was a Squadron 110 Vautour pilot from December 1963 until January 1971, from Vautour Conversion Course student to Deputy A. As a farewell gesture – aimed to stress his long service in the squadron – Bodinger was photographed as a "human statue" in the garden surrounding the squadron building. Bodinger's Vautour final flight was in 4321 on 7 January 1971, a nighttime air base attack practice over the Halutza Range. (PC)

The first shipment of A-4E Skyhawks for Squadron 110 arrived in July 1971; the official Vautour farewell fly-over was flown in a 4-ship formation on 14 July, but sporadic Squadron 110 Vautour flights continued; the final flight thus far unearthed was recorded on 10 August 1971, when Jonathan Shahar with Elisha Galon flew a reconnaissance Mission Order in Vautour 70.

Vautour 26 was damaged in an accident on 26 October 1970. (PC)

Final participation of the Vautour in the annual Ramat David Suez War Day turnaround competition was on 29 October 1970. (PC)

The Vautour's successor in Squadron 110 service was the A-4E Skyhawk. The first of these were delivered to the squadron in July 1971, when this photo was taken. (PC)

Ramat David Wing Commander Zorik Lev and Squadron 110 Commander Moti Yeshurun, surrounded by maintainers, at the start of IAF Year 1971 Term 1 in front of an A-4H Skyhawk loaned from Squadron 109, but already adorned with the badge of Squadron 110. (PC)

APPENDIX A

ISRAELI VAUTOUR DELIVERIES

The 32nd and last Vautour delivered to Israel was A8, which became 29. It was dismantled in France, airlifted to Israel, and assembled at Ramat David prior to entering service by May 1969, almost 12 years after A13 arrived at Hatzor. (PC)

The welcome party at Hatzor on 1 August 1957, left to right: Air Department Chief Ezer Weizman (tall person wearing a peaked cup just below the nose cone), IAF Commander Dan Tolkowsky (beret), and Prime Minister and Minister of Defense David Ben-Gurion (short person with gray hair). (PC)

OPERATION	AIRCRAFT	DEPARTURE	ROUTE	ARRIVAL	CREW
DANIELA 1	A13 4303	31 July 1957	St Nazaire-Bizerte-Hatzor	1 August 1957	Daniel Shapira
DANIELA 2	A12 4305	12 August 1957	St Nazaire-Bizerte-Hatzor	13 August 1957	Daniel Shapira
DANIELA 3	A14 4307	26 October 1957	St-Nazaire-Bizerte-Ramat David	26 October 1957	Daniel Shapira
DANIELA 3	A18 4308	26 October 1957	St-Nazaire-Bizerte-Ramat David	26 October 1957	Ezekiel Somech
SHULA 1	A15 4309	15 November 1957	Tours-Bizerte	15 November 1957	Daniel Shapira
		16 November 1957	Bizerte-Ramat David	16 November 1957	Ezekiel Somech
SHULA 1	A17 4311	15 November 1957	Tours-Bizerte	15 November 1957	Ezekiel Somech
		23 November 1957	Bizerte-Ramat David	23 November 1957	Daniel Shapira
SHULA 2	A19 4312	20 December 1957	Tours-Bizerte-Hatzor	20 December 1957	Ezekiel Somech
SHULA 2	A20 4314	20 December 1957	Tours-Bizerte-Hatzor	20 December 1957	Yoash Tsiddon
SHULA 2	A21 4315	20 December 1957	Tours-Bizerte-Hatzor	20 December 1957	Daniel Shapira
SHULA 3	A22 4317	9 January 1958	Tours-Bizerte-Hatzor	13 January 1958	Yoash Tsiddon
SHULA 3	A23 4318	9 January 1958	Tours-Bizerte-Hatzor	13 January 1958	Ezra Aharon
REBECCA 1	A16 4320	29 January 1958	Tours-Bizerte-Hatzor	30 January 1958	Yoash Tsiddon
REBECCA 1	A24 4321	29 January 1958	Tours-Bizerte-Hatzor	30 January 1958	Ezra Aharon
REBECCA 2	N322 4461	3 April 1958	Tours-Tebessa-Hatzor	4 April 1958	Tsiddon + Brosh
REBECCA 2	N325 4462	3 April 1958	Tours-Tebessa-Hatzor	4 April 1958	Aharon + French (Pozzo)
REBECCA 2	N326 4465	3 April 1958	Tours-Tebessa-Hatzor	4 April 1958	Somech + Erez
REBECCA 2	N327 4466	3 April 1958	Tours-Tebessa-Hatzor	4 April 1958	French + French
REBECCA 3	A27 4322	20 June 1958	Tours-Tebessa-Ramat David	21 June 1958	Reuben Harel
REBECCA 3	A28 4323	20 June 1958	Tours-Tebessa	20 June 1958	Simon Ash
		21 June 1958	Tebessa-Ramat David	21 June 1958	Ezekiel Somech
REBECCA 3	N328 4464	20 June 1958	Tours-Tebessa-Bizerte-Tel Nof	21 June 1958	Tsiddon + Brosh
REBECCA 3	N329 4467	20 June 1958	Tours-Tebessa	20 June 1958	Somech + Erez
		21 June 1958	Tebessa-Bizerte-Tel Nof	21 June 1958	Ash + Erez
ARIELA 1	A25 4325	25 July 1958	Tours-Bizerte-Ramat David	26 July 1958	Reuben Harel
ARIELA 1	A26 4326	25 July 1958	Tours-Bizerte-Ramat David	26 July 1958	Daniel Shapira
ARIELA 1	B616 4231	25 July 1958	Tours-Bizerte-Tel Nof	26 July 1958	Tsiddon + Brosh
ARIELA 1	N330 4469	25 July 1958	Tours-Bizerte-Tel Nof	26 July 1958	Agassi + Erez
ARIELA 2	B624 4233	23 December 1958	Cognac-Istres-Ramat David	23 December 1958	Somech + Kislev
ARIELA 2	B626 4235	23 December 1958	Cognac-Istres-Ramat David	23 December 1958	Tavor + Eyal
ARIELA 3	B628 4237	10 March 1959	Cognac-Istres-Ramat David	10 March 1959	Tsiddon + Talmor
	N301 4470			20 March 1964	
	07 4328				
	A7 4330		Chateaudun-Brindisi-Ramat David	21 December 1967	
	A8 4329			October 1968	

APPENDIX B

IAF VAUTOURS

Vautour 03 THE SILENCER leads Vautour 67 CYCLONE and Vautour 62 LILITH; 03 and 67 are already in the IAF post Six Day War lighter scheme, while 62 is still in the pre-Six Day War darker scheme. The Vautours are in different configurations: 03 with two 1,250 liter drop tanks, 62 with two 1,250 liter drop tanks and two Quadruple Bomb Racks, and 67 with four QBRs;. The two Vautour N aircraft are flown without navigators. (PC)

AIRCRAFT	NAME (1967 ONWARDS)	ACCIDENT/COMBAT LOSS/WFU	NOTES
4303	THE SILENCER	combat loss 22 April 1969	Naphtali Porat killed
4305	THE PERSUADER	combat loss 19 February 1970	Ami Alkalai ejected
4307		combat loss 5 June 1967	Abraham Vilan ejected
4308		combat loss 6 June 1967	Giora Goren ejected
4309	THE DESTROYER	WFU 1971	
4311		accident 25 January 1962	Amos Avni killed
4312	TORNADO	WFU 1971	
4314		combat loss 7 June 1967	Isaac Golan ejected
4315		accident 5 November 1962	Simon Zimon ejected
4317	THE PANTHER	WFU 1971	
4318	DEMON	accident 28 March 1971	David Zait ejected
4320		accident 28 June 1964, WFU 1971	Herzle Bodinger injured, aircraft repaired
4321	COBRA	WFU 1971	
4322		accident 21 July 1959	Gershon Lapid killed
4323		accident 22 July 1962	burnt out in HAS
4325		accident 21 July 1959	Gideon Magen ejected
4326	PYTHON	WFU 1971	
4328			Technical School instructional airframe
4329	POGROM	WFU 1971	
4330	EMBARGO	WFU 1971	
4231	THE LITTLE BROTHER	combat loss 1 December 1967	Aki Arzi and Elhanan Raz missing
4233	THE BIG BROTHER	WFU 1971	
4235		accident 5 October 1964	Yigal Yanay and Zvi Har-Nevo ejected
4237		accident 27 March 1961	Moshe Papo killed
4461	VAMPIRE	accident 5 July 1969	
4462	LILITH	WFU 1971	
4464		accident 11 January 1961	Squadron 119 crew (Amiran + Ramati) ejected
4465		combat loss 7 June 1967	Shlomo Keren and Alex Meltzer killed
4466	DRAGON	WFU 1971	
4467	CYCLONE	WFU 1971	
4469			returned to France in March 1964

Vautour 67 at the end of periodic depot maintenance. IAI was the D-Level provider for IAF Vautours. (PC)

Vautour 33 THE BIG BROTHER armed with a Shafrir air-to-air missile. (BIAF)

Vautour 17 THE PANTHER departs Ramat David with two 1,250 liter drop tanks and two napalm bombs; an impressive tally of kills (one Tu-16 and six fighters) and missions is painted on the right side of the vertical stabilizer. The Six Day War ribbon was painted on the left side of the vertical stabilizer, while 4317 was the only Vautour adorned with air-to-air kill markings that were painted below the windshield on the left side of the nose section. (PC)

APPENDIX C

SQUADRON 110 VAUTOUR COMMANDER CAREERS

EZEKIEL SOMECH, Squadron 110 Commander 1957 to 30 November 1959

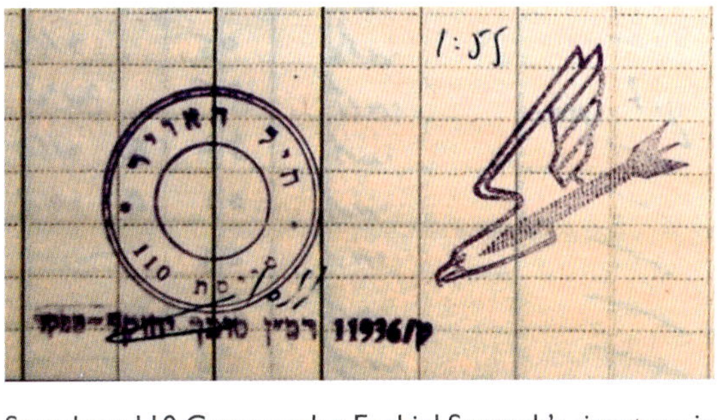

Squadron 110 Commander Ezekiel Somech's signature in a pilot's logbook. (PC)

Ramat David Wing Commander Ezekiel Somech (center, facing camera) presents Vautour's weaponry to visitors. (PC)

Ezekiel Somech was born in Iraq in 1927, and during 1935 his family immigrated to Palestine, where he joined the Palestine Aero Club in 1944. Selected in June 1948 for training as an IAF pilot, air cadet Somech trained in Italy from August until December 1948. When the IAF Flying School was opened in February 1950 Somech was assigned to Class 1, which graduated on 10 August 1950.

Somech remained in the Flying School from 1950 until 1955. Somech graduated from Flying Instructors Course and then instructed Class 3 students prior to progressing with Class 3 graduates to a Spitfire OTU Course – from May to July 1951 – for qualification as a fighter pilot. Returning to the Flying School, Somech instructed in the Flying Instructors Course from August 1951 and was Primary Flying Training Squadron 138 Commander (Acting) from October 1951 until June 1952. Somech then qualified as a multi-engine pilot (in a Flying School in house course) and served as Flying School Chief Training Officer until assignment to a Squadron 110 Mosquito Conversion Course from March to July 1954. Again returning to the Flying

School, Somech was Advance Flying Squadron Commander until assignment as commander of Squadron 109 Mosquito from 15 August 1955 to 1 June 1956.

Progress to jets followed with a Squadron 117 Meteor Conversion Course from June to August 1956 and a Jet Instructors Course in September 1956. When the reserve Squadron 110 Mosquito was mobilized in October 1956 for Suez War operations, Ezekiel Somech was assigned as squadron commander and flew four or five missions during the Sinai Campaign. By then, Somech was selected to lead the revival of Squadron 110 as a regular unit flying the French Vautour, so – in order to get acquainted with French fighters – he started an Ouragan Conversion Course in December 1956. Vautour Conversion Course in France followed from February 1957, and Somech commanded Squadron 110 Vautour until 30 November 1959.

In later years, Somech served as Flying School Commander from 1960 until 1962 and as Ramat David Commander from 1963 until 1967, flying eight missions during the June 1967 Six Day War in the Mirage (seven missions) and in the Vautour (one mission). Somech was IAF Air Department Chief during the 1969 to 1970 Attrition War and one of the finalists for assignment as IAF Commander from 1973, an assignment that was won by Benjamin Peled, the first IAF pilot to fly the Vautour back in 1956 and IAF Commander from 1973 until 1977.

Subsequently, Somech retired from IAF service in 1973 – that same year his daughter served in the operations cell of the Ramat David based F-4E Phantom equipped Squadron 69 – and started a second career as an El Al airline pilot, actually functioning as El Al Deputy Director for Operations from 1984. Ezekiel Somech passed away in April 2008.

Jacob Agassi took over from Ezekiel Somech as Ramat David Wing Commander on 30 June 1967. (PC)

JACOB AGASSI, Squadron 110 Commander 30 November 1959 to July 1963

Squadron 110 Commander Jacob Agassi's signature in a pilot's logbook. (PC)

Jacob Agassi was born in 1932 and graduated IAF Flying School Class 4 in August 1951. After a Squadron 105 Spitfire OTU course (during which he belly landed a Spitfire in a field after a collision with another Spitfire) Agassi returned to the Flying School, serving as instructor with emergency posting as a Squadron 105 Spitfire pilot. From 1954 until 1956 Agassi served at Ramat David, exploiting his Flying School instructor background at first as Squadron 105 Spitfire OTU instructor and then as Squadron 117 Meteor Jet Conversion Course instructor. After the RAF PAI Course, Agassi was assigned to Hatzor as Squadron 113 Ouragan Deputy A (in this capacity he flew nine missions during the Suez War) and Squadron 101 Mystere Deputy A until December 1957, when he was assigned to Squadron 110 as Deputy A until taking over from Ezekiel Somech – after completion of IDF Staff School Course – as Squadron 110 Commander from 30 November 1959 until July 1963.

Agassi's subsequent IAF service career included assignments as AIR5 Aerial Training Section Chief from 1963, Flying School Commander from 1964 to 1966 (during this time frame Agassi converted to the Mirage, switched emergency posting from 110 to 119, and then flew 15 missions during the Six Day War), Ramat David Wing Commander from 1967 until 1970 (ejecting from a Skyhawk during this time frame), and AIR3 Chief from 1970 with emergency posting as a Squadron 69 Phantom pilot. Agassi retired from IAF service as a Colonel and served as a Squadron 69 reserve pilot until 1982.

OVADIA NACHMAN, Squadron 110 Commander July 1963 to July 1964

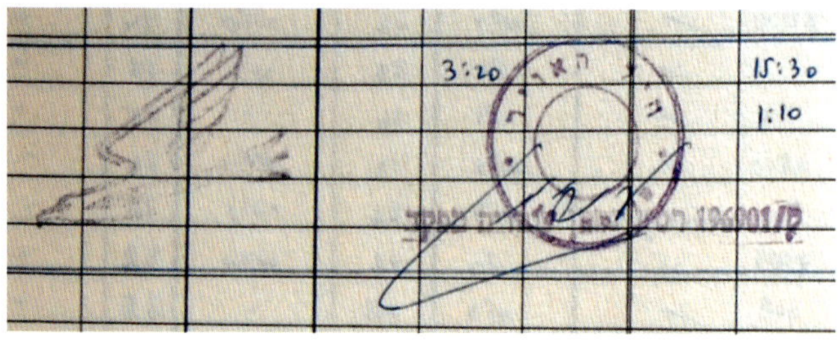

Squadron 110 Commander Ovadia Nachman's signature in a pilot's logbook. (PC)

Ovadia Nachman was born in 1932 and graduated IAF Flying School Class 5 in November 1951. Streamed to Mosquito, his training included a Squadron 141 Multi-Engine Course, a Squadron 109 Mosquito Conversion Course, a Squadron 105 Spitfire OTU (during which he had to land a Spitfire in a field due to an engine malfunction), and a Squadron 109 Mosquito OTU. Subsequent service as a Mosquito pilot (including bale out from a Mosquito that disintegrated in flight) was followed by an assignment to the Flying

Squadron 110 Commander Ovadia Nachman seated second from left in the first row as the pilots and navigators of Squadrons 109, 110 and 117 face President Shazar (note his hat) during his visit to Ramat David in October/November 1963. Sitting in the first row, left right: Squadron 109 Commander David Ivry, Squadron 110 Commander Ovadia Nachman, Ramat David Administration Squadron Commander Shaya Lazarson, and Squadron 117 Commander Samuel Shefer. (PC)

School from September 1954. As Squadron 140 Commander (Acting) Nachman flew a single mission during the Suez War. Continuing his service in the Flying School, Nachman converted to the Meteor, Ouragan, and Mystere – all during 1957 – prior to assignment to Squadron 119 Vautour from 1958. Having graduated from Staff School in 1960, Nachman returned to the Flying School as Basic Flying Training Squadron Commander, overseeing transition from Harvard to Fouga. From 22 May 1962 until 7 July 1963 Nachman was Squadron 119 Commander, moving north with the 119 Vautours to command Squadron 110 from July 1963 to July 1964.

AIR5 Aerial Training Section Chief from 1964 until 1966, Nachman continued to fly the Squadron 110 Vautours at first on an emergency posting service (flying 13 missions during the Six Day War), and then – having pursued a second career as an El Al pilot from 1967 to 1997 – as a reserve pilot, flying a Vautour for the last time on 1 April 1971.

YOSSI SARIG, Squadron 110 Commander July 1964 to July 1965

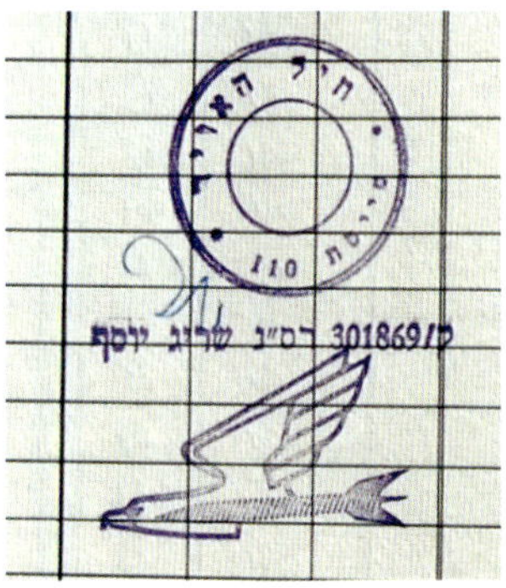

Squadron 110 Commander Yossi Sarig's signature in a pilot's logbook. (PC)

Yossi Sarig graduated IAF Flying School Class 19 in January 1956. After a Squadron 105 Spitfire OTU and a Squadron 117 Meteor Jet Conversion Course, Sarig returned to the Flying School as an instructor, flying eight missions during the Suez War as a Squadron 117 emergency posting pilot. Subsequently serving in Squadron 117 Meteor as an instructor, Sarig converted to the Mystere for emergency posting service with Squadron 109. In November 1958 he converted to the Vautour and subsequently assigned as Squadron 110 Deputy B, responsible for the reconnaissance effort of the squadron. During this time frame he attended, from August 1959, the first IAF Senior Leaders Course. Sarig was Squadron 107 Meteor Commander from April to July 1962, and then served in the Flying School as Instructors Squadron Commander and Advance Flying Training Squadron Commander until assignment as Squadron 110 Commander from July 1964 to July 1965.

Sarig served as AIR3 Attack Section Chief from 1965 until 1967, in charge of preparing the Operation FOCUS master plan. Following conversion to the Skyhawk in the USA during 1967, Sarig was Squadron 102 Commander until 1969. He was then assigned to AIR4 with intention to become AIR4 Chief in due course but retired from IAF service during 1971 in pursue of a different career as an El Al airline pilot.

Yossi Sarig in the cockpit of Vautour 14. (PC)

LEVY ZUR, Squadron 110 Commander July 1965 to 1 October 1967

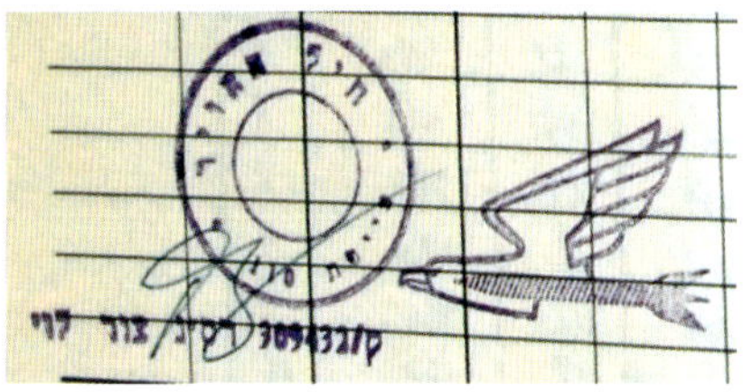

Squadron 110 Commander Levy Zur's signature in a pilot's logbook. (PC)

Squadron 110 Commander Levy Zur in front of the squadron's building in late June 1967. (PC)

Levy Zur was born in 1936 and in June 1956 graduated IAF Flying School Class 20 as the Outstanding Alumni. Zur attended a Squadron 105 Mustang OTU course and flew combat missions as a Squadron 105 Mustang pilot during the Suez War. After Meteor Jet Conversion Course and Ouragan OTU, Zur was assigned to Squadron 109 Mystere, including deployment for training in the Mystere at Dijon, France, from May to December 1957. Following service as Squadron 109 Deputy Commander and as Flying School Instructors Squadron Commander, he was assigned as Squadron 110 Commander from July 1965.

Zur flew 16 missions during the Six Day War. Subsequent service milestones included Flying School Chief Instructor, Hatzor Wing Flying Squadron Commander, head of special R&D projects and establishment of the then new Officers Department.

JACOB TERNER, Squadron 110 Commander
1 October 1967 to 13 July 1969

Jacob Terner was born in 1935 and graduated IAF Flying School Class 22 in July 1957. Terner attended a Squadron 113 OTU course and was assigned to Squadron 109 Mystere at Ramat David. Flying School service started with Flying Instructors Course 31 from April 1958, and Squadron 117 Mirage solo was flown in August 1962. Terner returned to the Flying School for a second tour prior to assignment as Squadron 107 Commander from July 1966 to July 1967 (flying 22 missions during the Six Day War).

Terner followed in the footsteps of Levy Zur as Flying School Chief Instructor from 1969 to 1971. Subsequent service included Tel Nof Flying Squadron Commander from 1971 to 1972, Flying School Commander from 1972 to 1974 (flying eight missions as a Squadron 107 Phantom emergency posting pilot during the October 1973 War), Training Department Chief from 1974 to 1977, Hatzerim Air Base Commander from 1977 to 1981, and Manpower Group Chief from 1981 to 1985 prior to retirement from IAF service as Brigadier General. Terner served in the Israeli Police from 1985, including tenure as the Police Chief Inspector from 1990 until 1993, and was the Mayor of Beer Sheba from 1998 until 2008. While serving as Hatzerim Air Base Commander, Terner entrepreneured the relocation from Ramat David to Hatzerim of Zorik Lev's collection of historic aircraft – including Vautour 09 – as the core of the IAF Museum that opened in 1991 with Terner as director ever since.

Squadron 110 Commander Jacob Terner (crouching second from left) with squadron service persons in front of Vautour 30 EMBARGO. (PC)

MOTI YESHURUN, Squadron 110 Commander
13 July 1969 to 29 March 1972

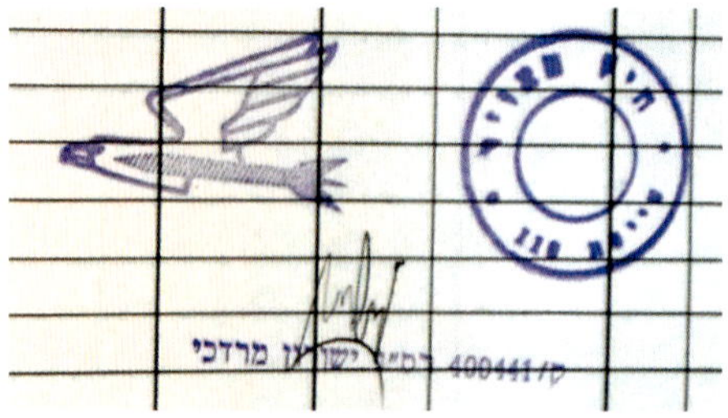

Squadron 110 Commander Moti Yeshurun's signature in a pilot's logbook. (PC)

Moti Yeshurun was born in 1937 and graduated IAF Flying School Class 26 in December 1958. Following a Squadron 113 Ouragan OTU, Yeshurun was assigned to Squadron 109 Mystere and served in the Flying School as instructor from 1960. Assigned to Squadron 101 Mirage from December 1962, Yeshurun was Squadron 116 Mystere Deputy B from August 1963 and then Squadron 116 Deputy A and Squadron 119 Mirage Deputy A (flying 17 missions during the Six Day War). After a tenure as Flying School Advance Training Squadron Commander – also flying as leader of the IAF Aerobatic Team – Yeshurun was assigned to command Squadron 110 from 13 July 1969.

Moti Yeshurun returns to the pen from his Vautour solo in 4309 THE DESTROYER. (PC)

Yeshurun commanded Squadron 110 until transition from Vautour to A-4E Skyhawk was fully accomplished. Sadly, Yeshurun was stricken with illness; he retired from IAF service as a Lieutenant Colonel and passed away in 1990.

APPENDIX D

SQUADRON 110 VAUTOUR FLIGHT COMMANDERS

TIMELINE	DEPUTY A	DEPUTY B	TECHNICAL FLIGHT COMMANDER
1957	Jacob Agassi	Simon Ash	Norik Harel
1959	Zeev Tavor	Reuben Harel	
1960	Reuben Harel	Yossi Sarig	
1962		Jonathan Shahar	
December 1962	Jonathan Shahar	Simi Saar	
September 1964	Simi Saar	Isaac Golan	
June 1965			Abraham Zahar
January 1966	Isaac Golan	Shlomo Keren	
February 1967			Nathan Segall
April 1967	Shlomo Keren		
June 1967		Jacob Gal	
October 1967	Aki Arzi		
December 1967	Jacob Gal	Naphtali Porat	
April 1969	Naphtali Porat	Herzle Bodinger	
April 1969	Herzle Bodinger	Gideon Magen	
September 1969		Giora Goren	
1971	Giora Goren	Amos Bar	

A Squadron 110 Technical Flight Commander reunion in the summer of 2005 in front of the "gate guardian" Vautour 26 PYTHON. (PC)

APPENDIX E

IAF VAUTOUR PILOTS AND NAVIGATORS

This tome chronicles Squadron 110 Vautour, rather than the complete history of the Vautour in Israeli service. However, the initial plans to operate four Vautour squadrons were scaled down to an actual force of two squadrons from 1958, with another scale down to a single squadron from 1963. All Squadron 119 Vautour pilots initiated Vautour flying in the Squadron 110 Vautour Conversion Course; many of the pilots and navigators assigned to Squadron 119 from 1958 until 1963 would continue their Vautour flying days with Squadron 110. Therefore, the scope of this appendix was broadened to include all known IAF pilots and navigators to fly the Vautour. Sorting is by date, then Flying School Class or Navigators Course Class, and finally name.

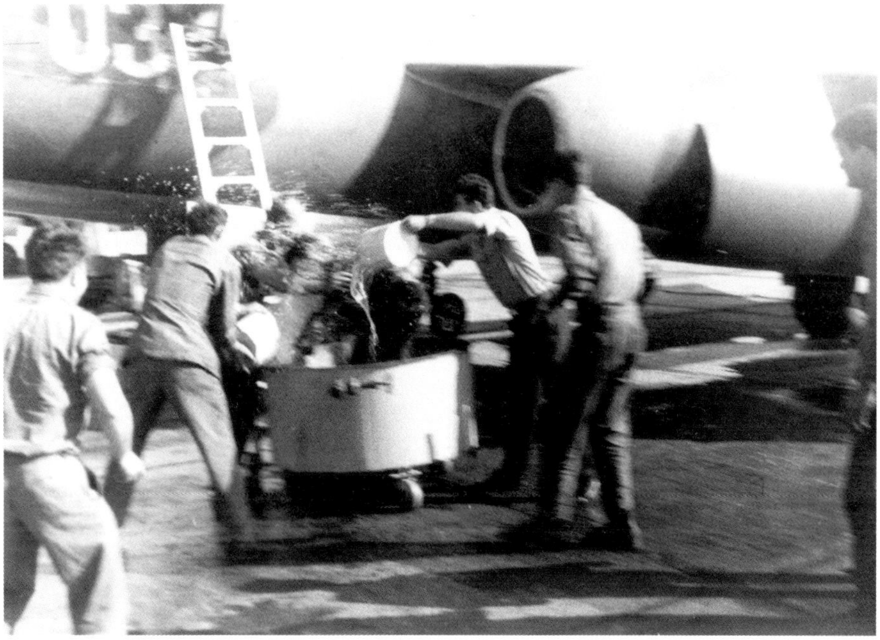

The ceremonial solo flight baptism was a ritual that was repeated at the start of almost every IAF Term. Pilots prepared for baptism by emptying their pockets and taking off their shoes prior to clambering out of the cockpit. The baptism included an initial shower from greasy buckets and a glorious dipping into an oily bath filled with water; always a cheerful experience, baptism in the summer was much more tolerable than in the winter, as illustrated by this IAF Year 1963 Term 3 baptism in front of Vautour 03. (PC)

DATE	FSC	NCC	NAME	A/C	VCC	NOTES
June 1956	2		Benjamin Peled	FAF 09		Evaluation Committee test pilot
June 1956			Shlomo Lahat	FAF 09		Evaluation Committee Chairman (back seat with Peled)
8 February 1957	- 2		Daniel Shapira	FAF A6		conversion in France
8 February 1957	1		Ezekiel Somech	FAF A6		conversion in France
4 October 1957	12		Ezra Aharon	FAF A6		Squadron 119 training with ECTT 3/30 at Tours
October 1957	2		Yoash Tsiddon			Squadron 119 training with ECTT 3/30 at Tours
November 1957		7	Shivi Brosh			Squadron 119 training with ECTT 3/30 at Tours
14 November 1957		10	Itche Erez	FAF N306		Squadron 119 training with ECTT 3/30 at Tours
9 April 1958	5		Zeev Tavor	4311	1	
9 April 1958	8		Simon Ash	4320	1	
9 April 1958	19		Reuben Harel	4308	1	
28 April 1958	5		Ovadia Nachman	4315	1	Squadron 119 pilot
April 1958	4		Jacob Agassi		1	
April 1958	15		Gershon Lapid		1	killed in Vautour 22 on 21 July 1959
5 May 1958	20		Gideon Magen	4311	1	
24 June 1958		10	Oded Erez	4465		Squadron 119 with Ezra Aharon
June 1958		10	Beni Ashkenazi			
6 August 1958		3	Achikar Eyal	FAF B609		CIB 328 Cognac with Lt Raimbault
7 August 1958	3		Adam Zivoni	4311	2	Squadron 119 pilot
8 August 1958	14		Meir Livneh	4323	2	Squadron 119 pilot
8 August 1958	22		Jonathan Shahar	4303	2	
8 August 1958	23		Eli Bahat	4303	2	
8 August 1958	23		Simi Saar	4323	2	

IAF Year 1963 Term 3 VCC student Micha Gil after his solo in 4320 on 18 December 1963; Gil flew the Vautour until 22 April 1969 and subsequently served as an A-4 Skyhawk pilot. (PC)

August 1958	15		Maoz Poraz		2	
August 1958	23		Oded Einav		2	
August 1958	23		Oded Marom		2	
August 1958	24		Beni Zohar		2	Squadron 119 pilot
August 1958		2	Samuel Kislev			CIB 328 Cognac
August 1958		9	Uri Talmor			CIB 328 Cognac
August 1958		11	Giora Bar-Nir			Squadron 119 navigator
August 1958		11	Menahem Pundak			Squadron 119 navigator
August 1958		11	Micha Ramati			Squadron 119 navigator
August 1958		11	Abraham Zfadya			Squadron 119 navigator
21 September 1958		10	Uri Ofer	4466		with Meir Livneh
24 November 1958	11		Yoel Dan		3	emergency posting pilot from Flying School
24 November 1958	19		Yossi Sarig	4323	3	emergency posting pilot from Flying School
November 1958	16		Zeev Sharon		3	emergency posting pilot from Flying School
November 1958	17		Jacob Rafaeli		3	emergency posting pilot from Flying School
11 December 1958			Ezer Weizman	4309		IAF Commander
1958 or 1959	- 1		Menahem Bar			Ramat David Wing Commander
1958 or 1959	14		Soli Amiran			Squadron 119 pilot
1 April 1959	1		Joe Alon	4326	1959 T1	
1 April 1959	25		Abraham Kimchi	4314	1959 T1	
April 1959	22		Baruch Haviv		1959 T1	
April 1959	25		Eli Mor		1959 T1	
April 1959	24		Moshe Ofer		1959 T1	
April 1959	25		Yair Ben-Shefer		1959 T1	
August 1959	24		Judah Fucs			
December 1959	25		Guri Palter		1959 T3	Squadron 119 pilot
1959		6	Shalom Dayan			Squadron 119 navigator

Year 1964 Term 1 VCC student Isaac Nir prepares for baptism – taking off his shoes in the cockpit of 4317 – after his Vautour solo flight on 12 April 1967; Nir flew the Vautour until assignment to a Year 1967 Term 2 Mirage Conversion Course, subsequently serving as a Mirage pilot and credited with two air-to-air kills during the Attrition War. (PC)

1959		7	Yigal Bar-Shalom			Squadron 119 navigator
1959		7	Elisha Galon			Squadron 119 navigator
1959		9	Johanan Levanon			Squadron 119 navigator
1959		13	Menahem Eini			Squadron 119 navigator
6 January 1960		5	Josef Shenar			with Ezekiel Somech (Shafrir test)
29 February 1960		10	Ephraim Ronen	4237		with Yossi Sarig
5 April 1960	28		Isaac Golan	4311	1960 T1	
April 1960	28		Moshe Papo		1960 T1	killed in Vautour 37 on 27 March 1961
4 December 1960	29		Pessach Shraga	4309	1960 T3	
December 1960	29		Uri Margalit		1960 T3	
1960		7	Arik Raviv			Squadron 110 emergency posting
April 1961	31		Giora Yoeli		1961 T1	Squadron 119 pilot
April 1961		14	Alex Meltzer		1961 T1	
August 1961	32		Jacob Gal		1961 T2	Squadron 119 pilot
August 1961	32		Shlomo Keren		1961 T2	Squadron 119 pilot, killed in Vautour 65 on 7 June 1967
December 1961	33		Amos Avni		1961 T3	killed in Vautour 11 on 25 January 1962
December 1961	33		Eitan Karmi		1961 T3	
December 1961	33		Mario Shaked		1961 T3	
December 1961	33		Menahem Yarden		1961 T3	
April 1962	34		Yigal Yanay		1962 T1	
April 1962	34		Simon Zimon		1962 T1	
August 1962	35		Gideon Dror		1963 T2	
August 1962	35		Boaz Gafni		1963 T2	killed in Meteor 14 on 28 November 1966
December 1962	36		Yossi Arazi		1962 T3	
April 1963		16	Zvi Har-Nevo		1963 T1	
18 December 1963	40		Herzle Bodinger	4314	1963 T3	last flew Vautour on 7 January 1971
18 December 1963	40		Micha Gil	4320	1963 T3	

Year 1968 Term 1 VCC student Isaac Gat after his solo flight in Vautour 21; Gat flew the Vautour until assignment to Year 1971 Term 2 Phantom Conversion Course. (PC)

19 December 1963	40		Ran Goren	4303	1963 T3	
December 1963	40		Zvi Shlomis		1963 T3	killed in car accident on 7 December 1964
12 April 1964	41		Isaac Nir	4317	1964 T1	
April 1964	41		Yaron Golan		1964 T1	
August 1964		17	Elhanan Raz		1964 T2	
December 1964	43		Menahem Eyal		1964 T3	
December 1964	43		Daniel Ilan		1964 T3	
April 1965	44		Michael Dvir		1965 T1	
April 1965	44		Giora Goren		1965 T1	
April 1965	44		Jacob Tal		1965 T1	
April 1965	44		Abraham Vilan		1965 T1	
16 July 1965	20		Levy Zur	4303		
December 1965	46		Absalom Fried-man		1965 T3	
December 1965	46		Ran Zur		1965 T3	
August 1966		20	Shabtai Ben-Shoa		1966 T2	
28 November 1966	49		Nahum Roman	—	1966 T3	killed in Meteor 14 prior to flying Vautour
December 1966	49		Ilan Lior		1966 T3	missing on Mont Blanc on 30 July 1968
10 April 1967	50		Jonathan Zorea	4318	1967 T1	killed in Mystere on 5 June 1967
April 1967	50		Moshe Teman		1967 T1	killed in Fouga on 24 February 1969
August 1967		21	Jacob Shmueli			
September 1967	22		Jacob Terner			pre assignment as squadron commander
October 1967	28		Aki Arzi			pre Deputy A assignment
29 December 1967	37		Naphtali Porat	4318		pre Deputy B assignment, killed in Vautour 03 on 22 April 1969
December 1967		21	Yair David			
December 1967		23	Zeev Zeevi			
April 1968	53		Isaac Gat	4321	1968 T1	
April 1968	53		Abraham Rodan	4326	1968 T1	

Year 1968 Term 3 VCC student Ami Alkalai after his Vautour solo flight; Ami Alkalai was killed in a Phantom accident on 22 April 1979. (PC)

29 December 1968	53		Ami Alkalai	4321	1968 T3	
29 December 1968	53		David Brisk	4303	1968 T3	
December 1968		24	Daniel Wittmann			
December 1968		24	Eli Tovel			
1968 or 1969		19	Jonathan Erel			
25 April 1969	56		Dani Haloutz		1969 T1	
April 1969	56		Hagai Tamir		1969 T1	
April 1969			Moshe Goldwasser			
May 1969	26		Moti Yeshurun	4309		pre assignment as squadron commander
December 1969	58		Joel Feldsho	4317	1969 T3	
December 1969	58		Ariel Cohen	4326	1969 T3	
April 1970	53		David Zait	4317	1970 T1	
April 1970	59		Daniel Lidai		1970 T1	
April 1970		19	Amos Eshet			
20 August 1970	60		Eldad Maayan	4329	1970 T2	
20 August 1970	60		Isaac Peleg		1970 T2	
August 1970	10		Zorik Lev	4320		Ramat David Wing Commander
August 1970	27		Zvi Lipchitz			
1971	49		Amos Bar			prior to assignment as Deputy B

Ramat David wing Commander Zorik Lev (right) shakes hands with Squadron 110 Deputy A Herzle Bodinger immediately after the wing commander flew his Vautour solo flight in 4320 during August 1970; Zorik Lev was instrumental in the future preservation of the Squadron 110 Vautour heritage. (PC)

APPENDIX F

SQUADRON 110 VAUTOUR HERITAGE

ISRAEL PURCHASED 32 VAUTOURS: 31 VAUTOURS ACTUALLY flew in IAF service; 17 Vautours were lost (ten in accidents and seven in combat); Vautour 69 returned to France, and 13 Vautours were withdrawn from IAF service in 1971. Luckily, from a future Vautour preservation perspective, Zorik Lev was Ramat David Wing Commander and Oded Erez was Air Defense Section Chief.

Zorik Lev organized a collection of historic combat aircraft at Ramat David that were preserved in perfectly serviceable status. The cornerstone of this collection was the Black Spitfire 57, and by October 1973 the collection also included Meteor 57, Ouragan 80, Mystere 60, and Vautour 09.

Oded Erez suggested to utilize the redundant Vautours stored at Ramat David as decoys in Sinai air bases. A typical order was issued on 11 May 1972 to cover the flight of four Vautours from Ramat David to Refidim on 12 May 1972; the schedule was: 05:30 departure from Ramat David to Refidim of Cessna with a team of maintainers, 07:30 take-off from Ramat David of the first two Vautours, 08:30 take-off from Refidim of Cessna with the two Vautour pilots, and 10:30 take-off of Cessna from Refidim to return to Ramat David the Vautour pilots and maintainers.

Jacob Terner entered office as Hatzerim Air Base Commander in 1977; shortly afterwards, "Zorik's Collection" was relocated from Ramat David to Hatzerim, and Terner started to expand the collection that eventually evolved into the IAF Museum, inaugurated on 20 June 1991 with three Vautours on display.

Moti Yeshurun flew Vautour 70 from Ramat David to Refidim in May 1972; this photo was taken over Ramat David. (PC)

Giora Goren prior to flying Vautour 66 DRAGON from Ramat David to Refidim in May 1972. (PC)

Vautour 66 "buzzes" Ramat David for the last time prior to flight to Refidim in May 1972. (PC)

Aircraft of "Zorik's Collection" were stored in a pen at Ramat David and were occasionally towed out for display, as for this June 1975 families day. (PC)

Vautour 09 inside the "Zorik's Collection" pen at Ramat David in 1978; all aircraft were still serviceable when this photo was taken, a short while prior to relocation to Hatzerim. (PC)

Vautour 33 was also retained at Ramat David and was photographed in 1979 prior to relocation to the Squadron. (PC)

Vautour 33 – still fitted with K-38 cover over the front suite of the bomb-bay – hoisted into position for service as the Squadron "gate guardian" from 1 March 1979. (PC)

Vautour 70 was relocated from Refidim (where it served as dummy from 1972) to Hatzerim (where it joined Terner's Collection) before IAF Air Base 3 was closed in December 1979 prior to Israeli withdrawal from Sinai; this photo of Vautour 70 at Hatzerim was taken in 1980. (BIAF)

Avihu Ben Nun commanded Tel Nof Air Base from September 1979 until February 1982, and it was during this time frame that a park of historic aircraft was opened at Tel Nof; among the exhibits was Vautour 12. (PC)

The IAF Technical School at Haifa also assembled a collection of historic aircraft that included Vautour 26. (PC)

Vautour 33 while serving as the Squadron "gate guardian" in June 1983; the K-38 bomb-bay modification is clearly visible in this view. THE BIG BROTHER was moved from Ramat David to Hatzerim in December 1983 to complete the Vautour variants collection at Hatzerim of Vautour A (4309), Vautour B (4233), and Vautour N (4470). (PC)

A Squadron 110 Commander reunion at Ramat David in January 1985 in front of Vautour 26. Standing left to right: Dave Orly (Squadron 110 Mosquito Commander 1954 to 1955), Yossi Sarig (Squadron 110 Vautour Commander 1964 to 1965), Jacob Agassi (Squadron 110 Vautour Commander 1959 to 1963), anonymous, anonymous, Hugo Marom (Squadron 110 Mosquito Commander 1953), and Ovadia Nachman (Squadron 110 Vautour Commander 1963 to 1964); crouching on the right is Jacob Terner (Squadron 110 Vautour Commander 1967 to 1969). (PC)

By 1989 the Hatzerim Collection was moved to a new location, where the IAF Museum would open on 20 June 1991. Vautour 09 was photographed in August 1989. (PC)

The convex leading edge wing of Vautour 33 is apparent in this 1989 image. (PC)

Vautour 70 in 1989, also with the convex leading edge wing and still fitted with the 1,250 liter drop tanks with which it flew from Ramat David to Refidim back in 1972. (PC)

By 1997, the three IAF Museum Vautours were repainted to represent the three IAF Vautour schemes: Vautour 33 was painted silver, Vautour 09 was painted in the darker pre-Six Day War scheme, and Vautour 70 was painted in the lighter post Six Day War scheme; a WART pod was placed in front of Vautour 70. (PC)

Vautour 29 POGROM – the convex leading edge wing is apparent in the lower image – was placed at Kfar Giladi (as photographed in 1999) in memory of Naphtali Porat. (PC)

Vautour 12 at the Technical School Collection in August 2003. (PC)

Vautour 26 PYTHON photographed in November 2003 as the new Squadron 110 "Gate Guardian". (PC)

The IAF Museum's Vautour 33 (in silver scheme) and Vautour 09 (in two tones darker pre-Six Day War scheme) on display at Hatzerim in March 2009. (PC)

APPENDIX G

ABBREVIATIONS AND ACRONYMS

" Inch
A-Level squadron (line) maintenance
AHQ Air HeadQuarters
AIR1 IAF Air Department Branch 1 (Organization)
AIR2 IAF Air Department Branch 2 (Weapon Systems)
AIR3 IAF Air Department Branch 3 (Operations)
AIR4 IAF Air Department Branch 4 (Intelligence)
AIR5 IAF Air Department Branch 5 (Training)

B-Level base/wing maintenance
BDA Battle Damage Assessment

cm centimeter

D-Level depot maintenance

FAF French Air Force
FTC Flight Test Center

G load factor
GCA Ground Control Approach
GHQ General Headquarters

HAS Hardened Aircraft Shelter
HMG Her/His Majesty Government (United Kingdom Government)

IAF Israel Air Force
IAI Israel Aerospace Industries
IFF Identification Friend or Foe
IFS IAF Flying School
IDF Israel Defense Force
IFF Identification Friend or Foe
IMOD Israel Ministry of Defense
IMI Israel Military Industries
IOC Initial Operational Capability

kg kilogram

LBR Light Bomb Rack

MAT1 IAF Material Department Branch 1 (Logistics – Organization)
MAT2 IAF Material Department Branch 2 (Engineering)
MAT3 IAF Material Department Branch 3 (Maintenance)
MAT4 IAF Material Department Branch 4 (Development)
MAT5 IAF Material Department Branch 5 (Ordnance)
MAT6 IAF Material Department Branch 6 (Logistics – General)
MAT7 IAF Material Department Branch 7 (Radar

Systems) established in 1964
MAT8 IAF Material Department Branch (Fire Control Systems) established in 1964
MAT9 IAF Material Department Branch (Electronic Warfare) established in 1966
mm millimeter
Mission Order Operational Order

NBR Navigators-Bombardiers-Radio Operators
NCO Non Commissioned Officer

OTU Operational Training Unit

PLO Palestine Liberation Organization
PC Personal Collection

QBR Quadruple Bomb Rack

RAF Royal Air Force
RPM Rounds per Minute

SAM Surface-to-Air Missile

TFC Technical Flight Commander

UHF Ultra High Frequency
UN United Nations
UNEF United Nations Emergency Force
USA United States of America

VHF Very High Frequency